Springer

Tokyo
Berlin
Heidelberg
New York
Barcelona
Budapest
Hong Kong
London
Milan
Paris
Santa Clara
Singapore

S. Nagakura (Ed.)

Functionality of Molecular Systems

Volume 1
From Molecules to
Molecular Systems

With 160 Figures

 Springer

Saburo Nagakura
Chairman
Kanagawa Academy of Science and Technology
3-2-1 Sakado, Takatsu-ku, Kawasaki
213 Japan

Springer-Verlag Tokyo Berlin Heidelberg New York

ISBN 978-4-431-66870-1   ISBN 978-4-431-66868-8 (eBook)
DOI 10.1007/978-4-431-66868-8

Preface

One of the most obvious trends in molecular science and engineering is the growing importance of molecular assemblies such as hydrogen-bonding systems, electron donor–acceptor complexes, van der Waals molecules, supermolecules, and molecular clusters. In particular, much importance is attached to molecular assemblies in which component molecules take regular orientations designed most suitably for desired functions.

These molecular assemblies, which are called "molecular systems," are known to play a major role in regulating various properties of molecular solutions and molecular solids, including biological systems. Many molecular scientists have succeeded in designing and preparing various kinds of molecular systems with desired properties. One of typical examples is molecular ferromagnet crystals in which component radicals like nitronyl nitroxide are regularly located in such a way that ferromagnetic coupling exists between them. The preparation of molecular devices, which may be regarded as the goal of minimizing electronic devices, is one of the final targets of the study of molecular systems.

In recognition of the importance of molecular systems, the New Project "Intelligent Molecular Systems with Controlled Functionality" was organized by Professor Hiroo Inokuchi in 1992 with the special support of a Grant-in-Aid for Scientific Research of New Project, Ministry of Education, Science, Sports, and Culture of Japan. Many molecular scientists joined the Project and took up the challenge of the new frontier of molecular science until the Project ended in 1995. Beautiful scientific fruits were produced in the study of structures and functions of molecules themselves; of molecular interactions in solutions and solids; of design, preparation, and functions of molecular systems; and of their application to molecular devices. In other words, the Project was very fruitful and significant in the sense that systematic and cooperative studies were carried out starting from molecules as unit components of molecular systems, elucidating the role of molecular systems as intermediates, and aiming at molecular devices as the goal.

It may be added that the Project was international, as revealed from the fact that a number of young foreign scientists participated as postdoctoral fellows and five international meetings were held as its activity. At present "molecular functionality" is one of the most active and promising research fields in Japan.

For the purpose of reviewing the present status of the study of molecular systems, paying special attention to the achievements of the Project, it was undertaken to publish *Functionality of Molecular Systems*, Vol 1, *From Molecules to Molecular Systems* and Vol 2, *From Molecular Systems to Molecular Devices*. Volume 1 consists of six parts. Part I is an introduction. Part II describes theories of electrons in molecules, molecular solutions, molecular solids, and specific molecular systems related to molecular ferromagnetism. Part III treats fundamental processes in molecules and molecular systems such as molecular recognition; self-regulation; electron, energy, and proton transfer; and chemical transformation. Part IV takes up optical, electric, magnetic, thermal, and mechanical properties of molecules and molecular systems, focusing on the relationship with their structures. Part V is concerned with environmental and external effects such as solvent effects, size and structure effects, polymer effects, and electric and magnetic field effects. Finally, the past, present, and future of molecular design and functionality of molecular systems are overviewed in Part VI.

The main purposes of this volume are to show fundamental aspects of the frontiers of the study of functionality of molecular systems and also to provide the readers with fundamental knowledge necessary for further development of the study of this important and promising research area. It is hoped that this volume will be useful for researchers and graduate students in molecular science and engineering, in materials science and engineering, in electronics, and in related fields.

The editor expresses his sincere thanks to the contributors for their kind cooperation and to Professor Yoshiyasu Matsumoto, Graduate University for Advanced Studies, and to Ms. Yuko Suzuki, Institute for Molecular Science for their kind editorial cooperation. His thanks are also due to Mr. Takeyuki Yonezawa, Ms. Motoko Fukuda, and Ms. Yumi Nishimura, Springer-Verlag Tokyo, for their invaluable editorial assistance.

S. Nagakura

Table of Contents

List of Authors

Part I

SABURO NAGAKURA
Chairman, Kanagawa Academy of Science and Technology, 3-2-1 Sakado, Takatsu-ku, Kawasaki, 213 Japan

Part II

KIMIHIKO HIRAO
Department of Applied Chemistry, Graduate School of Engineering, The University of Tokyo, 7-3-1 Hongo, Bunkyo-ku, Tokyo, 113 Japan

SHIGEKI KATO
Department of Chemistry, Graduate School of Science, Kyoto University, Kitashirakawa Oiwake-machi, Sakyo-ku, Kyoto, 606-01 Japan

KEIICHIRO NASU
Photon Factory, National Laboratory for High-Energy Physics, The Graduate University for Advanced Studies, 1-1 Oho, Tsukuba, Ibaraki, 305 Japan

KIZASHI YAMAGUCHI
Department of Chemistry, Faculty of Science, Osaka University, Toyonaka, Osaka, 560 Japan

Part III

KAZUO KITAURA
College of Integrated Arts and Sciences, Osaka Prefecture University, 1-1 Gakuen-cho, Sakai, Osaka, 593 Japan

OKITSUGU KAJIMOTO
Department of Chemistry, Graduate School of Science, Kyoto University, Kitashirakawa Oiwake-machi, Sakyo-ku, Kyoto, 606-01 Japan

KEITARO YOSHIHARA
Institute for Molecular Science, Okazaki National Research Institutes, 38 Aza-Nishigounaka, Myodaiji-cho, Okazaki, Aichi, 444 Japan

MICHIYA ITOH
Faculty of Pharmaceutical Sciences, Kanazawa University, 13-1 Takara-machi, Kanazawa, 920 Japan

TADASHI SUGAWARA
Department of Pure and Applied Sciences, College of Arts and Sciences, The University of Tokyo, 3-8-1 Komaba, Meguro-ku, Tokyo, 153 Japan

Part IV

YUSEI MARUYAMA
Faculty of Engineering, Hosei University, 3-7-2 Kajino-cho, Koganei, Tokyo, 184 Japan

YOSHINORI TOKURA
Department of Applied Physics, Faculty of Engineering, The University of Tokyo, 7-3-1 Hongo, Bunkyo-ku, Tokyo, 113 Japan

KAZUHIKO SEKI
Department of Chemistry, Nagoya University, Chikusa-ku, Nagoya, 464-01 Japan

TOSHIAKI ENOKI
Department of Chemistry, Tokyo Institute of Technology, Meguro-ku, Tokyo, 152 Japan

Part V

TADASHI OKADA
Department of Chemistry, Faculty of Engineering Science, Osaka University, Toyonaka, Osaka, 560 Japan

HIROSHI MIYASAKA
Department of Polymer Science and Engineering, Kyoto Institute of Technology, Matsugasaki, Sakyo-ku, Kyoto, 606 Japan

NOBUYUKI NISHI
Department of Chemistry, Faculty of Science, Kyushu University, 6-10-1 Hakozaki, Higashi-ku, Fukuoka, 812 Japan

TEIZO KITAGAWA
Institute for Molecular Science, Okazaki National Research Institutes, 38 Aza-Nishigounaka, Myodaiji-cho, Okazaki, Aichi, 444 Japan

KAZUHIKO SEKI
Department of Chemistry, Nagoya University, Chikusa-ku, Nagoya, 464-01 Japan

YOSHIFUMI TANIMOTO
Department of Chemistry, Faculty of Science, Hiroshima University, 1-3-1 Kagamiyama, Higashi-Hiroshima, Hiroshima, 739 Japan

Part VI

HIROO INOKUCHI
Institute for Molecular Science, Okazaki National Research Institutes, 38 Aza-Nishigounaka, Myodaiji-cho, Okazaki, Aichi, 444 Japan

Part I
Introduction

1
Introduction

SABURO NAGAKURA

1.1 Molecules and Molecular Systems

The number of molecules identified hitherto is over 12 million. Molecular materials composed of these molecules are diverse in their structures and properties and have many possibilities for future development. Generally speaking there are two fundamental aspects of investigating molecular materials. One is to study the structures and properties of molecules themselves as precisely and systematically as possible, focusing on the behavior of atoms and electrons within the molecules and to deduce fundamental concepts controlling molecular structures and properties. The other is to study structures and properties of molecular solids and liquids as molecular assemblies, paying special attention to the behavior of the component molecules and their interactions.

These microscopic and macroscopic aspects of molecular materials are closely related through molecular interaction. In other words, molecular interaction is the bridge connecting both aspects.

Rapid progress in molecular materials research from both the microscopic and macroscopic aspects and also progress in preparation techniques are now making it possible to design and prepare molecular assemblies with specified structures and desired functions. These molecular assemblies are called "molecular systems." Some typical examples are superconductive molecular crystals composed of specified electron donors and acceptors, molecular ferromagnetic crystals composed of radicals like nitronyl nitroxide, systematically stacked monomolecular films designed for specified functionality, and artificial polymers with regularly oriented functional groups.

The concept of molecular systems is important in materials science and engineering, particularly in the miniaturization of electronic devices. In Japan, studies of molecular systems and functionality have been made extensively and are now flourishing.

1.2 A New Trend in Molecular Science: From Molecules to Supermolecules and Molecular Systems

During the five decades since World War II, molecular science has made remarkable progress in elucidating structures and properties not only of stable molecules but also of excited and unstable molecules. In particular, the developments in molecular theory and molecular spectroscopy, including the combination of laser spectroscopy and molecular beam techniques, have broadened and enriched our knowledge of structures and dynamic behavior of unstable reaction intermediates and excited molecules such as radicals, clusters, excimers, exciplexes, and van der Waals molecules. Many molecular scientists have been attracted to the study of these chemical species because they have novel structures and properties and are important in interpreting reaction mechanisms.

Most of these species, which are assemblies of two or more molecules, may be called supermolecules in the sense that they have higher complexity than usual molecules and the bonds connecting the components go beyond the concept of the familiar covalent bond. Thus we can say there is a new trend in molecular science and in chemistry "from molecules to supermolecules." [1] This new trend is being accelerated by developments in the study of molecular cages and host–guest interactions in chemical and biological phenomena.

Materials science has recently made rapid progress, being stimulated by scientific interest to elucidate the structure–function relationship of various materials at the atomic and molecular levels. This progress has also been accelerated by industrial requirements to develop new materials with novel functions. Molecular materials are being targeted in materials science because of their huge diversity and potential as electronic materials suitable for the miniaturization of electronic devices, the final goal of which is considered to be molecular devices. These are the basic driving forces in the study of molecular systems. Progress in the preparation of organized molecular crystals, polymers, and thin films and also in direct observation of atoms and molecules by various kinds of electron and near-field optical microscopes are other important factors in promoting the study of molecular systems.

Supermolecules and molecular systems are not independent but are closely related. Both are composed of molecules, and molecular interactions are the common key for understanding their structures and properties. Some supermolecules may be regarded as intermediate stages between isolated molecules and molecular crystals or liquids. In this sense, knowledge of structures, properties, and molecular interactions of supermolecules is useful for the study of molecular systems.

1.3 Molecular Interactions and Functionality of Molecular Systems

Molecular interactions are usually of the van der Waals type, based mainly on electrostatic and dispersion forces, and are much weaker than the covalent bonds connecting atoms within a molecule. This means that molecules in the condensed phase essentially keep their inherent structures and properties and that electrons are usually localized within a molecule. Some types of molecular interactions, however, are considerably stronger than usual and exert significant effects on the structures and functionality of molecular systems. The electron donor–acceptor (EDA), or charge–transfer (CT), interaction is a typical example of strong molecular interactions.

1.3.1 EDA Complexes in Ground States

A large number of molecular complexes composed of electron donor (D) and acceptor (A) molecules have hitherto been studied with the aid of theoretical, spectroscopic, X-ray crystal analysis, and thermochemical methods among others. Aliphatic amine and iodine are known to form a strong EDA complex. The structure of the trimethylamine–iodine complex determined by X-ray crystal analysis [2] is shown in Fig. 1.3-1.

The N–I distance is 0.227 nm and is much smaller than the sum of the van der Waals radii of N and I, 0.353 nm. The N–I stretching band appears at 145 cm^{-1} [3]. The spectroscopic and thermochemical analyses show that the complex formation energy and the electron exchange energy between the component molecules are 50 kJ/mole and –2.09 eV, respectively [4]. This indicates that a special bond is formed between the component molecules. Furthermore the I–I distance of the complex, 0.284 nm, is longer than that of the I_2 molecule, 0.266 nm. These findings can be explained in terms of the resonance between the no-bond and dative structures proposed by Mulliken [5]:

$$\begin{array}{cc}
\text{R} & \text{R} \\
\text{R} \!\!\searrow\!\! \text{N} ----\text{I}-\text{I} \longleftrightarrow \text{R} \!\!\searrow\!\! \text{N}^+\!\!-\!\!\text{I}----\text{I}^- \\
\text{R} & \text{R}
\end{array}$$

no-bond structure   dative structure

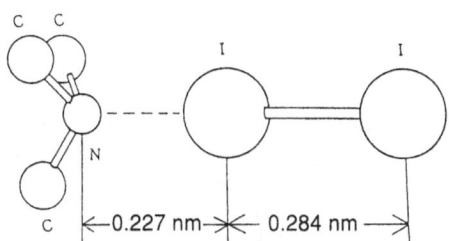

FIG. 1.3-1. The structure of trimethylamine–iodine complex determined by X-ray crystal analysis by Strømme [2]

This resonance scheme is strongly supported by the findings that the charge–transfer band characteristic of the resonance interaction appears at 266 nm for the trimethylamine complex [4] and the electric dipole moment of the trimethylamine complex is 6.5 Debye [5].

These data clearly show that a strong resonance interaction of almost 1:1 mixing occurs between the no-bond and dative structures. The nonbonding orbital electrons of the N atom are almost completely delocalized between the component molecules, and consequently the structures and optical and electric properties of the component molecules undergo considerable changes through the EDA interaction. In more polar environments, it is expected that the resonance contribution of the dative structure will become predominant and result in ionic dissociation.

$$\begin{matrix} R \\ R \\ R \end{matrix} \!\!\!\! > N^+ \!\!-\!\! I \text{- - - -} I^- \longrightarrow \begin{matrix} R \\ R \\ R \end{matrix} \!\!\!\! > N^+ \!\!-\!\! I + I^-$$

1.3.2 EDA Complexes in Excited States

The electronic structures and dynamic processes of excited EDA complexes have been extensively studied and have been elucidated for many EDA complexes [6].

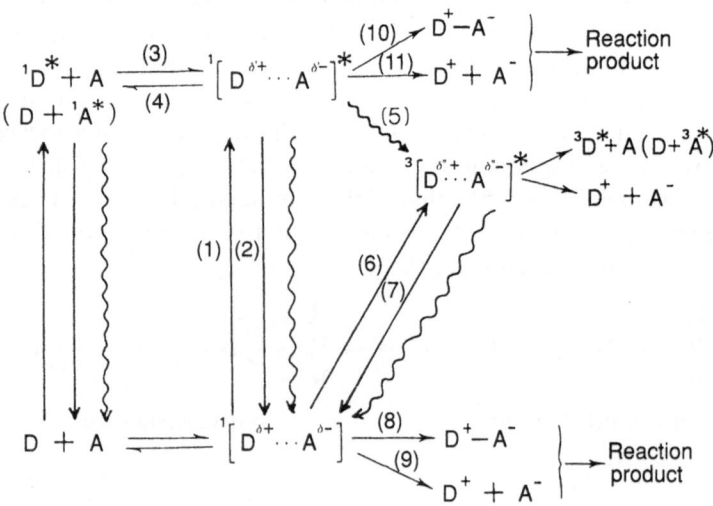

FIG. 1.3-2. Dynamic behavior of EDA complexes:

(1) Charge transfer (CT) absorption
(2) CT fluorescence
(3) Exciplex formation
(4) Exciplex degradation
(5) Intersystem crossing
(6) Singlet → triplet CT absorption
(7) CT phosphorescence
(8) , (10) Addition complex (reaction intermediate) formation
(9) , (11) Ion-radical (reaction intermediate) formation $0 < \delta, \delta', \delta'' < 1$

The observed excited species, including reaction intermediates, and dynamic processes are schematically shown in Fig. 1.3-2.

The degree of CT in the ground states δ is usually small except for strong EDA systems like the amine-iodine complex. On the other hand, the degree of CT in the excited states δ' and δ'' are almost equal to 1 in many cases. Consequently, we can observe CT absorption and emission bands and various nonradiative CT processes and can expect strong intermolecular bond formation depending on the relative orientation of components molecules. This means that photoexcitation from ground EDA complexes to their excited states causes major changes in the electron distribution and bond structures and frequently induces chemical reactions. It is well known that CT interaction plays an important role in photochemical reactions.

It is to be noted that the CT triplet state has been observed separately from the singlet one. This is direct evidence for intermolecular CT and electron delocalization through exchange interaction between D and A. It is of interest to note that the CT triplet exciton exists in some cation radical crystals and shows interesting magnetic behavior [7].

Because CT interaction is so sensitive to photoexcitation and because this causes large changes in the structures of EDA complexes and also in their optical, electric, and magnetic properties, it is expected that EDA complexes will play an essential role in the future development of materials science and in particular the development of opto-electronic materials, bridging molecules and molecular systems. It is hoped that the functionality of molecular systems will continue to grow as an active research field on the basis of wide-ranging, in-depth knowledge of both molecules and molecular interactions.

References

1. Lehn JM (1995) Supramolecular chemistry, VCH, New York
2. Strømme K (1959) X-ray analysis of the 1:1 compound; trimethylamine–iodine. Acta Chem Scand 13:268
3. Yada H, Tanaka J, Nagakura S (1962) Infrared absorption spectrum of charge–transfer complex between trimethylamine and iodine. J Mol Spectroscopy 9:461
4. Yada H, Tanaka J, Nagakura S (1960) Charge–transfer complexes between iodine and various aliphatic amines. Bull Chem Soc Jpn 33:323
5. Mulliken RS, Person WB (1969) Molecular complexes, Wiley, New York
6. Nagakura S (1975) Electron donor–acceptor complexes in their excited states. In: Lim EC (ed) Excited states, vol 2, Academic, New York, pp 321–383
7. Sakata T, Nagakura S (1970) Absorption spectrum of solid Wurster's blue perchlorate and its phase transition mechanism. Mol Phys 19:321

Part II
Theoretical Aspects of Electrons in Molecules and Molecular Systems

2.1
Electronic Structure Theory

Kimihiko Hirao

During recent years, *ab initio* molecular orbital theory has moved from a qualitative theory to a quantitative theory and has become available to nonexperts. The most frequently used *ab initio* methods such as Hartree–Fock (HF) and second-order perturbation theory (MP2) are not only the least expensive but also the easiest to use. Quantum mechanical calculations for molecules are now widely used as an instrument in studying problems in various fields of chemistry and molecular physics. Computational chemistry has certainly become an integral part of chemical research. Quantum molecular methods now have a predictive ability and increasing activity is evident in the design of electronic devices at the molecular level.

The conventional molecular theories are generally effective for medium-sized molecules in their ground state near equilibrium geometry. Typically, more than 98% of the full configuration interaction (CI) (exact) correlation energy is recovered for a given basis set. However, the difficulty with these theories is that as the number of electrons increases and the molecular bonds are stretched, the percentage of correlation energy recovered can decrease substantially. Real chemical processes such as multiple bond breaking, chemical reactions, excited states, and quasidegenerate problems are not easy to adequately describe using simple methods. For the design of chemical reactions and molecular devices, we need a theory that is quantitatively correct for all molecular states and all nuclear geometries.

Regarding basis sets, Gaussian-type functions (GTFs) have now evolved to become the predominant, almost the only, type of basis function used in *ab initio* molecular studies and their applications. Since the introduction of atomic natural orbital basis sets and correlation-consistent basis sets, GTFs have obviously reached a high level of sophistication and are capable of achieving excellent results, at least for relatively small molecules. While large to very large molecules can be treated with smaller basis sets (the double-zeta polarization basis set must be the smallest), such treatments sometimes produce erratic results, particularly for unusual molecules and structures. However, the bane of *ab initio* calculations is the great dependence of the computational effort on the number of the basis function n. This dependence is of the order of n^4 for self-consistent field (SCF) and MP2 calculations and is of the order n^6 and higher for correlated treatments beyond the MP2 level. Thus, there has been continuing interest in methods which reduce the dependence on n without utilizing the analytical basis. Basis set free (grid-based) methods and other approaches such as density functional theory and the pseudospectral method are likely to continue playing an important role in electronic structure theory for large to very large molecules. Such methods have not reached a level of applicability and reliability to challenge the traditional methods, but their relatively short history holds out hope that they will become much more useful in the future.

To qualify as a satisfactory theoretical model, a method ideally should satisfy a number of conditions and requirements [1]: (1) The model should be as simple as possible and demonstrate the required results within chemical accuracy. (2) It should be size consistent, i.e., should scale properly with the size of molecule. (3) It should be conventional. The method should be generally applicable to a wide class of problems and a wide variety of molecules within one framework. (4) It should be efficient and cost-effective with the amount of computer time not increasing too rapidly with the size of the system. (5) It should be able to describe properly the dissociation of a molecule into its fragments. (6) It should be applicable to both excited states and open shells. No single method proposed so far satisfies all these criteria; however, recent progress clearly indicates just what refinement is needed.

In this section, the significant progress in traditional electronic structure theory that has been made in the last few years will be reviewed and hopefully promising outlooks for the future will be discussed. The emphasis will be upon the most recent and important developments of the multireference-based (MR) theory. Topics addressed in Sect. 2.1.1 include multireference-based CI, perturbation, and cluster expansion theory. In Sect. 2.1.2, numerical examples illustrate results with MR-based theory for calculations of potential energy curves and excited states compared to those with single reference-based (SR) theory.

2.1.1 Electronic Structure Theory

2.1.1.1 Correlation Effects

Correlation energy is the difference between the exact eigenvalue of the Hamiltonian and its expectation value in the HF approximation for the state under consideration. Let us express the exact wave function of the system starting with the reference function Φ_{ref} and the correlation correction χ to the wave function as

$$\Psi_{exact} = \Phi_{ref} + \chi \tag{2.1-1}$$

If we introduce the intermediate normalization condition, $<\Phi_{ref}|\Phi_{ref}> = <\Phi_{ref}|\Psi_{exact}> = 1$, we get the exact energy projecting the Schrödinger equation onto the reference function

$$E = \left\langle \Phi_{ref} \middle| H \middle| \Phi_{ref} \right\rangle + \left\langle \Phi_{ref} \middle| H \middle| \chi \right\rangle = E_{ref} + E_{corr} \tag{2.1-2}$$

Usually we start with the HF approximation and E_{ref} is the HF energy. E_{corr} is the correlation energy. Thus, the correlation effect represents the deficiency of the independent electron models.

If the correlation correction χ is expressed in terms of the configuration interaction (CI) expansion as

$$\chi = \Phi_{ia} C_{ia} + \Phi_{ijab} C_{ijab} + \Phi_{ijkabc} C_{ijkabc} + \ldots \tag{2.1-3}$$

Where Φ_{ia}, Φ_{ijab}, . . . are singly, doubly, . . . excited configurations with respect to Φ_{ref} and C are respective expansion coefficients, then we have

$$E_{corr} = \left\langle \Phi_{ref} \middle| H \middle| \chi \right\rangle = \left\langle \Phi_{ref} \middle| H \middle| \Phi_{ijab} \right\rangle C_{ijab} \tag{2.1-4}$$

using the SCF orbitals. Only the doubly excited part of χ contributes to E_{corr}, the other terms such as triple and quadruple excitation contributions influence E_{corr} only in as much as their presence affects the optimum coefficients of the doubly excited configurations, C_{ijab}. The above relation is valid only for the exact wave function, but it suggests the physical nature of the correlation effects.

There are two origins of the correlation effects. The first comes from the correlation of the motion of the electrons. That is, two electrons with opposite spins are not prevented from occupying the same region of space at the same time in the HF approximation. The effect associated with the Coulomb hole is called the dynamical correlation. The dynamical correlation is not caused by the full Coulomb repulsion but by the sum of short-range fluctuation potentials [2]. Due to the short-range nature of the fluctuation potential, the dynamical correlation effects are nearly transferable from state to state and art independent of the nuclear charges and the number of electrons. The other correlation effect is referred to as the nondynamical correlation connected with the near-degeneracy effect. The nondynamical correlation is highly state specific and cannot be transferred from state to state. Some open-shell molecules and molecules in excited states often may not be described by a single HF determinant. Even the dissociation of such a simple system as H_2 cannot be properly described by a single HF wave function over the entire potential curve. This deficiency is connected with the near-degeneracy problem where the rearrangement of the electron and spin couplings becomes very important.

In general, the separation of the correlation energy into the dynamical and nondynamical components is not possible. It should be noted, however, that there are two different physical origins of the correlation effects. This implies that we should use a *different* prescription for *different* correlation effects.

2.1.1.2 Variational Approach

There are three approaches to the electron correlation. The first method is based on the variational principle. The most commonly used variational method is the method of configuration interaction, the so-called CI method [3]. The wave function in the CI method is expressed as a linear combination of the configuration state functions (CSFs)

$$\Psi_{CI} = \sum_R C_R \Phi_R + \sum_I C_I \Phi_I \qquad (2.1\text{-}5)$$

where Φ_R denotes the reference configuration functions. If the first sum contains just one term, it is called SRCI or simply CI. In this case the reference function is usually the SCF function. The SCF reference function would certainly not be expected to give a balanced description for both the equilibrium and distorted geometries of a molecule. Thus, the SRCI is valid only near the equilibrium geometries. If there is more than one reference function, the expansion is called a multireference (MR) CI expansion, where the reference function is usually determined by the multiconfigurational SCF (MCSCF) or complete active space SCF (CASSCF) methods. In most cases, the expansion is truncated after all single and double excitations with respect to one or several reference configurations. Application of the linear variational method to the function Ψ_{CI} leads to the simple eigenvalue equation. The group theoretical (unitary and symmetry group) approach [4] has been developed to evaluate the necessary matrix elements.

The conventional CI algorithm consists of three steps: selection of the number of electron spaces for the CSFs, evaluation of the matrix elements projecting the Hamiltonian to these spaces, and solution of the matrix eigenvalue problems. The advantages of this algorithm are that numerical selection of important CSFs is possible and that non-Abelian symmetry can be used. The most time-consuming step is the calculation of matrix elements, but they are computed only once. Thus CI has conceptual simplicity and generality. The conventional algorithm can usually handle up to 10^6 CSFs. The complexity in the CI including single and double excitations (CISD) roughly equals $m^2 n^4$, where m and n are the numbers of occupied and virtual orbitals, respectively. CISD is approximately equivalent in accuracy as well as in computational complexity to the third-order perturbation theory (PT). The CI energy is variational and converges uniformly to the upper bound for the absolute energy. CI can also be applied to the excited states; however, it suffers from one significant weakness: the slow convergence of the wave function. The CI expansion becomes progressively less compact and less efficient as the number of electrons in the system grows.

There is an alternative algorithm, so-called *direct* CI, originally due to Roos and Siegbahn [5]. Direct CI has stimulated the development of the CI algorithm, enabling a very long expansion of CSFs. Fast methods have been given for the evaluation of the vector $\sigma = (HC)_J$, where C is an estimate for the CI vector. The evaluation of σ is essential in any iterative scheme for the lower eigenvalues of a large matrix. This algorithm can handle large-scale SRCI and MRCI with 10^6 to

10^8 CSFs with little storage. However, the graphical approach requires fixed structure of CSF space. The selection of reference space is rather tedious although many CSFs have negligible CI coefficients and may be discarded safely from the calculation. The Abelian point group can be used, but it is difficult to take into account the non-Abelian symmetry in the direct CI. To avoid the calculation of the general two-electron matrix elements, an intermediate projection CI method has been proposed for large-scale calculations. This algorithm requires only elements of single-shift operators or special products of two operators. However, it is crucial to address random CSFs in this algorithm, and no general programs exist except for full CI.

One of the disadvantages of the restricted CI is a size-inconsistency problem although the MRCI is approximately size consistent. The size-consistency problem is significant to achieve a balanced description of a molecule and its fragments and is not easily addressed within the variational framework. Davidson's correction for SRCI restores size consistency up to the fourth order in perturbation series. Generalization of Davidson's correction to MRCI also gives good results although there is no solid justification in this case. Several modified CI methods have been proposed to correct the size inconsistency. One of them is the averaged coupled-pair functional approximation [6], which optimizes the energy expression through the use of partial renormalization denominators. Pople et al. proposed a quadratic CI method [7] to correct the size-consistency error. This is one of the variants of the cluster expansion method. Various types of MR coupled electron pair approximations [8] also comprise the CI method modified to remedy the size-consistency error.

The main bottleneck of the conventional MRCI method is the fact that the size of the configuration expansion and the computational effort rapidly increase with the number of reference configurations. This makes it necessary to apply configuration selection schemes using perturbation theory (PT) or to select a small number of dominant reference configurations. The method most commonly used is a contracted CI, which is an approximation to the CI that aims to reduce the length of CI expansion. The externally and internally contracted CIs also give better results and can be generalized to the MR-type wave functions. A similar idea has been employed in the superdirect CI by Duch and Meller [9], where the Hamiltonian is projected onto a space of fewer functions approximating the solution. However, it is rather difficult to code the general program due to many complicated formulae.

The CI method has undergone a long stormy development during these last two decades, and calculations with a million CSFs have now become quite feasible. However, application of MRCI is still limited to medium-size molecules.

2.1.1.3 Perturbative Approach

As an alternative to the variational approach, we may use the perturbation theory (PT) [10] to solve the Schrödinger equation. The second-, third-, and fourth-order many-body PT is very successful. This is a well-defined theory that can be applied unambiguously. PT can be applied safely for the ground states of closed-shell molecules around equilibrium geometry.

The SR-based PT is a size-consistent theory. But the energy does not have an upper-bound nature, so the application is limited only to the ground state. The introduction of diagrammatic analysis is a powerful way of handling and summing various types of terms in a perturbation expansion. However, the use of a diagram is only an aid; it does not essentially alter the fact that we are doing a CI calculation whose convergence is basis-set dependent and is normally rather slow. Moreover it is often not appropriate to assume that the usual fluctuation potential is small, i.e., to assume convergence of the perturbation series. In addition, there is some ambiguity as to the definition of the zero-order Hamiltonian, which determines the speed of the convergence. Thus, the success of PT depends on the proper zero-order description of the system. There is no a priori analysis prescribing the order at which one should stop to get a reliable answer. Because of this, perturbation methods were not widely accepted for the treatment of correlation effects. Systematic study by the Pople group and the availability of the Gaussian code [11] were instrumental in establishing the reliability of perturbation techniques. We can conclude from the experience gained that the convergence of the perturbation series with Møller-Plesset (MP) partitioning is reasonable on the whole, but it is very poor for describing the nondynamical correlation. For dynamical correlation, decoupling of the first-order electron pairs has proved to be a fairly good approximation.

The quasidegenerate perturbation approach is now being intensively developed. The common feature of this approach and of many others is the concept of the effective Hamiltonian. When diagonalized in the model space, the effective Hamiltonian gives the part of the spectrum of the exact Hamiltonian defined in the complete Hilbert space. However, there are many problems with the implementation of the generally applicable theory. This type is a "perturb and then diagonalize" approach involving the construction of an effective Hamiltonian. An alternative approach is a "diagonalize and then perturb" method. The PT is applied to each state individually. The two approaches will give the same results at infinite order but may yield different results at low order. Our MRMP [12] and CASPT2 [13] methods use the latter approach.

The essential feature of the MRMP theory is that the MR technique is used as a means of recognizing the nondynamical correlation. Once the state-specific nondynamical correlation is taken into account, the rest is primarily composed of a dynamical pair correlation. Individual pair correlations can be calculated independently using the second-order PT. The performance of PT depends critically on the choice of the zero-order Hamiltonian. For closed shells, the best results are obtained with MP partitioning. Thus, the MR version is formulated with a close analogue of MP partitioning. The key idea underlying the theory is that the dynamical and nondynamical correlations can be obtained by separate calculations and that the correlation energy is given predominantly by the sum of pair contributions if the nondynamical correlation is removed.

We start with an optimized MCSCF or CASSCF wave function in a reference space of m CSFs, $\Xi_1, \ldots \Xi_m$

$$\Phi_0 = \sum_i C_i \Xi_i \qquad (2.1\text{-}6)$$

The active space is spanned by the basis functions that have a filled core and the remaining active electrons distributed over a set of active orbitals. The orthogonal complete space incorporates all other possible basis functions which are characterized by having at least one vacancy in a core orbital and/or at least one electron in a virtual orbital. The $(m-1)$ orthogonal-complement functions in the reference space will be $\Phi_1, \Phi_2, \ldots \Phi_{m-1}$ while all other external CSFs will be Φ_m, $\Phi_{m+1}, \ldots$. The orthogonal-complement functions as well as all external functions higher than those doubly excited relative to Φ_0 will not contribute to the second- and third-order energies. Let us define all zero-order energies in terms of a set of generalized orbital energies

$$E_i = \sum_p \varepsilon_p \left\langle \Phi_i \middle| E_{pp} \middle| \Phi_i \right\rangle \qquad (2.1\text{-}7)$$

where E_{pp} is a unitary group operator. The matrix element of E_{pp} gives the occupancy of the p-th orbital of Φ_i. Therefore, the E_{pp} matrix element for the MR function is just a one-electron density matrix of $<\Phi_0|E_{pp}|\Phi_0> = D_{pp}$. E_0 is given by

$$E_0 = \sum_p \varepsilon_p D_{pp} \qquad (2.1\text{-}8)$$

An alternative approach would use

$$E_0 = \sum_{pq} \varepsilon_{pq} \left\langle \Phi_0 \middle| E_{pq} \middle| \Phi_0 \right\rangle = \sum_{pq} \varepsilon_{pq} D_{pq} \qquad (2.1\text{-}9)$$

In this case, the orbital ambiguity is resolved to make the ε_{pq} as diagonal as possible. The generalized orbital energies are commonly defined as

$$\varepsilon_{pq} = \left(p|q\right) + \sum_{rs} \left\{ \left(pq|rs\right) - 1/2 \left(pr|qs\right) \right\} D_{rs} \qquad (2.1\text{-}10)$$

The virtual orbital energies are defined in a similar manner. This definition reduces to the ordinary MP partitioning in the SR closed-shell case. The definition of an active space, the choices of active orbitals, and the specification of the zero-order Hamiltonian completely determine the perturbation approximation.

The second-order MRMP method is fairly reliable and retains the attractive features of the SRMP method. The MRMP theory has conceptual simplicity due to the independent electron pair model. It is approximately size-consistent. It is very efficient and cost-effective. The complexity in the computation is $m^2 n^2$ if m is the number of orbitals in the reference space and n is that of virtual space. Neither iteration nor diagonalization is necessary in the calculation of the first-order correction to the wave function. The MR technique can dissociate a molecule correctly into its fragments. It is applicable to open shells and excited states.

Clearly, a reliable MR formulation of third-order MP would be desirable. This would be worth studying further and work is currently being carried out in this area.

2.1.1.4 Cluster Expansion Approach

The third approach to electron correlation is a theory based on the cluster expansion of the exact wave function. It is neither variational nor perturbational; it possesses some features of both and is a size-consistent theory. The cluster expansion of the wave function converges more rapidly than the CI expansion through the exponential. The essential characteristics of this approach are to express the exact wave function Ψ_g for the ground state as a cluster expansion in the neighborhood of an independent particle wave function Φ_0

$$\Psi_g = \exp(S)\Phi_0, \quad S = \sum S_i \qquad (2.1\text{-}11)$$

Here S_i is a cluster operator which produces i-fold symmetry adapted excited configurations when operating on Φ_0. Thouless' theorem states that the optimized wave function of the single particle cluster expansion of the form

$$\Psi_{HF} = \exp(S_1)\Phi_0 \qquad (2.1\text{-}12)$$

is the HF wave function. The correlated wave function including correlation effects thus could be given by

$$\Psi_{SAC} = \exp(S_1 + S_2 + \ldots)\Phi_0 \qquad (2.1\text{-}13)$$

This is our symmetry adapted cluster (SAC) wave function for the ground state [14]. The SAC formalism is essential for open-shell systems. The coupled cluster singles and doubles (CCSD) [15] is a special case of the SAC theory. Usually HF approximation recovers more than 99.5% of the total electronic energy and SAC theory recovers more than 98% of the total correlation energy.

To make some connection between the cluster expansion and the more conventional CI expansion, let us expand $\exp(S)$ and collect terms of the common excitation level. By grouping the terms of a given excitation level together we see that

$$\begin{aligned}
\Psi_{SAC} &= \exp(S_1 + S_2 + \ldots)\Phi_0 \\
&= S_1 + (S_2 + S_1^2/2!) + (S_3 + S_1^3/3! + S_1 S_2) \\
&\quad + (S_4 + S_1^4/4! + S_2^2/2! + S_3 S_1 + S_2 S_1^2/2!) + \ldots
\end{aligned} \qquad (2.1\text{-}14)$$

We thus see that the quadruple excitation, for example, that would be obtained in a CI treatment can be viewed within the cluster framework as consisting of five separate terms. Some contributions are very important but others are not. In other words, the CI expansion is a mixed expansion of important and less important terms. This is why the convergence of a CI expansion is rather slow. In quantum mechanics the energy is additively separable. Hence the energy contains only linked clusters. The unlinked clusters are formally not present with the use of the exponential ansatz. Both CI and cluster expansions are representations of the exact wave function. There is nothing wrong in the CI expansion from the purely mathematical point of view, but the cluster expansion of the wave function is more strongly based on physics.

The singly excited cluster S_1 represents the orbital correction or orbital optimization and we can expect S_1 to be negligible in view of the Brillouin theorem. Thus, single cluster contributions to unlinked clusters in higher order can be made small by using the SCF orbitals. However, S_1 is commonly included in the calculations since it represents the SCF effect due to Thouless' theorem. The most important role in this expansion is played by the doubly excited cluster S_2. It represents the electron pair correlation. The next important term is the S_2 by S_2 component, which represents the simultaneous interactions of two distinct pairs of electrons. The S_4 term is expected to be quite small since it describes the simultaneous interactions of four electrons. The S_3 term represents the simultaneous interactions of three electrons and is usually considered to be small; however, when the nondynamical correlation is significant, the theory often requires S_3. The method including S_3 terms is known as CCSDT. Although the CCSD is quite tractable, the accurate calculation of S_1, S_2, and S_3 in CCSDT is rather difficult. Various approximations have been introduced to simplify CCSDT.

The cluster expansion theory can also be applied to the electronic excited states, ionized states, and electron attached states of molecules. As long as these excitations or ionizations involve only one or two electrons directly, it is reasonable to assume that the majority of the dynamical electron correlations in the ground state will not be drastically changed. Thus, it will be more convenient to start from the correlated ground state function Ψ_{SAC} and calculate the change with respect to this state. The correlated excited state wave function Ψ_{SAC-CI} is expressed in terms of the SAC ground state function and the excitation operator T,

$$\Psi_{SAC-CI} = T\Psi_{SAC} = T\exp(S)\Phi_0 = \exp(S)T\Phi_0 \quad \text{with} \quad T = \sum_i T_i \quad (2.1\text{-}15)$$

The symmetry-adapted excitation operator T generates the zero-order wave function for the excited states when acting on Φ_0 and also describes the changes in the correlation relative to the ground state. Expanding $T\exp(S)$, we have

$$T\exp(S) = T_1 + \left(T_2 + T_1 S_1\right) + \left(T_3 + T_2 S_1 + T_1 S_1^2/2! + T_1 S_2\right)$$
$$+ \left(T_4 + T_3 S_1 + T_2 S_1^2/2! + T_2 S_2 + T_1 S_1^3/3! + T_1 S_2 S_1^2/2! + T_1 S_3\right) + \dots$$

$$(2.1\text{-}16)$$

If the ground state SAC wave function is constructed by Brueckner orbitals $(S_1 = 0)$, the above expansion is simplified as

$$T\exp(S) = T_1 + T_2 + (T_3 + T_1 S_2) + (T_4 + T_2 S_2 + T_1 S_3) + \ldots \qquad (2.1\text{-}17)$$

$T_1, T_2, \ldots$ describe the one-electron, two-electron, $\ldots$ excitations. In the ground state, the most important terms are the pair clusters of S_2 and their interactions. For excited states, it is assumed that pair clusters in the closed-shell part are not significantly changed by the transition. The higher excited states thus contribute through disjoint clusters simulated by the products of pair clusters S_2 and the excitation operators T_i such as $T_1 S_2$, $T_2 S_2$, etc. Other terms are either neglected or treated less rigorously. Because S and T are defined in the same restriction, T commutes with S. T is linear and the theory is called SAC–CI [16, 17]. The equation of motion coupled cluster (EOMCC) theory, which has recently been proposed, is a variant of the SAC–CI theory.

This approach makes it possible to calculate the direct determination of the pertinent energy differences, i.e., excitation energies, ionization potentials, and electron affinities [18]. The success of the calculation of these transition energies depends on the extent to which differences in correlation energy between the two states involved in the transition are accounted for. The SAC–CI theory can thereby account for the proper balance of the energy difference between the two states.

While the SR cluster expansion approach has been exploited during the past two decades, the extension of this formalism to the MR case cannot be considered to be complete even today, despite significant theoretical progress made in recent years. This is undoubtedly due to the inherent complexity of the problem as well as to our lack of understanding of the cluster structure of the general MR wave function. One of the difficulties lies with the validity of the commutation property of the excitation operators. Noncommutative algebra makes the theory very complicated and conceptual simplicity is lost.

A general formalism of the MR cluster expansion theory has been given by many authors. Hirao [19] has proposed a generalized SAC theory starting with an MR function,

$$\Phi_{\text{ref}} = \sum_i C_i \Phi_i \qquad (2.1\text{-}18)$$

The generalized SAC wave function is then defined by

$$\Psi_{\text{GSAC}} = \exp(R)\,\Phi_{\text{ref}}, \quad \Phi_{\text{ref}} = T\phi_{\text{core}} \qquad (2.1\text{-}19)$$

The Φ_{core} is the doubly occupied part of Φ_{ref}. The T operator generates all the configurations appearing in the starting function Φ_{ref} and also includes the internal and semi-internal correlation effects, which are specific in their magnitudes and depend to a great extent on the symmetry of the state. The R operator represents the dynamical (all external) correlation. T and R can be chosen uniquely and commute each other. The theory is reduced to the single-reference SAC theory in the absence of near-degeneracy. However, the application of the theory is still limited to relatively small systems.

The extension of the closed-shell formulation of the SAC and CC theories to more general cases, including those requiring a MR space, has turned out to be a very tough problem. At present, the development of the MR-based SAC or CC theories has not yet reached a level comparable to that of the corresponding SR theory.

2.1.2 Some Illustrative Calculations

2.1.2.1 Potential Energy Curves

The performance of the various methods discussed above will be demonstrated by numerical examples. The focus is on the nondynamical correlation included in the MR-based theory. Correlation energies of the H_2O calculations are summarized in Table 2.1-1. Calculations are performed with a double-zeta quality with three geometries, corresponding to stretching of the two OH bonds to r_e (equilibrium bond distance), $1.5*r_e$ and $2.0*r_e$ [20–24]. In general the SCF reference function would certainly not be expected to give a balanced description for both the equilibrium and distorted geometries of a molecule. Thus, examining the validity of the SR-based methods for quasidegeneracy problems provides a good example. Overall, an agreement of the highest versions of the perturbation and CC (SAC) methods with full CI is very good when the SR wave function is an appropriate starting point. It is seen that the convergence of the cluster expansion is more rapid than that of the linear CI analogues. The performance of the SR-based methods is made worse for H_2O with stretched bonds as expected. The perturbative method is applicable up to about $1.5*r_e$. Relatively, the most stable method is CC (SAC). The CC with single, double, and triple excitations is still very good at $1.5*r_e$ but it overestimates correlation effects at $2.0*r_e$. Thus, SR-based methods cannot overcome the deficiency of the poorness of the zero-order wave function. The coefficient of the SCF configuration in the full CI at r_e is 0.979, which drops to 0.764 at $2.0*r_e$. Table 2.1-1 also lists results computed with MR-based methods. The reference function is CASSCF generated by four active orbitals with four electrons. This is the smallest active space that gives a qualitatively correct description of the two OH bonds dissociation process. Contrary to SR-based methods, MR theories give excellent results. The errors are almost constant over the entire range of bond lengths. In the calculation of potential energy surfaces, it is important to keep the errors to full CI as constant as possible, to obtain an accurate surface. With equilibrium geometry, the energy error of the MRSAC relative to the full CI is only 0.04 mhartree. Even at $2.0*r_e$, the error is no more than 0.7 millihartree. The potential energy curves are very close to that of the full CI. It is obvious that SR-based methods, in general, cannot be expected to satisfy this requirement. Appropriate MR-based methods, on the other hand, can give a more balanced treatment and nearly constant errors.

The ground state potential energy curve for N_2 has also been studied [25–26] by a variety of standard *ab initio* techniques and is summarized in Figs. 2.1-1 and

TABLE 2.1-1. Errors to the full configuration interaction (CI) of H_2O with double-zeta basis (in millihartree)

	r_e	$1.5*r_e$	$2.0*r_e$
SCF	148.03	210.99	310.07
Single-reference-based methods			
CISD	7.85	22.41	249.67
CISDT	6.71	18.68	49.72
CISDTQ	0.26	1.10	4.35
MP2	8.55	19.94	52.79
MP3	7.16	25.13	70.45
MP4	0.99	6.13	16.40
SAC(CCSD)	1.79	5.59	9.34
CCSDT	0.45	1.47	−4.09
Multireference-based methods[a]			
CASSCF	60.80	61.88	61.20
MRMP	5.67	1.19	1.33
MRCISD	2.06	2.25	1.10
MRSAC	0.05	0.55	0.75

SCF, self-consistent field; CISDTQ, CI with singles, doubles, triples, and quadruples; MPn, nth-order Møller–Plesset perturbation; SAC, symmetry adapted cluster; CCSDT, coupled cluster with singles, doubles, and triples; CASSCF, complete active space SCF; MRMP, multireference Møller–Plesset perturbation; MRCISD, multireference CI with singles and doubles; MRSAC, multireference SAC.
Full CI energies: $r_e = -76.157866$; $1.5*r_e = -76.014521$; $2.0*r_e = -75.905247$ Å. CI and full CI, from [20–24].
[a]The reference function is the CASSCF with four electrons in four active orbitals.

2.1-2. The basis set used is the double-zeta polarization plus diffuse functions. The reference space for the MRMP method was of the CASSCF type. The CASSCF wave functions were obtained by distributing six electrons among six $2p$ active orbitals, corresponding to 52 reference configurations. This is the smallest active space that leads to the qualitatively correct description of the triple bond dissociation. Results using the MRCISD and MR linearized coupled-cluster method (MRLCCM) were obtained based on the CASSCF function with 176 larger reference functions. The dissociation energies and equilibrium distances are listed in Table 2.1-2.

The SCF potential well is over three times as deep as the experimental values of 9.91 eV and the equilibrium distance is nearly 0.03 Å shorter. The SR-based finite-order perturbation series was found to diverge beyond approximately 3.0 bohr. Even in the minimum region the perturbation series is oscillatory. The

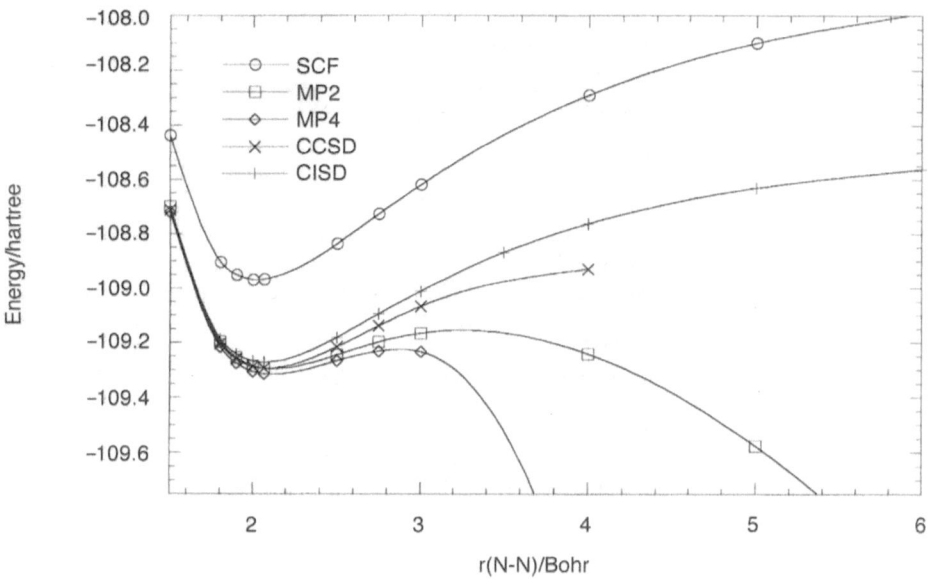

FIG. 2.1-1. Potential energy curves for N_2. *SCF*, self-consistent field; *MPn*, *n*-th order Møller–Plesset perturbation; *CISD*, configuration interaction with singles and doubles; *CCSD*, coupled cluster with singles and doubles

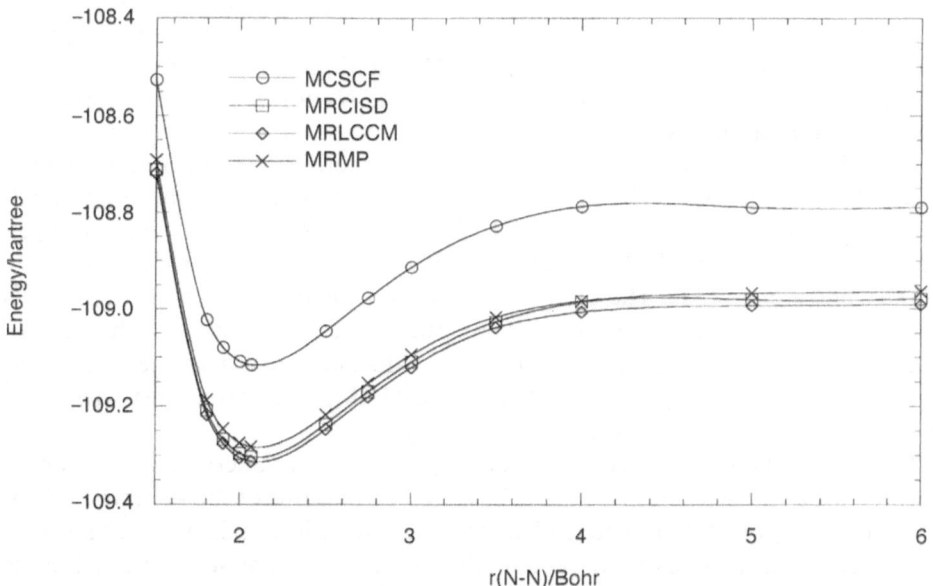

FIG. 2.1-2. Potential energy curves for N_2. *MCSCF*, multiconfigurational SCF; *MRCISD*, multireference CI with singles and doubles; *MRLCCM*, multireference linearized coupled cluster method; *MRMP*, multireference Møller–Plesset perturbation

TABLE 2.1-2. Equilibrium bond length and dissociation energy for N_2

Method	r_e (Å)	D_e (eV)
SCF	1.0703	34.57
CISD[a]	1.0958	23.15
CCSD[a]	1.1037	—
MCSCF-52[b]	1.1087	8.560
MCSCF-176[a]	1.1098	8.903
MRCISD[a]	1.1120	8.857
MRLCCM[a]	1.1128	8.778
MRMP[b]	1.1131	8.674
Experiment	1.0977	9.91

MCSCF, multiconfigurational SCF; MRLCCM, multireference linearized coupled-cluster.
[a] From [25].
[b] From [26].

CISD dissociation energy is still too high by a factor of more than 2 compared to experimental values. The CISD and experimental r_e values differ by only 0.002 Å. The full CI r_e on this basis is estimated as being around 1.113 Å. If this is the case, the CISD r_e is nearly 0.02 Å too short. The CCSD appears to give a much better estimate of r_e. Also the CCSD curve is accurate to 4.0 bohr. However, it was reported that the CCSD curve beyond 4.5 bohr cannot be obtained due to the convergence difficulty of the iterative CCSD equations. Thus, none of these SR-based methods can describe all the regions of the N_2 potential energy curve to a high begree of accuracy.

The dissociation energies and equilibrium geometries have been improved in the MR-based methods. The MCSCF surface itself contains no substantial qualitative defects. Like MRCISD and MRLCCM, the MRMP energy curve dissociates correctly and the three curves are nearly parallel. The MRMP gives D_e = 8.67 eV and r_e = 1.113 Å. The D_e is computed as being only 12% in error and r_e is within 0.015 Å of the experimental value and is nearly identical with the estimated full CI r_e. Although the MRCISD curve deviates most from the MRLCCM surface in the region surrounding 4 bohr, the MRMP curve is almost parallel to the MRLCCM curve for the entire bond lengths.

In terms of the Fermi sea determined by the reference function, the first-order corrections to the wave function may be classified in terms of the number (0, 1, or 2) of external orbitals introduced as internal, semi-internal, or external. The internal correlation was found to be small, within 0.01 Å. The semi-internal terms include significant single excitations that arise from the failure of the reference function to satisfy the Brillouin theorem. The external terms resemble the pair correlation of the closed-shell theory. The external terms are rather insensitive to changes in the bond length. Thus, the total correlation curve is almost parallel to the semi-internal curve, i.e., the balanced description of the potential curves cannot be obtained before the semi-internal terms are correctly taken into account.

To clarify the influence of electron correlation effects on the transition states, we will discuss recent results of the activation energy or energy barriers. In general, the transition state is often a loosely bound structure having stretched and deformed bonds and including atoms with hypervalent states. Thus, one can expect an important role for the correlation effects, both dynamical and nondynamical, in calculating the barrier heights. One of the most theoretically studied elimination reactions is $H_2CO \rightarrow H_2 + CO$ [27–31]. The reaction is Woodward–Hoffmann forbidden and the highly asymmetric transition structure is shown in Fig. 2.1-3. During the dissociation process, two electrons, one from each C—H bond, pair up to form the H—H bond while the other two form a lone pair on C in CO. Thus, the MCSCF wave function with four electrons in four active orbitals can give a qualitatively correct description of the dissociation process. The optimized geometries of the transition state are listed in Table 2.1-3. Since the H—H bond is stretched to almost double its value in the H_2 molecule, the reactant diabatic surface is not well represented at the SCF level with respect

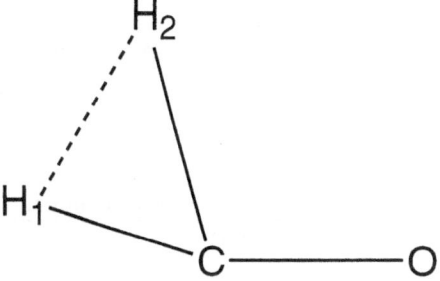

FIG. 2.1-3. The transition state geometry for molecular dissociation to $H_2 + CO$

TABLE 2.1-3. Transition structure parameters for H_2CO to $H_2 + CO$

Method[a]	R(CO) (Å)	R(CH$_1$) (Å)	R(HH) (Å)	<OCH$_1$ (degrees)
SCF	1.151	1.104	1.208	161.9
CASSCF	1.160	1.098	1.350	162.25
MP2	1.180	1.092	1.356	164.0
CISD	1.153	1.094	1.263	162.7
CCSD	1.163	1.093	1.307	163.1
MRCI	1.183	1.096	1.301	162.65

[a] CASSCF and MRCI, from [29]; MP2, from [28]; CCSD, from [31].

to the H—H stretch. The CASSCF H—H bond is significantly larger than the SCF value. This means that the transition state occurs at an earlier stage in the reaction path at the SCF level. Although the SCF wave function gives qualitatively correct results, an MR wave function is required to obtain the same level of accuracy for the transition structure as for the SCF equilibrium geometry. Moore and co-workers [32] determined the barrier for dissociation from careful experimental studies and concluded that the experimental observations are best fit by a potential barrier height (including zero-point energy) of 78.0–81.1 kcal mol^{-1}. Table 2.1-4 shows the results of recent calculations of an activation barrier at various level of theory. The zero-point vibrational correction is not included, which is estimated to be 5.1–5.4 kcal mol^{-1}. While H_2CO is a stable closed-shell molecule, the transition structure has partly formed bonds and is of a diradical character. The correlation effects cannot be expected to cancel in the same manner as occurs in closed-shell systems. Although the correlation contribution is very sensitive to the basis-set quality, the electron correlation certainly lowers the barrier. The differential correlation energy between the equilibrium geometry and the transition structure comes mainly from the nondynamical correlation. The MP perturbation series with the same basis set gives rise to oscillation in the barrier height. The triple excitations in the CC theory lower the barrier height by 3–4 kcal mol^{-1}. These SR-based CI and CC theories require the inclusion of more than double excitations. In contrast, MRMP results are within the range of the observed values. Again, MR methods are usually indicated.

In summary, a few examples given in this section should show that the computation of electron correlation effects, especially the nondynamical correlation

TABLE 2.1-4. Calculated barrier height (kcal mol^{-1}) in the H_2CO to H_2 + CO elimination

Basis set[a]	DZ	DZP	TZP	QZP
SCF	113.7	107.8	103.3	101.8
	105.2			
MP2	96.2	96.0	90.3	90.3
MP3	98.8	97.9	92.7	92.2
MP4	90.7	91.7	85.8	85.9
CISD	98.5	95.0		
CISD(Q)	94.6	90.8		
CCSD	94.4	90.4		
CCSDT	91.0	86.8		
MRCI/CASSCF(4/4)		91.3		
MRCI/CASSCF(6/6)		91.0		
MRMP/CASSCF(10/10)		84.3	84.8	84.7

DZ, double-zeta basis set; DZP, double-zeta plus polarization basis set; TZP, triple-zeta plus polarization basis set; QZP, quadruple-zeta plus polarization.
[a] MPn, from [28]; MRCI, from [29]; CC, from [31] ; MRMP, from Nakano, Nakayama, Hirao, and Dupuis, 1996, unpublished data.

effect, is a problem that requires special attention. It is apparent that potential energy surfaces can be obtained reliably over the entire range of geometrical variables only by the MR-based theory.

2.1.2.2 Excitation Energies

As an example of the calculation of excited states, we will discuss results of the valence and Rydberg excitation states of benzene [33]. Benzene is often used, being an intermediate case between large and small molecules. The calculation was carried out for ground and low-lying electronic excited states of benzene. All calculations were performed with a double-zeta plus polarization quality or better. For the calculation of Rydberg states, appropriate Rydberg functions were added to the valence basis set. The ground state of benzene is well described in a single HF configuration but the single configurational description is made worsen for excited states, particularly for valence excited states, due to the severe quasidegeneracy problems. The reference space should be chosen to be large enough such that all near-degeneracy effects are included. For MR theory, valence six π-electrons are treated as active electrons and distributed among π- (a_{2u} and e_{1g}) and π^*- (e_{2u} and b_{2g}) orbitals for valence states. The contributions from σ-electrons are included as dynamical correlation through the second-order perturbation in the MRMP treatment. For the calculation of Rydberg excited states, an extended active space is used by adding appropriate Rydberg orbitals in addition to valence π- and π^*-orbitals.

Results for singlet and triplet valence π–π^* excitation energies are summarized in Table 2.1-5 and Fig. 2.1-4. The results of SAC/SAC–CI are cited from the study done by Kitao and Nakatsuji [34]. The SAC/SAC–CI reproduced the experimental excitation energies to within 0.5 eV. There are several CI works on the transition energies of benzene but we have cited the study of MRD–CI results by

TABLE 2.1-5. Valence π–π^* excitation energies (eV) of benzene[a]

State	CASSCF	MRMP (error)	Exp.	CASPT2	SAC-CI	MRD-CI	NO-CI
Singlet							
1^1B_{2u}	5.07	4.77 (−0.13)	4.90	4.70	5.25	5.19	5.17
1^1B_{1u}	8.10	6.28 (0.08)	6.20	6.10	6.60	6.67	6.38
1^1E_{1u}	9.37	6.98 (0.04)	6.94	7.06	7.47	7.55	7.32
1^1E_{2g}	8.17	7.88 (0.08)	7.80	7.77		8.17	
Triplet							
1^3B_{1u}	4.58	4.11 (0.16)	3.95	3.89	4.06		3.90
1^3E_{1u}	5.62	4.81 (0.05)	4.76	4.50	5.02		4.57
1^3B_{2u}	7.41	5.34 (−0.26)	5.60	5.44	6.02		5.42
1^3E_{2g}	7.23	7.04 (0.21)	6.83	7.03			

Exp., experiment; CASPT2, second-order perturbation theory based on CASSCF; MRD-CI, multireference CI with doubles; NO-CI, CI with natural orbitals.
[a] CASSCF and MRMP, from [33]; CASPT2, from [37]; SAC-CI, from [34]; MRD-CI, from [35]; NO-CI, from [36].

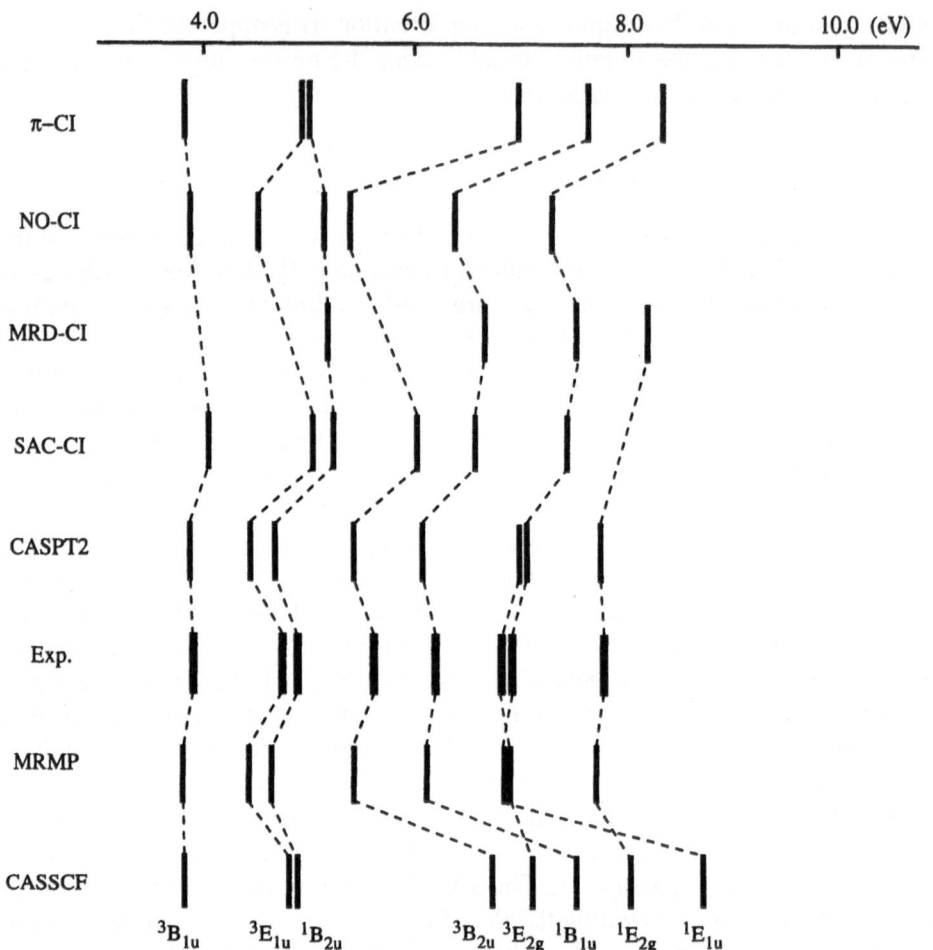

FIG. 2.1-4. Schematic summary of the calculated results of valence π–π^* excitation energies of benzene. π-CI, CI within π-space; NO-CI, CI with natural orbitals; MRD-CI, multireference CI with doubles; SAC-CI, CI with symmetry-adapted clusters; CASPT2, second-order perturbation based on complete active space based on SCF (CASSCF)

Palmer and Walker [35] and the most recent work by Yamamoto et al. [36]. The former predicted the line position for the singlet valence states with an accuracy of 0.61 eV. The latter is a single reference CI based on the composite natural orbitals with the extended basis set and gave results within 0.38 eV of the experimental values. CASPT2 results by Roos et al. [37] are also listed for comparison. CASPT2 achieved an accuracy of 0.26 eV or better. CASPT2 is the perturbation treatment based on the CASSCF reference function and is very similar to our MRMP method.

We observe that the CASSCF analysis tends to overestimate the excitation energies compared with the experimental values. Figure 2.1-4 shows that the largest errors are found in the states with the largest ionic character: $^1B_{1u}$, $^1E_{1u}$, and $^1B_{2u}$. For instance, the CASSCF excitation energy for the strongly ionic state, $^1E_{1u}$, is 9.37 eV, which is 2.4 eV too high compared with the experimental value of 6.94 eV. The CASSCF active space contains only π-orbitals and is not adequate for the description of the ionic states, so-called V states. Adequate description of V states can be obtained only when the σ-σ and σ-π interactions are correctly taken into account. These interactions can be included through the second-order correlation. For the covalent excited states, CASSCF transition energies are still more than 0.9 eV too large.

The MRMP theory corrects the deficiency and represents a great improvement over CASSCF. Inclusion of the dynamical correlation changes the situation dramatically, and the MRMP excitation energies become quite close to the experimental values both for ionic and covalent states. The calculated valence singlet π-π^* excitation energies with MRMP (experimental values in parentheses) are $^1B_{2u}$, 4.77 (4.90), $^1B_{1u}$, 6.28 (6.20), $^1E_{1u}$, 6.98 (6.94), and $^1E_{2g}$, 7.88 (7.80) eV, respectively. The calculated values of the excitation energy are within 0.13 eV of experimental values for all the singlet excited states. It is worth nothing that the second-order correlation effects are very important especially for the ionic states. Results of similar accuracy have been obtained for triplet excited states.

Calculated Rydberg excitation energies are listed in Table 2.1-6. Rydberg excitation energies are usually calculated to a reasonable accuracy even at a low level of theory if the basis set is chosen adequately. For Rydberg excitation energies, CASSCF shows a trend that is the reverse of that in the valence excitation energies. That is, CASSCF produces excitation energies that are too small compared with the experiment although the deviation is much smaller than that of the valence excitation energies. The addition of dynamical correlation effects through the perturbation treatment remedies this, and MRMP excitation energies follow fairly closely to the experiment. The MRMP Rydberg excitation energies agree with the experimental values to within 0.18 eV. The SAC/SAC–CI also reproduced the Rydberg excitation energies with an accuracy of 0.3 eV. The difference between the singlet and triplet excitation energies for each Rydberg state is expected to be small. The MRMP predicts that the triplet state is lower than the corresponding singlet state for all the Rydberg excited states listed in the table. MRMP theory has been applied to larger systems such as cyclopentadiene, pyrrole, furan, and naphthalene, and has proved to be very successful in predicting details of molecular electronic spectra [38–40].

In general, there are three types of excited states: covalent excited states, ionic excited states, and Rydberg-type excited states. The degree of nondynamical correlation effects is significantly different in these excited states. To get a balanced description for both the ground and excited states, all nondynamical correlation effects should be fully included in the wave function. The CASSCF overestimates the excitation energies for ionic and covalent excited states and underestimates those for Rydberg excited states, while the dynamical correlation effects increase in the order of ionic excited states, covalent excited states, the ground state, and Rydberg excited states. The MR-based CI and PT theories

TABLE 2.1-6. Rydberg excitation energies (eV) of benzene[a]

State	CASSCF	MRMP (error)	Exp.	SAC-CI
1^1E_{1g} (3s)	6.01	6.39 (0.06)	6.33	6.31
1^3E_{1g} (3s)	5.98	6.36		6.28
1^1A_{2u} (3pσ)	6.48	6.84 (−0.09)	6.93	6.88
1^3A_{2u} (3pσ)	6.45	6.81		6.82
1^1E_{2u} (3pσ)	6.54	6.92 (−0.03)	6.95	6.99
1^3E_{2u} (3pσ)	6.53	6.87		7.02
1^1A_{1u} (3pσ)	6.61	6.93		7.10
1^3A_{1u} (3pσ)	6.61	6.93		7.15
2^1E_{1u} (3pπ)	7.01	7.27 (−0.14)	7.41	6.91
1^1B_{1g} (3dσ)	7.16	7.51 (0.05)	7.46	7.42
1^3B_{1g} (3dσ)	7.15	7.51		7.43
2^3E_{1u} (3pπ)	7.00	7.27		6.89
1^1B_{2g} (3dσ)	7.00	7.53 (0.07)	7.46	7.42
1^3B_{2g} (3dσ)	6.99	7.53		7.42
2^1E_{1g} (3dδ)	6.99	7.56		7.44
2^3E_{1g} (3dδ)	6.99	7.55		7.42
3^1E_{1g} (3dδ)	7.05	7.61 (0.07)	7.54	7.35
3^3E_{1g} (3dδ)	7.04	7.61		7.33
2^1A_{1g} (3dπ)	7.26	7.62 (−0.18)	7.80	7.64
2^3A_{1g} (3dπ)	7.23	7.62		7.66
2^1E_{2g} (3dπ)	7.24	7.63 (−0.18)	7.81	7.64
2^3E_{2g} (3dπ)	7.23	7.61		7.71
1^1A_{2g} (3dπ)	7.27	7.66		7.57
1^3A_{2g} (3dπ)	7.27	7.66		7.57

[a] MRMP, from [33]; SAC-CI, from [34].

correct the CASSCF results by taking account of the dynamical correlation effects and give very accurate excitation energies for all types of excitations.

References

1. Urban M, Cernusak I, Kello V, Noga J (1987) Electron correlation in molecules. In: Wilson S (ed) Methods in computational chemistry, vol 1. Plenum, pp 117–250
2. Sinanoglu O (1961) Many-electron theory of atoms and molecules I. Shells, electron pairs vs many-electron correlations. J Chem Phys 36:706–717
3. Shavitt I (1977) The method of configuration interaction. In: Schaefer HF (ed) Methods of electronic structure theory, vol 3. Plenum, pp 189–275
4. Paldus J (1974) Group theoretical approach to the configuration interaction and perturbation theory calculations for atoms and molecular systems. J Chem Phys 61:5321–5330
5. Roos BO, Siegbahn PEM (1977) The direct configuration interaction method from molecular integrals. In: Schaefer HF (ed) Methods of electronic structure theory, vol 3. Plenum, pp 277–318

6. Ahlichs R, Scharf P, Ehrhardt C (1985) The coupled pair functional (CPF). A size-consistent modification of the CI (SD) based on an energy functional. J Chem Phys 82:890–898

7. Pople JA, Head-Gordon M, Raghavachari (1987) Quadratic configuration interaction. A general technique for determining electron correlation energies. J Chem Phys 87:5968–5975

8. Mayer W (1973) PNO–CI studies of electron correlation effects. I. Configuration expansion by means of nonorthogonal orbitals and application to the ground state and ionized states of methane. J Chem Phys 58:1017–1035

9. Duch W, Meller J (1994) On multireference superdirect configuration interaction in third order. Int J Quantum Chem 50:243–271

10. Kelly HP (1963) Correlation effects in atoms. Phys Rev 131:684–699

11. Frisch AMJ, Trucks GW, Head-Gordon M, Gill PMW, Wong MW, Foresman JB, Johnson BG, Schlegel HB, Robb MA, Replogle ES, Gomperts R, Andres JL, Raghavacari K, Binkley JS, Gonzalez C, Martin RL, Fox DJ, Defrees DJ, Baker J, Stewert JJP, Pople JA (1992) Gaussian 92. Gaussian, Pittsburgh

12. Hirao K (1992) Multireference Møller–Plesset method. Chem Phys Lett 190:374–380

13. Andersson K, Malmqvist P, Roos BO, Sadlej AJ, Wolinski K (1990) Second-order perturbation theory with a CASSCF reference function. J Phys Chem 94:5483–5488

14. Nakatsuji H, Hirao K (1978) Cluster expansion of the wavefunction. Symmetry-adapted cluster expansion, its variational determination, and extension to the open shell orbital theory. J Chem Phys 68:2053–2065

15. Purvis GD, Bartlett RJ (1982) A full coupled-cluster singles and doubles model: The inclusion of disconnected triples. J Chem Phys 76:1910–1918

16. Nakatsuji H (1978) Cluster expansion of the wavefunction. Excited states. Chem Phys Lett 59:362–364

17. Nakatsuji H (1979) Cluster expansion of the wavefunction. Electron correlations in ground and excited states by SAC (symmetry-adapted cluster) and SAC CI theories. Chem Phys Lett 67:329–333

18. Hirao K (1983) Direct cluster expansion method. Application to glyoxal. J Chem Phys 79:5000–5010

19. Hirao K (1991) The generalized symmetry adapted cluster theory. J Chem Phys 95:3589–3595

20. Saxe P, Schaefer HF, Handy N (1981) Exact solution (within a double-zeta basis set) of the Schrödinger electronic equation for water. Chem Phys Lett 79:202–204

21. Harrison RJ, Handy N (1983) Full CI calculations on BH, H_2O, NH_3 and HF. Chem Phys Lett 95:386–391

22. Hirao K, Hatano Y (1993) Cluster expansion of the wave function. Comparison with full CI results. Chem Phys Lett 100:519

23. Hirao K, Hatano Y (1993) Full CI and SAC CI calculations for ionized states, electron-attached states and triplet excited states of H_2O. Chem Phys Lett 111:533

24. Yamamoto S, Hirao K (1993) Molecular application of the generalized symmetry-adapted cluster theory. Chem Phys Lett 204:315–319

25. Laidig WD, Saxe P, Bartlett RJ (1987) The description of N_2 and F_2 potential energy surfaces using multireference coupled-cluster theory. J Chem Phys 86:887–907

26. Hirao K (1992) Multireference Møller–Plesset perturbation treatment of potential energy curve of N_2. Int J Quantum Chem Symp 26:517–526

27. Goddard JD, Yamaguchi Y, Schaefer HF (1981) Features of the H_2CO potential energy hypersurface pertinent to formaldehyde photodissociation. J Chem Phys 75:3459–3465
28. Frisch MJ, Kirtman R, Pople JA (1981) The lowest singlet potential surface of CH_2O. J Phys Chem 85:1467–1468
29. Dupuis M, Lester WA, Lengsfield BH, Liu B (1983) Formaldehyde: *Ab initio* MCSCF+ CI transition state for $H_2CO \rightarrow CO + H_2$ on the S_0 surface. J Chem Phys 79:6167–6173
30. Frisch MJ, Binkley JS, Schaefer HF (1984) *Ab initio* calculation of reaction energies. III. Basis-set dependence of relative energies on the FH_2 and H_2CO potential energy surfaces. J Chem Phys 81:1882–1893
31. Scuseria GE, Schaefer HF (1989) The photodissociation of formaldehyde: A coupled-cluster study including connected triple excitations of the transition state barrier height for $H_2CO \rightarrow H_2 + CO$. J Chem Phys 90:3629–3636
32. Chuang M-C, Foltz MF, Moore CB (1987) T_1 barrier height, S_1–T_1 intersystem crossing rate, and S_0 radical dissociation threshold for H_2CO, D_2CO and HDCO. J Chem Phys 87:3855–3864
33. Hirao K, Nakano H, Hashimoto T (1995) Multireference Møller–Plesset perturbation treatment for valence and Rydberg excited states of benzene. Chem Phys Lett 235:430–435
34. Kitao O, Nakatsuji H (1987) Cluster expansion of the wavefunction. Valence and Rydberg excitations and ionizations of benzene. J Chem Phys 87:1169–1182
35. Palmer MH, Walker IC (1989) The electronic states of benzene and the azines. I. The parent compound benzene. Correlation of vacuum UV and electron scattering data with *ab initio* CI studies. Chem Phys 133:113–121
36. Yamamoto Y, Noro T, Ohno K (1994) *Ab initio* CI calculations on benzene with an extended basis set. Int J Quantum Chem 51:27
37. Roos BO, Andersson K, Fulscher MP (1992) Towards an accurate molecular orbital theory for excited states: the benzene molecule. Chem Phys Lett 192:5–13
38. Tsuneda T, Nakano H, Hirao K (1995) Study of low-lying electronic states of ozone by multireference Møller–Plesset perturbation method. J Chem Phys 103:6520–6528
39. Nakano H, Tsuneda T, Hashimoto T, Hirao K (1996) Theoretical study of the excitation spectra of five-membered ring compounds: Cyclopentadiene, furan, and pyrrole. J Chem Phys 104:2312–2320
40. Hashimoto T, Nakano H, Hirao K (1996) Theoretical study of the valence $\pi \rightarrow \pi^*$ excited states of polyacenes: Benzene and naphthalene. J Chem Phys 104:6244–6258

2.2
Electrons in Molecular Liquids

SHIGEKI KATO

The development of theoretical methods to describe the electronic structures of molecules in the liquid phase is one of the most important issues of quantum chemistry because most chemical reactions are observed in solution. In this section, several topics are discussed concerning electrons in molecular liquids that are fundamental problems in describing chemical reactions in solution. First, intermolecular potentials are examined. It is shown that theoretical characterization of the energy components constituting the intermolecular potential is important to construct potential functions for use in simulation studies of molecular liquids. Second, the methods for calculating the electronic structures of solute molecules in solution are introduced. Several theoretical models based on the dielectric continuum approximation for solvents are compared. Further, a new method is discussed combining the *ab initio* methods for the electronic structures of solute molecules and the integral equation in the liquid state theory to obtain the distribution function of solvent molecules around the solute. Last, the structural and dynamic properties of solvated electrons in polar solvents are discussed.

2.2.1 Intermolecular Potential

Knowledge of intermolecular potentials is important in the theoretical understanding of dynamic and structural properties of molecular liquids. In particular, the analytical form of intermolecular potentials, the so-called potential functions, is required in computer simulation studies based on Monte Carlo (MC) and molecular dynamics (MD) techniques.

The origin of intermolecular interaction has been investigated from an early stage of quantum mechanics. It is convenient to use perturbation theories to analyze various components of the interaction energy. Murrel, et al. [1] obtained the energy expanded as a power series in the intermolecular potential U and the overlap integral S. The Hamiltonian is represented by

$$H = H_a(i) + H_b(j) + U \tag{2.2-1}$$

where $H_a(i)$ and $H_b(j)$ correspond to the Hamiltonians of isolated molecules A and B satisfying the Schrödinger equations,

$$H_a(i)A_r(i) = E_r A_r(i) \quad (r = 0,\ 1,\ 2,\ \ldots) \tag{2.2-2a}$$

$$H_b(j)B_s(j) = E_s B_s(j) \quad (s = 0,\ 1,\ 2,\ \ldots). \tag{2.2-2b}$$

The wave function of the composite system is written by

$$\Phi = \Psi_0 + \sum_t C_t \Psi_t \tag{2.2-3}$$

where Ψ_0 is the zero-order wave function

$$\Psi_0 = \mathscr{A} A_0(i)B_0(j) \tag{2.2-4}$$

and Ψ_t represents the excited configuration

$$\mathscr{A} A_r(i)B_s(j) \tag{2.2-5}$$

and the charge transfer configurations

$$\mathscr{A} A_k^+(i')B_l^-(j'), \qquad A_m^-(i')B_n^+(j'). \tag{2.2-6}$$

It is straightforward to derive the energy expression up to the order $U^2 S^2$ and the result is given by

$$\Delta E = E^{10} + E^{12} + E^{20} + E^{22} \tag{2.2-7}$$

where the first superscript refers to the power of U and the second to the power of S.

The first term in Eq. (2.2-7), E^{10}, is the electrostatic energy due to Coulomb forces between the two charged clouds. This can be evaluated by calculating the field acting on one molecule due to the other. In many cases, the field is approximated by multipole contributions such as the dipole–dipole and the dipole–quadrupole interaction terms. For systems with neutral charge, the dipole–dipole interaction is important to characterize the properties of the liquid, and the energy between two molecules separated by R is expressed by

$$E_{\text{dipole–dipole}} = \frac{\mu_a \mu_b \left\{ \sin\theta_a\ \sin\theta_b\ \cos(\phi_a - \phi_b) - 2\cos\theta_a\ \cos\theta_b \right\}}{R^3} \tag{2.2-8}$$

where μ_a and μ_b are the magnitudes of the dipole moments of molecules A and B, respectively, and the angles, (θ_a, ϕ_a) and (θ_b, ϕ_b), represent the directions of the dipole moments.

The second term, E^{12}, denotes the exchange energy. This arises because electrons must obey the Pauli exclusion principle. When the electron clouds of two molecules overlap, electrons with parallel spins avoid each other thus causing a repulsive interaction between the molecules. In the 12–6 Lennard–Jones potential, the exchange energy is approximated by the R^{-12} term.

E^{20} is the sum of induction and dispersion energies. The induction energy represents the attraction between the potential field of one molecule and the induced moment of the other molecule. Using the dipole moments, μ_a and μ_b, and polarizabilities, α_a and α_b, the induction energy is given by

$$E_{ind} = -\frac{\alpha_a \mu_b^2}{R^6} - \frac{\alpha_b \mu_a^2}{R^6}.$$

(2.2-9)

It is noted that the induction energy is also called the polarization energy because this term originates from the polarization of the electron cloud. The dispersion energy is due to the attractive force between the instantaneous polarizations of the electron clouds of two molecules. This term is of quantum origin and plays an important role for nonpolar molecules. London proposed an approximate formula to calculate the dispersion energy

$$E_{disp} = \frac{3}{2}\frac{I_a I_b}{I_a + I_b} \cdot \frac{\alpha_a \alpha_b}{R^6}$$

(2.2-10)

where I_a and I_b are the ionization potentials of molecules A and B.

The last term, E^{22}, is the sum of the charge transfer and exchange polarization energies. The charge transfer energy acts to stabilize the donor–acceptor complex. The contribution of exchange polarization energy is usually small compared to other terms.

Jeziorski and van Hemert [2] have calculated several components of the interaction energy, Eq. (2.2-7), for the water dimer $(H_2O)_2$ in a linear form. Figure 2.2-1 shows the variation of each term as a function of O–O distance, where the

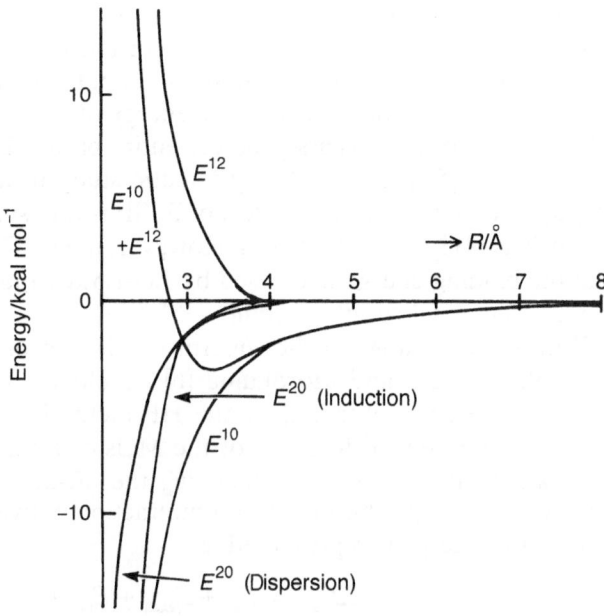

FIG. 2.2-1. Energy components for linear $(H_2O)_2$ as functions of O–O distance R

electrostatic, induction, and dispersion energies are attractive while the exchange energy is repulsive. It was also found that the sum of the electrostatic, exchange, and induction energies is very close to the self-consistent field (SCF) binding energy in the range of $R_{oo} > 2.6$ Å. The dispersion energy increased the binding energy by $2 \, \text{kcal} \, \text{mol}^{-1}$.

Although the calculations of intermolecular potentials based on perturbation theories provide physically meaningful insight in to the origin of intermolecular interactions, these require complicated procedures to evaluate each energy component because the molecular orbitals (MOs) of the two molecules are nonorthogonal and the convergence of the perturbation series is rather slow for strongly interacting systems. In this respect, variational calculations such as Hartree–Fock (HF) and configuration interaction (CI) methods are more practical where interacting molecules, A . . . B, are regarded as one molecular system, the so-called supermolecule. The interaction energy is thus obtained by

$$\Delta E = E(A \ldots B) - E(A) - E(B) \tag{2.2-11}$$

where $E(A \ldots B)$ is the energy of the supermolecule and $E(A)$ and $E(B)$ are of isolated monomers, respectively.

Morokuma and Pedersen [3] were the first to apply the HF method to predict the geometry of $(H_2O)_2$ and found that the linear configuration is most stable. This result was surprising at that time, but was confirmed by microwave experiments later. After the pioneering work by Morokuma and Pedersen, many *ab initio* calculations have been carried out on a variety of hydrogen-bonding systems, and the effect of the basis set and electron correlation on the geometries and interaction energies has been examined. It is well known that the basis set superposition error (BSSE) overestimates the binding energy and underestimates the equilibrium intermolecular distance when small basis sets are employed. The magnitude of the BSSE is estimated by the counterpoise correction, which is the difference between the energy of the isolated monomer and that in which each monomer is assigned the entire basis set of the complex [4]. Newton and Kestner [5] performed a systematic study to analyze the BSSE effect on $(H_2O)_2$. At the HF level, the effect of BSSE is very small when a rather large basis set such as $[4s3p1d/2s1p]$ is used. However, the O–O distance increased by 0.1 Å and the binding energy decreased by 1.0–$1.5 \, \text{kcal} \, \text{mol}^{-1}$ when the electron correlation effect is taken into account.

With the advantage of the supermolecule approach based on the variational principle, Kitaura and Morokuma [6] developed a method to decompose the interaction energy obtained by the HF method. In their procedure, the Fock matrix is first defined in terms of the MOs of isolated monomers and the SCF calculations are repeated by including the off-diagonal elements in a stepwise manner by considering the orbital interaction between the molecules. Thus the interaction energy is represented as

$$\Delta E_{SCF} = E_{es} + E_{exch} + E_{pol} + E_{ct} + E_{mix} \tag{2.2-12}$$

where E_{es} and E_{exch} are the electrostatic and exchange energies as in Eq. (2.2-7), and E_{pol} and E_{ct} are the polarizantion (induction) and charge transfer energies. Noted that E_{pol} and E_{ct} include higher than second-order terms because they are calculated by the SCF procedure. E_{mix} represents contributions that cannot be assigned to the first four terms of Eq. (2.2-12).

Figure 2.2-2 shows the variation of each component in Eq. (2.2-12) for a linear-form $(H_2O)_2$ dimer as a function of θ, the angle between the hydrogen-bond axis, O ... H—O, and the bisector of the proton acceptor's HOH angle. It is note-worthy that the θ-dependence of E_{es} is very similar to that of ΔE_{SCF}.

Analytical potential functions have been developed for use in computer simu-lations of liquids and solutions. For water, the SPC, ST2, and TIP4P models have been successfully applied to calculate the structural and thermodynamic proper-ties of liquid water and aqueous solutions. These model potentials are composed of the Coulombic interactions between all intermolecular pairs of charges along with a single Lennard–Jones potential between oxygen atoms. The orientation dependence of the intermolecular interaction is thus solely determined by the Coulombic interaction. This might be justified by the observation mentioned above that the angular dependence of the electrostatic term resembles that of the total interaction energy. For many organic solvents such as alcohols and nitriles, Jorgensen and Swenson [7] proposed simple analytical potential functions.

By referring to the component analysis of the interaction energy, Honda and Kitaura [8] proposed a new form of potential function. The interaction energy between molecules A and B is given as a sum of bond–bond (or lone-pair) and Coulombic interactions

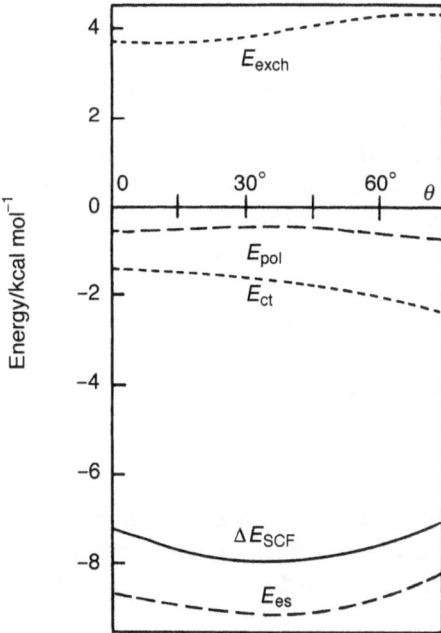

FIG. 2.2-2. Energy components for linear $(H_2O)_2$ as functions of angle θ at $R = 2.98$ Å. See Eq. (2.2-12) for E_{es}, E_{exch}, E_{pol} and E_{ct}

$$\Delta E = \sum_{I}^{A}\sum_{J}^{B} V_{IJ} + \sum_{I}^{A}\sum_{K}^{B} V_{IK} + \sum_{K}^{A}\sum_{L}^{B} V_{KL} + \sum_{r}^{A}\sum_{s}^{B} q_r q_s / R_{rs} \qquad (2.2\text{-}13)$$

where I and J refer to bonds and K and L to lone-pairs. q_r is the effective charge on nucleus r and R_{rs} the interatomic distance. The bond–bond interaction is represented by

$$V_{IJ} = C_{IJ} S_{ij}^2 + C'_{IJ} S_i *_j^2 + C''_{IJ} S_{ij}^2 * \qquad (2.2\text{-}14)$$

where S_{ij} is the overlap integral between the localized molecular orbitals (LMOs), and i and $i*$ denote the bonding and antibonding LMOs belonging to the bond I. In Eq. (2.2-14), the term including S_{ij} corresponds to the exchange energy and those with S_i*_j and $S_{ij}*$ to the charge transfer energy. The parameters C_{IJ}, C'_{IJ}, and C''_{IJ} were determined for H_2O, NH_3, and HF by the least-square fitting of interaction energies calculated by *ab initio* MO methods.

2.2.2 Electronic Structure of Solute Molecules in Solvents

In many simulation studies of solutions, the potential functions and the atomic effective charges on solute molecules are determined from the electronic structure calculations for the isolated solute molecule or solute–solvent 1:1 complex. Such an approach is valid only for systems in which the solute electronic structure is insensitive to the electric field from the solvent molecules. However, there are many systems in which the solute electronic structure undergoes large modification by the solvent, and it is important to carry out the calculations for solute molecules under the influence of the solvent field in describing these systems.

Several approaches incorporating the solvent effect in quantum chemical calculations of solute moleules have been proposed. One of the most widely used methods is to describe the solvent by a polarizable dielectric continuum medium. In this approach, the solute molecule is located in a vacuum cavity embedded in a continuum characterized by a dielectric constant ε, and the reaction field induced by the solute charge distribution influences the solute electronic structure.

The polarization of the dielectric media can be represented by the virtual charge density $\sigma(s)$ induced on the cavity surface S. The density $\sigma(s)$ can be determined from the total electrostatic potential

$$V(r) = V_\rho(r) + V_\sigma(r) \qquad (2.2\text{-}15)$$

where V_ρ and V_σ are the contributions from the solute charge distribution and the surface virtual charge distribution, respectively. Using the boundary conditions at the cavity surface

$$V(s)_{-} = V(s)_{+} \tag{2.2-16a}$$

$$\left(\frac{\partial V(s)}{\partial n}\right)_{-} = \varepsilon\left(\frac{\partial V(s)}{\partial n}\right)_{+} \tag{2.2-16b}$$

where the indexes − and + indicate points lying immediately inside and outside the cavity and n being normal to the surface. The virtual charge distribution is calculated by

$$\sigma = -\left[P(r)\cdot n(r)\right]_{+} = -\frac{\varepsilon-1}{4\pi}E_{n}(s)_{+} = -\frac{\varepsilon-1}{4\pi\varepsilon}E_{n}(s)_{-} \tag{2.2-17}$$

$P(r)$ is the solvent polarization and E_n the total electric field along n. The electronic Hamiltonian of the solute molecule is thus written as the sum of the Hamiltonians of the isolated solute molecule H_0 and the electrostatic potential arising from $\sigma(s)$

$$H = H_0 + V_\sigma(r) \tag{2.2-18}$$

and the free energy of the system is given by

$$G = \left\langle \Psi \left| H_0 + V_\sigma \right| \Psi \right\rangle - \frac{1}{2}\int_s \sigma(r)V_\rho(r)dr \tag{2.2-19}$$

where the wave function Ψ is for the Hamiltonian, Eq. (2.2-18). Miertus et al. [9] developed a practical method to calculate $\sigma(s)$ for a cavity with an arbitrary shape.

Although it takes an iterative procedure to calculate the surface charge $\sigma(s)$ and then the reaction field $V_\sigma(r)$ with the use of Eq. (2.2-17), simple expressions of the reaction field are available for a spherical cavity. Tapia and Goscinski [10] proposed the self-consistent reaction field (SCRF) method in which the solute molecule is regarded as a point dipole located at the center of the cavity and the reaction field due to the solute dipole moment is calculated using Onsager's model. The effective Hamiltonian for the solute molecule is written as

$$H = H_0 - {}^t\boldsymbol{\mu}\cdot\mathbf{g}\cdot\left\langle \Psi \left| \boldsymbol{\mu} \right| \Psi \right\rangle \tag{2.2-20}$$

where $\boldsymbol{\mu}$ is the total dipole moment operator and the superscript t means the transpose. The coupling tensor g is given by

$$g = \frac{2(\varepsilon-1)}{(2\varepsilon+1)a^3}\mathbf{1} \tag{2.2-21}$$

where a is the cavity radius and $\mathbf{1}$ is the unit tensor.

The SCRF method has been widely used as a convenient way to calculate the potential surfaces of polyatomic molecules in polar solvents because the derivatives of energy with respect to the nuclear coordinates can be readily obtained analytically. Another simple method for including the solvent effect in the electronic structures of solute molecules has been derived from the image charge model proposed by Friedman [11]. With this approximation, the free energy of the system is given by

$$G = \langle \Psi | H_0 | \Psi \rangle + \frac{1}{2} \int V_\rho(\mathbf{r}) \hat{I}_m \rho(\mathbf{r}) d\mathbf{r} + V_{ex} \qquad (2.2\text{-}22)$$

where $\hat{I}_m$ is the image operator which generates the image charge of solute charge density outside the cavity with radius a centered at N

$$\hat{I}_m \rho(\mathbf{r}) = -\frac{\varepsilon - 1}{\varepsilon + 1} \int d\mathbf{r}' \frac{a}{|\mathbf{r}' - N|} \cdot \frac{\rho(\mathbf{r}')}{\left| \mathbf{r}' - a^2(\mathbf{r}' - N)/|\mathbf{r}' - N| \right|^3} \qquad (2.2\text{-}23)$$

V_{ex} is the exclusion potential arising from the exchange interaction with the electrons of solvent molecules.

The image charge model has been used for the HF and MCSCF wave functions [12]. The analytical energy gradient method has also been developed [13]. Figure 2.2-3 shows the potential energy profiles for the S_N2 reaction, $Cl^- + CH_3Cl \rightarrow ClCH_3 + Cl^-$, both in the gas phase and aqueous solution. The difference between the C–Cl distances is chosen as the reaction coordinate. In the gas phase, the ion–dipole complex exists and the calculated energy of the transition state relative to this ion complex is 15.5 kcal mol^{-1}. In the solution phase, the complex disappears and both reactant and product with localized charge are stabilized due to the solvation. The barrier height is calculated to be 20.4 kcal mol^{-1}.

Although the dielectric continuum model has been utilized as a practical method to take account of the solvation effect in the electronic structure calculations of solute molecules, the use of the macroscopic dielectric constant is too crude to represent the microscopic solvation structure in the vicinity of solute molecules. The most serious problem has been the interpretation of the solute–solvent hydrogen bonding for proton donor solvents. There is also some ambiguity in defining the size and shape of cavity.

Several approaches have been proposed to include microscopic solvation structures. The most promising method is fully quantum chemical calculations on the systems including solute and solvent molecules. However, such a method is impractical to apply to a system including a large number of solvent molecules. Alternative approaches have been developed to combine the quantum chemical calculations for solute molecules with statistical mechanical calculations for solvent molecules employing molecular-mechanics-based models.

The integral equation of Ornstein and Zernike in the liquid state theory has proved to be a powerful tool in calculating the pair correlation function for liquids. For molecular liquids, the reference interaction site model (RISM) [14]

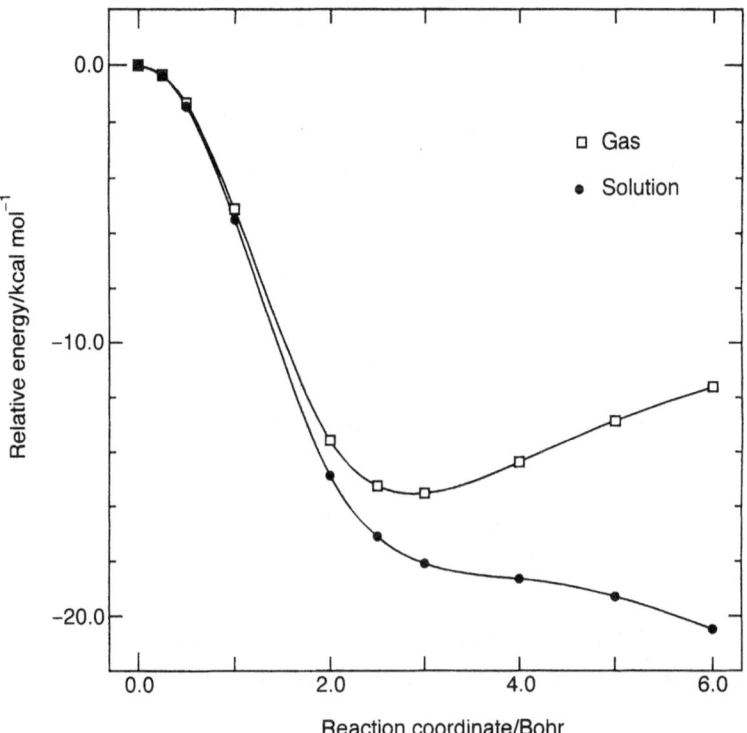

FIG. 2.2-3. Potential energy curves for $S_N 2$ reaction in the gas and solution phase

has been applied successfully to a variety of problems in solution chemistry. The hybrid approach combining *ab initio* MO theories with RISM enables us to calculate the electronic structure of solute molecules in solution and the statistical distribution of solvent molecules in a self-consistent manner [15]. The solute electronic Hamiltonian may be written as

$$H = H_0 + \sum_\lambda b_\lambda V_\lambda \qquad (2.2\text{-}24)$$

where b_λ is the population operator defining the effective charge q_λ on the site λ of the solute molecule. The electrostatic potential acting on the site λ is given by

$$V_\lambda = \rho \sum_a q_a \int_0^\infty 4\pi r^2 \frac{h_{\lambda a}(r)}{r} dr \qquad (2.2\text{-}25)$$

where $h_{\lambda a}$ is the total correlation function, q_α is the effective charge on the site α of the solvent molecule, and ρ is the number density of the solvent. The total correlation function is defined in terms of the RISM Ornstein–Zernike (OZ) equations for the solute–solvent system, in which a solute molecule is at infinite dilution in the solvent

$$h^{vv} = \omega^v * c^{vv} * \omega^v + \rho \omega^v * c^{vv} * h^{vv} \qquad (2.2\text{-}26a)$$

$$h^{uv} = \omega^u * c^{uv} * \omega^v + \rho \omega^u * c^{uv} * h^{vv} \qquad (2.2\text{-}26b)$$

where the asterisk denotes the spatial convolution integral, and h and c stand for the matrices whose elements are the site–site intermolecular pair correlation functions and the direct correlation functions. The indices u and v represent the quantities concerning the solute and solvent molecules, respectively. The intramolecular correlation matrix, ω, defines the molecular geometry in terms of distance constraints,

$$\omega_{\alpha\gamma}(r) = \delta_{\alpha\gamma}\delta(r) + (1 - \delta_{\alpha\gamma})s_{\alpha\gamma}(r) \qquad (2.2\text{-}27)$$

where $\delta_{\alpha\gamma}$ is a Kronecker delta and $s_{\alpha\gamma}$ is the distance between the site α and γ belonging to the solute or solvent molecule. In order to solve the OZ equations, Eq. (2.2-26), the closure relation connecting the total and direct correlation function is required. The hypernetted chain (HNC)-like relation

$$c_{\alpha\gamma}^{jv} = \exp\left(-\beta u_{\alpha\gamma}^{jv} + h_{\alpha\gamma}^{jv} - c_{\alpha\gamma}^{jv}\right) - \left(h_{\alpha\gamma}^{jv} - c_{\alpha\gamma}^{jv}\right) - 1 \qquad (2.2\text{-}28)$$

where $j = u, v$, has been utilized for polar solvents. The site–site intermolecular potential $u_{\alpha\gamma}^{jv}$ is expressed as a sum of short-range and long-range contributions,

$$u_{\alpha\gamma}(r) = u_{\alpha\gamma}^*(r) + \frac{q_\alpha q_\gamma}{r} \qquad (2.2\text{-}29)$$

where $u_{\alpha\gamma}^*(r)$ is the short-range part conveniently given by a Lennard–Jones-type potential.

With the use of total correlation functions obtained by solving the RISM OZ equation, the solvated Fock poerator is defined as

$$F_i^{\text{solv}} = F_i - f_i \sum_\lambda V_\lambda b_\lambda \qquad (2.2\text{-}30)$$

for the HF method. F_i is the Fock operator for the intramolecular contribution of the solute molecule and the population operator is related to the effective charge as

$$q_\lambda = q_\lambda^{(N)} - \sum_i f_i \langle \phi_i | b_\lambda | \phi_i \rangle \qquad (2.2\text{-}31)$$

where $q_\lambda^{(N)}$ is the nuclear charge and f_i is the occupation number of orbital ϕ_i. Although there are several ways to define the population operator, the least-square relation provides a simple but reliable procedure to generate the effective charge, q_λ. In this procedure, the electrostatic potential $V_\rho(r)$ arising from the solute charge distribution is fitted to the Coulombic potential, $\Sigma_\lambda q_\lambda / |r - R_\lambda|$.

The RISM–SCF method was applied to examine the solvation effect on absorption spectra of carbonyl compounds, formaldehyde, acetaldehyde, and acetone in aqueous solution [16]. For the singlet $n \to \pi^*$ excitation, formaldehyde shows a blue shift of 0.16 eV in water. This shift increases with an increase in the number of methyl groups, mainly due to the inductive effect of the electron-donating nature of the methyl group. Figure 2.2-4 shows the carbonyl(O)–water(H) and carbonyl(O)–water(O) pair correlation functions in the electronic ground state. The first peak at around 1.7 Å in the O–H correlation functions grows as the number of methyl groups increases, indicating that the hydrogen bonding becomes stronger with the substitution of the methyl group. The pair correlation functions in O–O are almost invariant, which indicates the dominance of the O–O contact in these functions. However, some development of the structure can be observed in the second peak. This is due to the propagation of the enhanced hydrogen bond around the carbonyl oxygen through the water–water hydrogen bond.

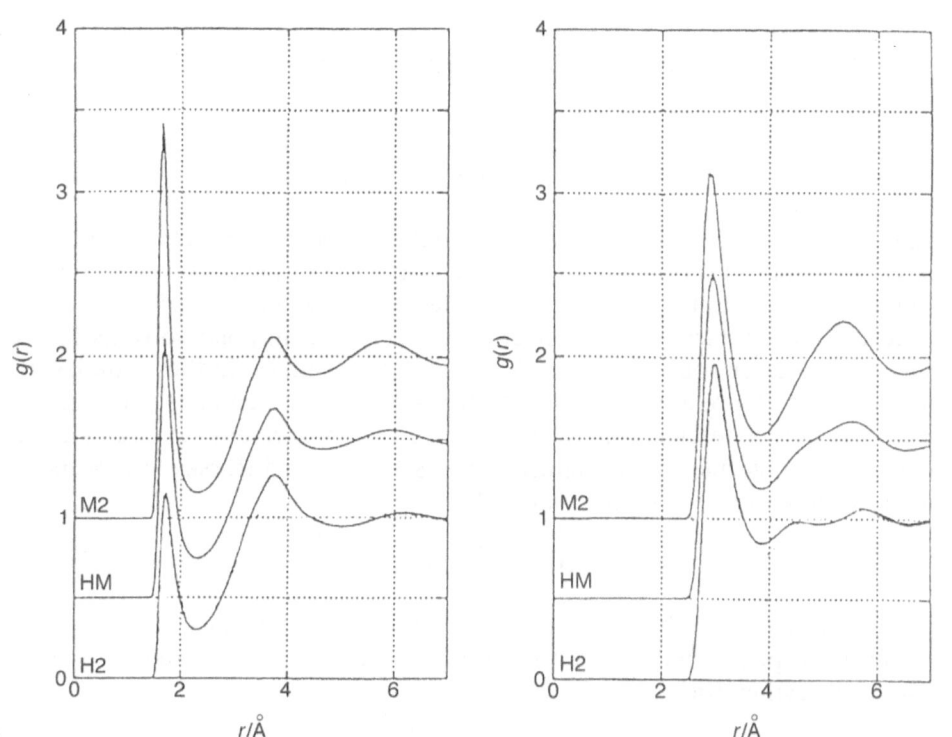

FIG. 2.2-4. Carbonyl-water pair correlation functions. **a** Carbonyl(O)-water(H); **b** carbonyl(O)-water(O). *M2*, *HM*, and *H2* denote acetone, acetaldehyde, and formaldehyde, respectively

2.2.3 Solvated Electrons

The nature of the excess electrons in polar solvents has been the subject of intense theoretical study in the past 50 years. The cavity models for electrons in liquids were first developed to account for the spectroscopic and thermodynamic data. *Ab initio* MO calculations were later carried out to take account of the importance of solvent molecules in the first solvation shell. Newton [17] proposed a semi-continuum model in which the cluster containing the excess electrons and the first solvation shell molecules are embedded in a polarizable idelectric continuum. The energy of the system is given by

$$E = E_0 + U \tag{2.2-32}$$

where E_0 is the electronic energy of the cluster and U is the classical polarization energy of a dielectric continuum approximated by

$$U = \frac{1}{2}\left(1 - \frac{1}{\varepsilon}\right)\int\int \rho(r_1) g(r_1, r_2) \rho(r_2) dr_1 dr_2 \tag{2.2-33}$$

In Eq. (2.2-33), ρ is the charge density of cluster and ε is the dielectric constant. The truncated Coulomb interaction associated with the finite cavity is defined as

$$g(1,2) = -\frac{1}{\max(r_1, r_2, r_c)} \tag{2.2-34}$$

where r_c is the cavity radius. By employing HF method, this model was applied to excess electrons in water and ammonia and the role of the long-range polarization of the media in trapping the excess electrons was emphasized.

Recently the relaxation process of excess electrons in polar solvents has received much attention in quantum MD simulation studies. In these studies, the development of a relatively simple electron–molecule potential is necessary because the complete electron treatment for a system including excess electrons and solvent molecules remains intractable. Schnitker and Rossky [18] proposed an electron–water pseudopotential in the form

$$V(r) = V_s(r) + V_{pl}(r) + V_R(r) + V_E(r). \tag{2.2-35}$$

The first term is a purely electrostatic contribution in which the effective charges of the single point charge (SPC) water model are taken. The second term is a spherically symmetric polarization term,

$$V_{pl}(r) = (\alpha_0/2r^4)\left\{1 - \exp\left[-(r/r_c)^6\right]\right\} \tag{2.2-36}$$

where α_0 is the isotropic part of the molecular polarizability and r_0 is the cutoff distance. The third term is an effective repulsive term introduced to account for the orthogonality between the wave function of excess electron and those of water molecules. This potential is represented by

$$V_R(r) = \sum_i \sum_{j\in i} B_j r^{n_j} \exp(-\rho_j r)$$ (2.2-37)

where i denotes the nuclear centers and j denotes the basis functions centered at the nuclei i. The last term is a local exchange potential derived from a free-electrongas approximation and is usually omitted because its contribution is quite small compared to those of the other terms.

The Feynman path integral technique was applied to study the structure of solvated electrons in water at room temperature [19, 20]. The electron–water pseudopotentials were used in these studies. It was found that there is no sharp geometrical coordination number for water and that the pair correlation functions of solvent molecules in the vicinity of the electrons are largely modified from those in the bulk.

Nonadiabatic quantum dynamic simulations have been performed to explore the relaxation of excess electrons initially prepared in the bulk water [21]. Figure 2.2-5 shows the dynamic history of the adiabatic eigenstates for two typical trajectories. One is an example to show a rapid relaxation to the ground state through the excited states. The time to reach the ground state was calculated to be in the range of 50–150 fs. Another example shows that the electron is

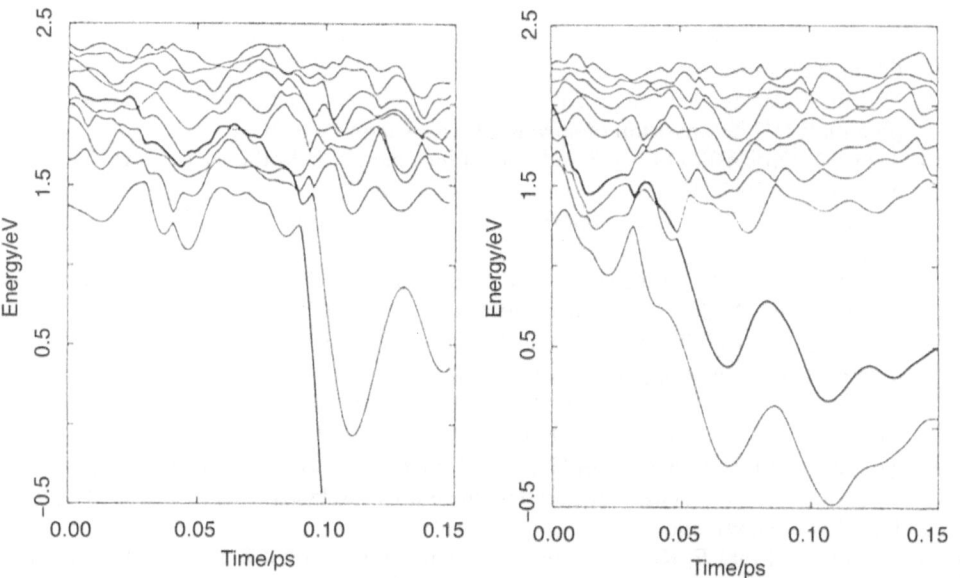

FIG. 2.2-5. Dynamical history of adiabatic energies of solvated electrons for a trajectory, **a** ground state, **b** excited state. The *bold line* denotes the occupied eigenstate. (From [21], with permission)

characterized by a well-defined isolated excited state and the time scale for the appearance of this state is comparable to the complementary relaxation time to reach the ground state. The lifetime of the isolated excited state was estimated to be about 1 ps. These studies reveal that a solvated electron is formed at the site of a preexisting void in the solvent.

References

1. Murrell JN, Randic M, Williams DR (1964) The theory of intermolecular forces in the region of small orbital overlap. Proc Roy Soc (London) A284:566–581
2. Jeziorsk B, van Hemert M (1976) Variation–perturbation treatment of the hydrogen bond between water moleculaes. Mol Phys 31:713–729
3. Morokuma K, Pedersen LG (1968) Molecular–orbital studies of hydrogen bonds. An *ab initio* calculation for dimeric H_2O. J Chem Phys 48:3275–3282
4. Boys SF, Bernardi F (1970) The calculation of small molecular interaction by the differences of separate total emergies. Some procedures with reduced errors. Mol Phys 19:553–566
5. Newton MD, Kestner NR (1983) The water dimer. Theory versus experiment. Chem Phys Lett 94:198–201
6. Kitaura K, Morokuma K (1976) A new energy decomposition scheme of molecular interactions within the Hartree–Fock approximation. Int J Quantum Chem 10:325–340
7. Jorgensen WL, Swenson CJ (1985) Optimized intermolecular potential functions for amides and peptides. Structure and properties of liquid amides. J Am Chem Soc 107:569–578
8. Honda K, Kitaura K (1987) A new form for intermolecular potential energy functions. Chem Phys Lett 140:53–56
9. Miertus S, Scrocco E, Tomasi J (1981) Electrostatic interaction of *ab initio* molecular potentials for the prevision of solvent effects. Chem Phys 55:117–129
10. Tapia O, Goscinski O (1975) Self-consistent reaction field theory of solvent effects. Mol Phys 29:1653–1661
11. Friedman MI (1975) Image approximation to the reaction field. Mol Phys 29:1533–1543
12. Karlström G (1989) Electronic Structure of HF^- and HCl^- in condensed phases studied by a CASSCF dielectric cavity model. J Phys Chem 93:4952–4955
13. Sato S, Kato S (1994) Potential surfaces of chemical reactions in solution by the dielectric continuum method. J Mol Struct 310:67–75
14. Chandler D, Andersen HC (1972) Optimized cluster expansions for classical fluids. II. Theory of molecular liquids. J Chem Phys 57:1930–1937
15. Ten-no S, Hirata F, Kato S (1993) A hybrid approach for the solvent effect on the electronic structure of a solute based on the RISM and Hartree-Fock equations. Chem Phys Lett 214:391–396
16. Ten-no S, Hirata F, Kato S (1994) Reference interaction site model self-consistent study for solvation effect on carbonyl compounds in aqueous solution. J Chem Phys 100:7443–7453
16. Newton MD (1975) The role of *ab initio* calculations in elucidating properties of hydrated and ammoniated electrons. J Phys Chem 79:2795–2808

18. Schnitker J, Rossky PJ (1987) An electron–water pseudopotential for condensed phase simulation. J Chem Phys 86:3462–3470

19. Schnitker J, Rossky PJ (1987) Quantum simulation study of the hydrated electron. J Chem Phys 86:3471–3486

20. Wallqvist A, Thirumalai D, Berne BJ (1987) Path integral Monte Carlo study of the hydrated electron. J Chem Phys 86:6404–6418

21. Webster FJ, Schnitker J, Friedrichs MS, Friesner RA, Rossky PJ (1991) Solvation dynamics of the hydrated electron. A nonadiabatic quantum simulation. Phys Rev Lett 66:3172–3175

2.3
Electrons in Molecular Solids

KEIICHIRO NASU

As is well known, there are various types of molecules in the world. However, it is also well known that when molecules are condensed macroscopically, they can make even more numerous varieties of crystals with periodic lattice structures. In this case, the electrons associated with each molecule can move all over the crystall and result in various new macroscopic characteristics. In some cases, these new characteristies can be directly attributed to the nature of a single constituent molecule. However, we often encounter entirely new characteristics that can scarcely be imagined from the original properties of each isolated molecule, and these are attributed to the macroscopic condensation itself. In this section, we will be mainly concerned with theories explaining these new characteristics exhibited by electrons in various molecular crystals.

2.3.1 Excitons in Molecular Solids

2.3.1.1 Frenkel, Wannier, and Charge Transfer Excitons

Let us start from insulating molecular crystals. There are many molecules that have a wide energy gap between their highest occupied molecular orbital (HOMO) and their lowest unoccupied molecular orbital (LUMO). When these wide-gap molecules have been condensed macroscopically, the resultant crystals become well-defined insulators or semiconductors with a wide energy gap.

In the case of these insulating molecular crystals, the most basic and important properties are their optical characteristics, especially the nature of their lowest optical excitation. When a photon with an energy $h\nu$ is absorbed by a crystal, an electron in the HOMO goes up to the LUMO, as schematically show in Fig. 2.3-1. An electron vacancy created in the HOMO is usually called a hole, and it has a positive charge +e relative to the unexcited molecules. This hole and the excited electron, thus created by the photon, can move from one molecule to another through the Huckel-type resonance transfer energy between molecular orbitals. Thus, they finally become two plane wave states extending over the crystal, being quite independent of each other.

Frenkel exciton

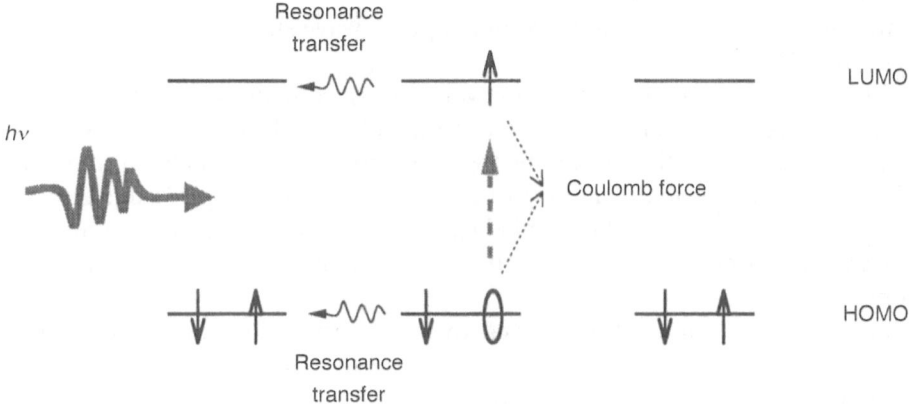

Charge transfer exciton

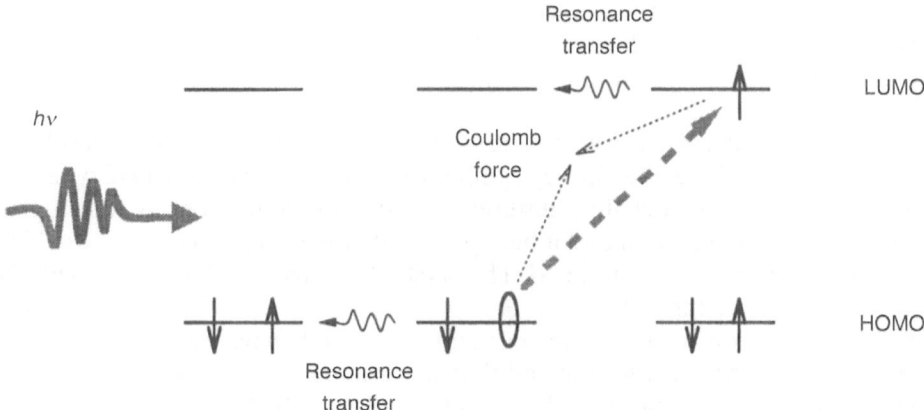

FIG. 2.3-1. The Frenkel and charge transfer (CT) excitons. *Small arrows* correspond to electrons and *circles* correspond to holes. Thick-dashed arrows correspond to excitations of electrons by lights $h\nu$

The interelectron Coulombic interaction greatly modulates this independent motion. The excited electron and the hole attract each other through the Coulombic force, and their relative motion falls into a bound state, keeping their center of mass motion still in a stationary plane wave state. This is usually called an exciton [1].

When this Coulombic attraction is very strong compared with the resonance transfer energy, the excited electron and the hole are always confined within the same molecule, and they itinerate together from one molecule to another as schematically shown in the upper part of Fig. 2.3-1. This strongly bound electron–hole pair is usually called the Frenkel exciton or the intramolecular exciton, and its essential properties are almost the same as that of the intramolecular electronic excitation.

Since the Coulombic attraction acts over a relatively long range, the hole can form a bound state with an electron not only in a same molecule but also in neighboring molecules as shown in the lower part of Fig. 2.3-1. This bound state of a distant electron–hole is usually called the Wannier exciton. Since this excitation is nothing but a charge transfer excitation between two molecules in the crystal, it is also called the charge transfer (CT) exciton. The appearance of this CT exciton is the result of the condensation itself, and is not attributable to a single molecule.

2.3.1.2 Excitons in C_{60} Crystals

These Frenkel and CT excitons always coexist and appear in the light absorption spectrum of insulating molecular crystals as characteristic peaks. Let us consider this coexistence problem in much more detail, taking a C_{60} crystal as a typical example.

The C_{60} molecule, discovered by Kroto et al. [2] in 1985, is almost like a sphere as shown in Fig. 2.3-2, but has I_h symmetry. This molecule has two types of intercarbon bonds with slightly different lengths; the longer one is the single-bond shared by a pentagon and a hexagon, while the shorter one is the double-bond shared by two hexagons [3, 4]. The crystal has a face-centered cubic (FCC) structure at room temperature.

Optical properties of C_{60} molecules and C_{60} crystals have been studied by a number of researchers, who showed that three main peaks appear in the light absorption spectrum of both, as shown in Fig. 2.3-3 [5–7]. They are all located in the region above 3 eV, and are very similar in the molecule and the crystal. This is consistent with the fact that the C_{60} crystal is a well-defined molecular crystal. In the region below 3 eV, a new peak appears only in the case of the crystal, which is not found in the spectrum of the molecule. This difference seems to be very important in clarifying the nature of excited states in the C_{60} system.

Keeping these points in mind, let us consider the C_{60} crystal theoretically. Using an extended Hubbard model for this π-electron system [8], we will explain a theory for calculating the light absorption spectrum, and will show how the Frenkel and CT excitons appear in this spectrum.

Fig. 2.3-2. A C_{60} molecule

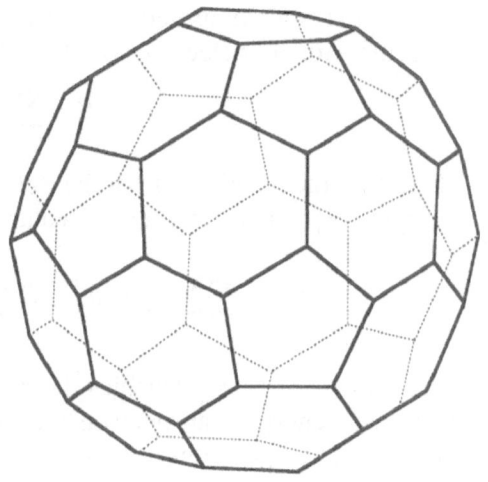

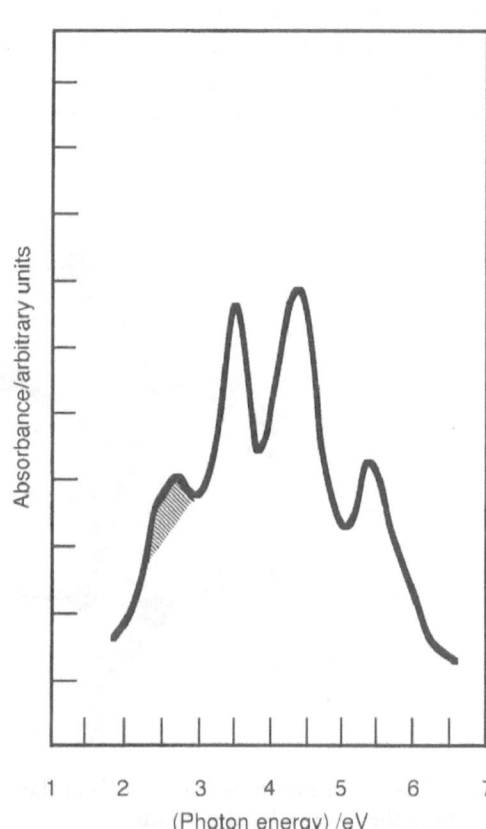

Fig. 2.3-3. The light absorption spectrum
of the FCC-type C_{60} crystal. The *shaded
area* is a peak that is not present in the
spectrum of a C_{60} molecule schematic

Take a crystal composed of N lattice sites ($N \gg 1$), each occupied by a C_{60} molecule with 60 π-electrons. Its Hamiltonian, H, is defined as

$$H \equiv \sum_{\ell} h_{\ell}^m + H_{\text{inter}}, \qquad (2.3\text{-}1)$$

where h_{ℓ}^m is the Hamiltonian of the isolated molecule at a lattice site $\ell (= 1, 2, \ldots,$ N), while H_{inter} denotes intermolecular interaction. h_{ℓ}^m is defined as

$$h_{\ell}^m = -\sum_{\langle i,j \rangle} \sum_{\sigma} T_{i,j} a_{\ell i \sigma}^+ a_{\ell j \sigma} + U \sum_i n_{\ell i \alpha} n_{\ell i \beta} + \frac{V_0}{2} \sum_{\langle i,j \rangle} \sum_{\sigma, \sigma'} n_{\ell i \sigma} n_{\ell i \sigma'}. \qquad (2.3\text{-}2)$$

Here, $i (= 1, 2, \ldots, 60)$ specifies the π-orbitals within a molecule, and $T_{i,j} (>0)$ is the transfer energy of an electron between two neighboring π-orbitals in the same molecule. Their positions are specified by indices i and j. The symbol $\sum_{\langle i,j \rangle}$ denotes that the summation over i and j should be taken only between two neighboring π-orbitals. As mentioned before, there are two kinds of bonds in the molecule. The value of $T_{i,j}$ for the long bond is taken to be T_0, while the short one is T_1. $a_{\ell i \sigma}^+$ denotes the creation operator of an electron at a π-orbital specified by (ℓ, i) and spin $\sigma (= \alpha, \beta)$. The molecules are arranged at each FCC lattice site. The orientation of the three-fold symmetry axis of each molecule is assumed to be the same as proposed by David et al. [9]. This arrangement leads to a pentagon of one molecule facing a double-bond of its neighboring molecule as shown in Fig. 2.3-4. U in Eq. (2.3-2) denotes the intraorbital Coulombic repulsive energy, and $n_{\ell i \sigma} \equiv a_{\ell i \sigma}^+ a_{\ell i \sigma}$. V_0 is the repulsive energy between the neighboring orbitals.

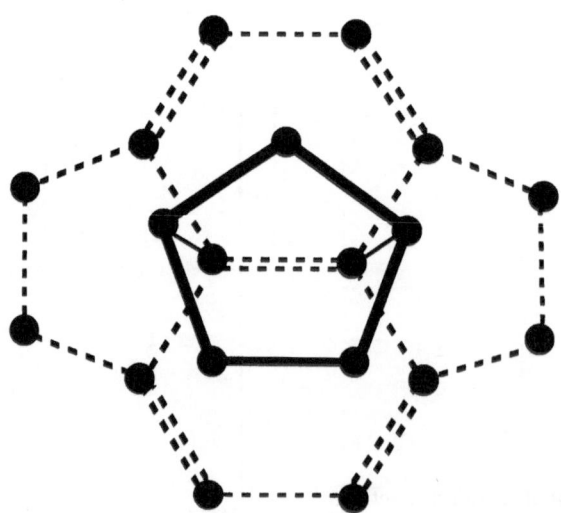

FIG. 2.3-4. The intermolecular bonds of the FCC-type C_{60} crystal. The two *thin solid lines* denote the intermolecular bonds

H_{inter} is defined as

$$H_{inter} \equiv -T_2 \sum_{[\ell i, \ell' j]} \sum_{\sigma} a^+_{\ell i \sigma} a_{\ell' j \sigma} + \frac{V_1}{2} \sum_{[\ell i, \ell' j]} \sum_{\sigma, \sigma'} n_{\ell i \sigma} n_{\ell' i \sigma'}. \tag{2.3-3}$$

Here, T_2 (>0) is the intermolecular transfer energy of an electron. This transfer energy is assumed to be nonzero only between the two nearest π-orbitals as shown in Fig. 2.3-4. Consequently, a molecule has 24 intermolecular bonds. The symbol $\Sigma_{[\ell i, \ell' j]}$ denotes the conditional summation over these intermolecular bonds between ℓi and $\ell' j$. V_1 is the intermolecular Coulombic repulsive energy between ℓi and $\ell' j$, which is assumed not to be zero only in the bonds mentioned above.

Using this model Hamiltonian H, we first determine the ground state of a single molecule h^m_ℓ within the Hartree–Fock (HF) approximation. In this approximation, we can decouple the Coulombic repulsions between two electrons, and h^m_ℓ is reduced to the HF Hamiltonian ($\equiv h^m_{HF}$) which has only single-body terms as

$$h^m_\ell \rightarrow h^m_{HF} = -\sum_{\langle i,j \rangle} \sum_{\sigma} t_{i,j} a^+_{i\sigma} a_{j\sigma} + u \sum_i \left\{ \langle n_{i\alpha} \rangle n_{i\beta} + \langle n_{i\beta} \rangle n_{i\alpha} - \langle n_{i\alpha} \rangle \langle n_{i\beta} \rangle \right\}$$

$$+ \frac{V_0}{2} \sum_{\langle i,j \rangle} \sum_{\sigma, \sigma'} \left\{ \langle n_{i\sigma} \rangle n_{i\sigma'} + \langle n_{i\sigma'} \rangle n_{i\sigma} - \langle n_{i\sigma} \rangle \langle n_{i\sigma'} \rangle \right\}$$

$$- \frac{V_0}{2} \sum_{\langle i,j \rangle} \sum_{\sigma} \left\{ m_{ij\sigma} a^+_{j\sigma} a_{i\sigma} + m_{ij\sigma} a^+_{j\sigma} a_{i\sigma} - m_{ij\sigma} m_{ji\sigma} \right\}. \tag{2.3-4}$$

Here, the site index l is omitted because there is only one molecule. $m_{ij\sigma}$ is defined as $m_{ij\sigma} \equiv \langle a^+_{i\sigma} a_{j\sigma} \rangle$, and <...> denotes the average with respect to the HF ground state($\equiv |g>$):

$$\langle \ldots \rangle \equiv \langle g | \ldots | g \rangle. \tag{2.3-5}$$

At this stage, however, the ground state is not known, but it will be determined self-consistently.

In the HF ground state, it is natural to assume

$$\langle n_{i\alpha} \rangle = \langle n_{i\beta} \rangle = 1/2. \tag{2.3-6}$$

That is, in the ground state $|g>$, all the carbon atoms are neutral with no spin. In this case, the Hubbard-type repulsion U in Eq. (2.3-2) becomes irrelevant. To determine single-body eigenvalues and eigenfunctions of h^m_{HF}, we introduce the following new operator, $b^+_{\lambda\sigma}$:

$$b^+_{\lambda\sigma} \equiv \sum_i f_{\lambda\sigma}(i) a^+_{i\sigma}, \tag{2.3-7}$$

so that it can diagonalize h^m_{HF} as

$$h_{\mathrm{HF}}^m \equiv \sum_{\lambda\sigma} \varepsilon_{\lambda\sigma} b_{\lambda\sigma}^+ b_{\lambda\sigma} + \text{constant terms.} \qquad (2.3\text{-}8)$$

Here, $\varepsilon_{\lambda\sigma}$ is the λ th eigenvalue of h_{HF}^m, and $f_{\lambda\sigma}(i)$ is its eigenfunction. Eigenvalues $\varepsilon_{\lambda\sigma}$ are numbered according to their energies from lower to higher, $\lambda = 1, 2, \ldots$, 60. Using this eigenfunction, the ground state of a single molecule is now determined as

$$|g\rangle = \prod_{\lambda}^{\mathrm{occ}} b_{\lambda\alpha}^+ b_{\lambda\beta}^+ |0\rangle, \quad |0\rangle \equiv \text{true electron vacuum.} \qquad (2.3\text{-}9)$$

The single-body states with $\lambda = 1$ to 30 are all occupied states, while the states $\lambda = 31$ to 60 are unoccupied.

As mentioned before, the C_{60} crystal is a molecular crystal in which the original characteristics of the molecule are well peserved. Therefore we assume that H_{inter} is small compared with h_{ℓ}^m. We also assume that the ground state ($\equiv |G\rangle$) of the whole crystal is the direct product of $|g\rangle$'s of each molecule. Thus, the ground state $|G\rangle$ can be written as

$$|G\rangle = \prod_{\ell}\prod_{\lambda}^{\mathrm{occ}} b_{\ell\lambda\alpha}^+ b_{\ell\lambda\beta}^+ |0\rangle, \qquad (2.3\text{-}10)$$

where

$$b_{\ell\lambda\sigma}^+ \equiv \sum_i f_{\lambda\sigma}(i) a_{\ell i\sigma}^+. \qquad (2.3\text{-}11)$$

Using this ground state $|G\rangle$, we can now describe excited states with an electron and a hole. We assume that the wave function of the relative motion of the electron and the hole is nonzero only when they are in two neighboring molecules or within the same molecule. Since we are going to calculate the light absorption spectrum, only the excited states whose total momentum ($\equiv k$) is zero are relevant. For this reason, we take the following excited state as our base

$$|\nu\,\mu;\Delta\ell\rangle \equiv \frac{1}{\sqrt{N}} \sum_{\ell} e^{-ik\cdot\ell} b_{\ell+\Delta\ell,\nu,\sigma}^+ b_{\ell\mu\sigma} |G\rangle, \quad k = 0, \qquad (2.3\text{-}12)$$

where ν and μ specifies the unoccupied and occupied molecular states, respectively, that is, $\nu = 31, 32, \ldots, 60$ and $\mu = 1, 2, \ldots, 30$. In this equation, $\Delta\ell$ means the electron–hole distance in the scale of the lattice constant.

In practical calculations, we first rewrite the total Hamiltonian of crystal H in terms of $b_{\ell i\sigma}^+$ and $b_{\ell i\sigma}$. Then we calculate the following matrix element between the excited states

$$\langle \nu\,\mu;\Delta\ell | H | \nu'\,\mu';\Delta\ell' \rangle. \qquad (2.3\text{-}13)$$

Finally, we diagonalize this energy matrix and obtain final eigenvalues and eigenfunctions of the excited states ($\equiv |\zeta\rangle$; $\zeta = 1, 2, \ldots$).

To calculate the absorption spectral shape, we introduce a polarization operator P

$$P \equiv iT' \sum_i \sum_{\langle i,j \rangle} \sum_\sigma \left(e_{ij} \cdot p\right)\left\{a_{li\sigma}^+ a_{lj\sigma} - a_{li\sigma}^+ a_{lj\sigma}\right\}$$

$$+ iT'' \sum_{[li,l'j]} \sum_\sigma \left(e_{ij}^{ll'} \cdot p\right)\left\{a_{li\sigma}^+ a_{l'j\sigma} - a_{l'j\sigma}^+ a_{li\sigma}\right\}. \qquad (2.3\text{-}14)$$

Here, T' is the dipole matrix element of the intramolecular transition, which is assumed to be nonzero only between neighboring orbitals i and j. e_{ij} denotes a unit vector from i to j, and p is the unit vector of the polarization of light. T'' is the dipole matrix element of the intermolecular transition. $e_{ij}^{ll'}$ denotes the unit vector from li to $l'j$. The ratio T''/T' is assumed to be equal to T_2/T_0. After calculating

$$\left|\langle \zeta | P | G \rangle\right|^2 \Big/ \left(\text{photon energy}\right), \qquad (2.3\text{-}15)$$

we can finally get the relative intensity of the optical absorption. The value of T_0 is taken to be 2.3 eV, and the value of T_1/T_0 is taken o be 1.1 from the Harrison relation [10].

In Fig. 2.3-5, we have shown the spectral shape calculated using $T_2/T_0 = 0.2$, $V_0/T_0 = 0.4$, and $V_1/T_0 = 0.17$. We can see three large peaks in the region above 3 eV.

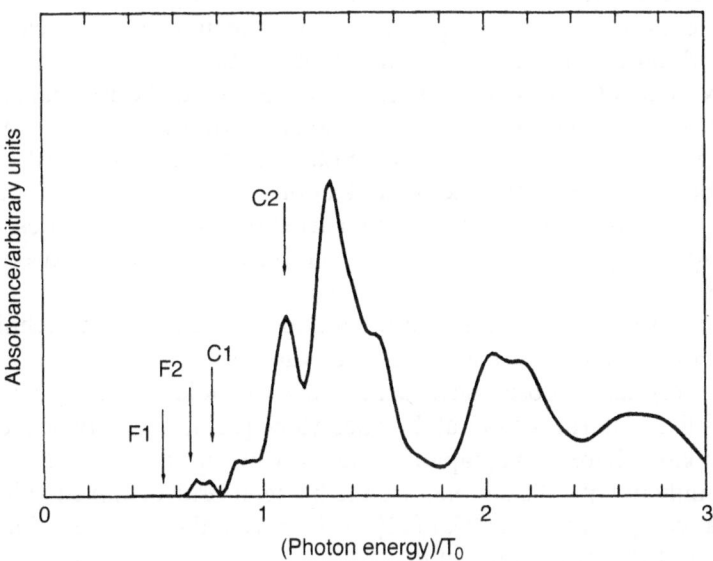

FIG. 2.3-5. The calculated light absorption spectrum of the C_{60} crystal. $T_0 = 2.3$ eV. F1, F2, C1, and C2 are explained in the text

These three correspond to the three large peaks in Fig. 2.3-3, although the calculated peak energies differ a little from the experimental values. These three calculated peaks are mainly due to the intramolecular transitions. We can also see another strong peak in the region labeled C2. This peak corresponds to the peak at 2.5~2.8 eV in Fig. 2.3-3. We can also calculate the absorption spectrum of a single C_{60} molecule, and in this case the lower energy peak is absent, although the three high-energy peaks are present [8].

In connection with the difference between the Frenkel exciton and the CT one mentioned before, the density distribution of the excited electron seen from the hole is very important. For this reason, we have analyzed the excited state eigenfunctions of F1, F2, C1, and C2 marked in Fig. 2.3-5. The calculated probabilities that the electron and the hole are in the same molecule are 88% and 17.7% in the case of F1 and F2, respectively, while the values for C1 and C2 are 5.5% and 3.3%. Therefore, we can now conclude that the experimentally observed peak [6] at around 2.5~2.8 eV is mainly due to the CT exciton (C2).

We have now seen that the C_{60} crystal is a trpical example in which both the Frenkel excitons and the CT excitons coexist.

2.3.2 Peierls Transitions and Solitons in Molecular Solids

2.3.2.1 Peierls Instability

Let us now proceed to the problems of atoms or molecules that have an unpaired electron in their outermost molecular orbital. When these atoms or molecules are condensed as a crystal, we can expect various new properties to arise from the unpaired electrons. As mentioned before, an unpaired electron can itinerate from one orbital to another though the hybridizations between these orbitals. The matrix element due to this hybridization is called the resonance transfer energy between orbitals. Through this resonance transfer energy, an electron finally becomes a plane wave state extending over the crystal with a definite wave-number ($\equiv k$), ($\hbar = 1$). In order to look at this situation in detail, let us restrict ourselves to the case of a one-dimensional array of N ($\gg 1$) molecules, assuming that the resonance tranfer energy ($\equiv -T$) is nonzero only between two neighboring orbitals.

Recent progress, in chemical synthesis and characterization techniques for molecule-based materials, has produced various new kinds of quasi-one-dimensional crystals, whose constituent atoms or molecules have an unpaired electron in their outermost orbital. Polyacetylene [11] is one of typical examples, and in this case there is an unpaired electron in the π-orbital of the carbon. Another family of typical examples are the halogen-bridged mixed-valence transition metal complexes (HMMCs) [12]. In HMMCs, the unpaired electron is in the d-orbital of the transition metal.

Generally, the energy eigenvalues ($\equiv E(k)$) of the plane wave state make a continuum in the wave-number space k, as shown in Fig. 2.3-6. This continuum

FIG. 2.3-6. The electron
energy band $E(k)$ in the
metallic state (*dashed line*),
and that after the Peierls
transition (*solid line*). E_f is
the Fermi level

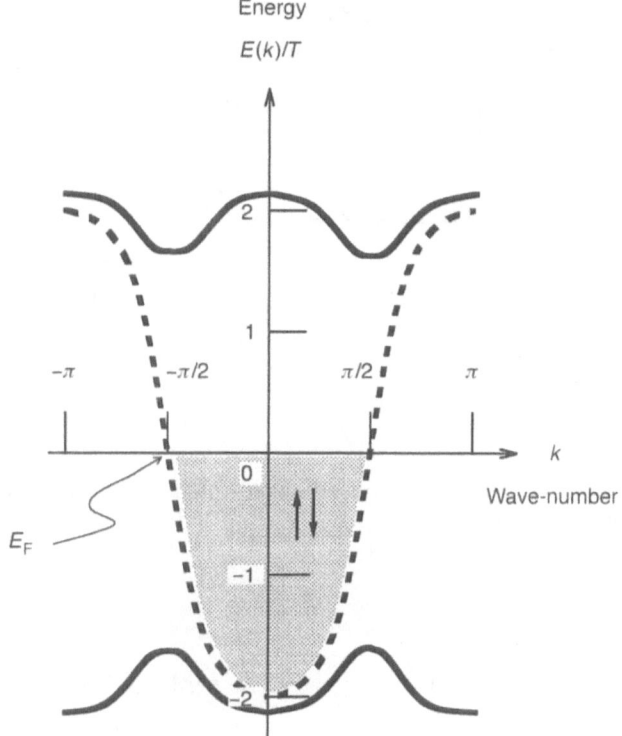

$E(k)$ is called an electron energy band. Since there are N molecular orbitals, the total number of eigenstates is N, while in the case of the nonmagnetic ground state, the total number of electrons with up- or down-spin is $N/2$. Thus, the lower half of the energy band $E(k)$ is filled by up- and down-spin electrons, equally, and the Fermi level ($\equiv E_F$) is just at the center of this band. This is a metallic state in the sense that the minimum energy required to excite electrons from this ground state is zero.

However, the one-dimensional metallic state is always unstable against the interaction between electrons and phonons. When this electron–phonon (e–ph) interaction is not zero, the one-dimensional metallic state always undergoes a phase transition and becomes an insulator with a finite energy gap between the ground and the excited states [13].

To analyze this problem in detail, let us consider the HMMC as a typical example. The HMMCs are composed of trivalent transition metal ions ($\equiv M^{3+}$, M = Pt, Pd, Ni), bridged by halogen ions ($\equiv X^-$, X = Cl, Br, I), as shown in Fig. 2.3-7. M^{3+} is coordinated by a ligand ($\equiv AA$), a planar organic molecule such as ethylenediamine. These AA ligands are bound together through counter ions ($\equiv Y$), where Y is ClO_4^- or BF_4^-.

In the d_{z^2} orbital of M^{3+} (where the z-axis is parallel to the chain), there is an unpaired electron, and this electron can itinerate from one d_{z^2} orbitals to another

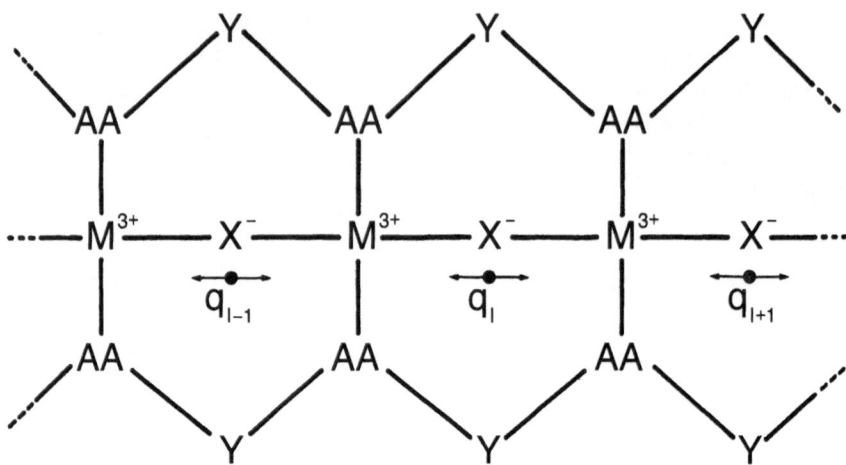

FIG. 2.3-7. Hydrogen-bridged mixed-valence transition metal complexes (HMMC). $M =$ Pt, Pd, or Ni; $X = $ Cl, Br, or I. AA and Y denote a ligand and a counter-ion, respectively. q_ℓ denotes the coordinate of X

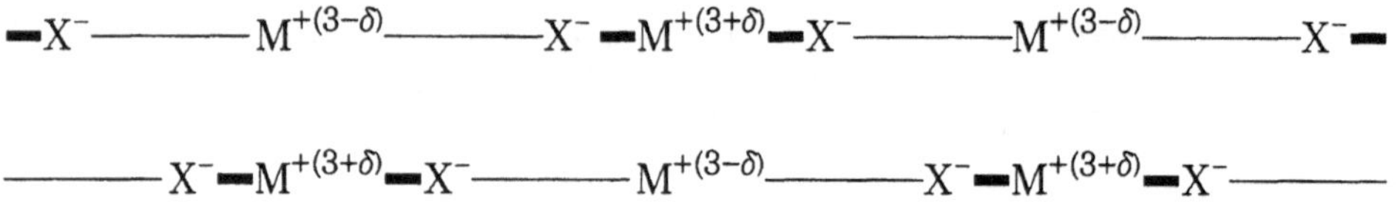

$$-X^-\text{——————}M^{+(3-\delta)}\text{—————}X^- -M^{+(3+\delta)}-X^-\text{————}M^{+(3-\delta)}\text{————}X^- -$$

$$\text{————}X^- -M^{+(3+\delta)}-X^-\text{————}M^{+(3-\delta)}\text{————}X^- -M^{+(3+\delta)}-X^-\text{————}$$

FIG. 2.3-8. Two degenerate mixed-valence states of HMMC. δ denotes the degree of charge transfer

through the resonance transfer energy. In the present case, however, this transfer between neighboring d_{z^2} orbitals can occur by virtually passing through the p_z orbital of X$^-$, which has a closed-shell structure. This is usually called supertransfer. Because of this supertransfer, the d_{z^2} orbitals make a quasi-one-dimensional energy band, the lower half of which is occupied by electrons as shown in Fig. 2.3-6.

However, this metallic state is unstable against e–ph interactions, and it results in a Peierls-type structural phase transition [13]. In the case of HMMCs, the unpaired electron in the d_{z^2} orbital of the M^{3+} ion strongly couples with the phonon modes of the X$^-$ ion through mutual electrostatic repulsion, and this e–ph interaction results in the Peierls-type phase transition. In fact, it has been confirmed by various experiments [12] that the charge transfer occurs between two neighboring M ions to give the mixed valence state shown in Fig. 2.3-8. In this figure, X$^-$ ions have been distorted with a new periodicity that is twice the period of the original lattice. δ denotes the degree of charge transfer.

This mixed-valence state can also be called the charge density wave (CDW) state, since the charge density in the d_{z^2} orbital undulates with twice the period. This double period corresponds to the wave-number $\pm \pi$ (where unit of length is the lattice constant), and this new periodicity causes mixing between the two one-electron states $|k>$ and $|k \pm \pi>$, where k is the wave-number of the electron. Since $E(k)$ and $E(k \pm \pi)$ become equal at the Fermi level with $k = \pm \pi/2$, this structural change opens up an energy gap at $k = \pm \pi/2$, as shown in Fig. 2.3-6. This is a kind of Jahn–Teller effect, and the system has now changed from a metal to an insulator.

2.3.2.2 Nonlinear Lattice Relaxation and Solitons

However, this insulator is quite exotic because its ground state is doubly degenerate due to the two possible phases of the Peierls distortion, as shown in Fig. 2.3-8. This double-degeneracy brings about various exotic nonlinear phenomena in the HMMC. Especially, the optical excited states and the lattice relaxation process become quite exotic compared with ordinary insulators with no degeneracy.

Let us consider this problem in detail. As mentioned before, the most fundamental optical excitation of an insulator is the exciton. In general, the exciton created by a photon in an insulator interacts with phonons, and this *e–ph* interaction brings about various lattice relaxation processes, which finally induce local lattice distortion around the exciton [14]. In the case of an ordinary insulator with a wide energy gap, however, the total number of excitons remains unchanged during this relaxation. In this sense, the relaxation of excitons in ordinary insulators can be called a linear lattice relaxation.

In contrast, the ground state of the CDW has a degeneracy or a multistability as mentioned before. In this multistable state, an exciton created by a photon from one of the ground states relaxes to low-lying nonlinear collective excited states, such as solitons, wherein many excitons have been condensed. This new characteristic is called the nonlinear relaxation of the exciton.

Let us look at this nonlinearity, much more intuitively. In the HMMC ground state, the energy levels of the d_{z^2} orbitals undulate as X$^-$ approaches or leaves the neighboring M, because of the electrostatic repulsion of the X$^-$ ion. The d_{z^2} electrons are influenced by this undulation as shown in Fig. 2.3-9a.

The HMMC has a strong light absorption band in the visible region [15] that corresponds to the CT excitation of an electron from the occupied d_{z^2} orbital to the vacant ones as shown in Fig. 2.3-9b. Thus, an electron and a hole are created, and they attract each other through the Coulombic force, resulting in the CT exciton.

This photoexcitation is a backward charge transfer, and the electron number per orbital becomes almost equal in this region (Fig. 2.3-9b). The Peierls distortion therefore has lost its reason for presence, and tends to disappear, as seen in Fig. 2.3-9c. Thus, the exciton self-induces a local lattice distortion, and is trapped in it. This is usually called a self-trapped exciton (STE) [14].

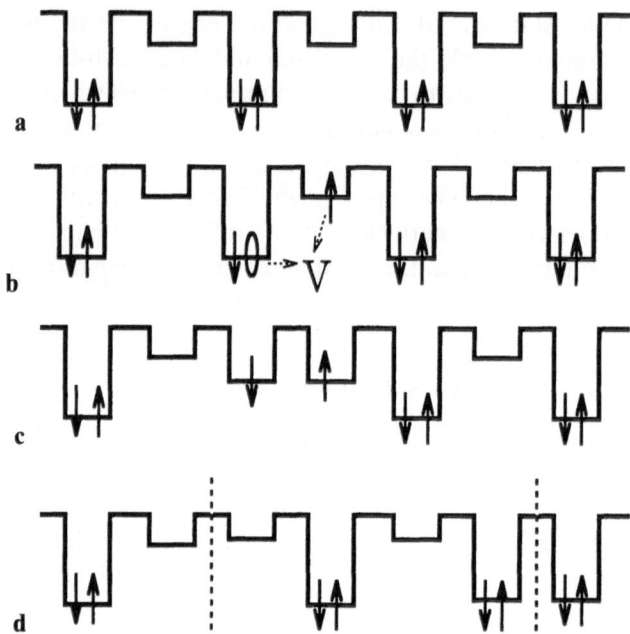

FIG. 2.3-9. Potential energy for the d_{z^2} electrons. **a** Charge density wave (CDW) state, **b** CT exciton, **c** self-trapped exciton (STE), **d** solitons. *Small arrows* correspond to electrons with corresponding spin, and a *circle* corresponds to the hole. The two *dashed lines* denote the centers of the solitons

In the case of ordinary insulators, this STE is the final state of the relaxation [14], while, in the present case, it is not the final state, but further relaxations are expected. In the CDW, as mentioned before, there is a low-lying excited state with a collective nature. That is, one phase of the ground state can appear locally in the other phase of the ground state at the expense of creating boundaries between the two phases. This boundary is usually called a soliton, and the exciton is expected to relax finally to the state consisting of a pair of solitons, as shown in Fig. 2.3-9d. In contrast to the initial single-charge transfer which occurred from the CDW ground state, a large number of charges have been transferred to create this soliton pair. This is the essence of the nonlinear lattice relaxation.

2.3.2.3 Quasi-One-Dimensional Extended Peierls–Hubbard Model

To clarify this nonlinearity from a theoretical point of view, let us study the following quasi-one-dimensional extended Peierls–Hubbard model ($\equiv H$) [16, 17], ($\hbar = 1$),

$$H = -T \sum_{\ell,\ell',\sigma} \left(a_{\ell\sigma}^{+} a_{\ell'\sigma} + \text{h.c.} \right) + \omega \sum_{\ell} q_{\ell}^2 / 2 + \left(S\omega \right)^{1/2} \sum_{\ell,\ell',\sigma} \left(q_{\ell} - q_{\ell'} \right) n_{\ell\sigma}$$

$$+ U \sum_{\ell} n_{\ell\alpha} n_{\ell\beta} + V \sum_{\ell,\ell',\sigma} n_{\ell,\sigma} n_{\ell',\sigma'}, \tag{2.3-16}$$

where ℓ' denotes the nearest neighboring site of ℓ. $a_{\ell\sigma}^+$ is the creation operator of an electron localized at a metallic site ℓ with spin σ ($= \alpha, \beta$). q_ℓ is the dimensionless coordinate of the displacement of X$^-$ as shown in Fig. 2.3-7, and ω is the energy of this phonon mode. The kinetic energy is neglected under the adiabatic approximation. S is the site-diagonal e–ph coupling energy, and $n_{\ell\sigma} \equiv a_{\ell\sigma}^+ a_{\ell\sigma}$. U and V denote the intrasite and intersite Coulombic repulsive energies, respectively. The dimensionless coordinate $Q_\ell [\equiv (\omega/S)^{1/2} q_\ell]$ is used to obtain a convenient form of the Hamiltonian.

Let us now calculate the ground and excited states. To calculate the ground state, we use the mean-field theory for interelectron Coulombic interactions and the adiabatic approximation for phonons. Within the mean-field theory H can be approximated by the mean-field Hamiltonian ($\equiv H_{HF}$), in which $n_{\ell\sigma}$ and $(a_{\ell+1\sigma}^+ a_{\ell\sigma})$ are replaced by their averages so that

$$n_{\ell\sigma} \Rightarrow \langle n_{\ell\sigma} \rangle \quad \text{and} \quad (a_{\ell+1\sigma}^+ a_{\ell\sigma}) \Rightarrow \langle a_{\ell+1\sigma}^+ a_{\ell\sigma} \rangle. \tag{2.3-17}$$

These $\langle n_{\ell\sigma} \rangle$ and $\langle a_{\ell+1\sigma}^+ a_{\ell\sigma} \rangle$ are unknown parameters to be self-consistently determined later. By diagonalizing this H_{HF}, we can obtain the energies of the ground and excited states. To take the electron–hole correlation into account in the excited state, we define the difference ($\equiv \Delta H$) between the mean-field Hamiltonian H_{HF} and the true Hamiltonian H as

$$\Delta H = H - H_{HF}, \tag{2.3-18}$$

and diagonalize it within the basis of the one-electron excited states obtained from H_{HF}. This is the first-order perturbation theory for the electron-hole correlation. The new energies of the excited states and their wave functions, including the exciton effects can then be determined. By completing this process for various lattice configurations, we can obtain adiabatic potential energy surfaces.

As mentioned before, this system is highly nonlinear in the sense that the electrons and phonons are strongly correlated with each other. To study this nonlinearity from a quantitative point of view, let us now define an index that expresses the degree of nonlinearity of relaxation. The number of electrons ($\equiv N_e$) that are forced to go out from the CDW ground state by optical excitation and the subsequent lattice distortions is given as

$$N_e = \sum_{j=1}^{N/2} \langle \varphi_j | \bar{N}_e | \varphi_j \rangle, \tag{2.3-19}$$

where the operator $\bar{N}_e$ is defined by

$$\bar{N}_e \equiv \sum_{i=N/2+1}^{N} |\phi_i\rangle\langle\phi_i|. \tag{2.3-20}$$

Here, N is the total number of electrons in the system. $|\phi_i\rangle$ is the i-th one-electron wave function of the CDW grand state, while $|\varphi_j\rangle$ is the j-th wave function of the relaxed excited state. The summation for j is taken over all the occupied states, while i is taken over all the unoccupied states. Thus, N_e gives the total number of excited electrons.

Although N_e is almost unity in the Franck–Condon state, it increases as the lattice relaxation proceeds. This increase represents the nonlinearity of the relaxation. We can also examine the variation of N_e from Fig. 2.3-9, as follows. N_e is equal to unity in the case of the CT exciton, as shown in Fig. 2.3-9b. In the case of the soliton pair described in Fig. 2.3-9d, however, there are four electrons transferred from the CDW ground state; i.e., N_e increases from unity to four as a result of the lattice relaxation. Thus, N_e is the index that can properly express the degree of nonlinearity of the relaxation.

Let us now consider the nonlinear lattice relaxation of the exciton in a one-dimensional CDW. To describe the relaxation path from the free exciton to the soliton pair through the STE, we use the following variational function for Q_ℓ [16, 17]

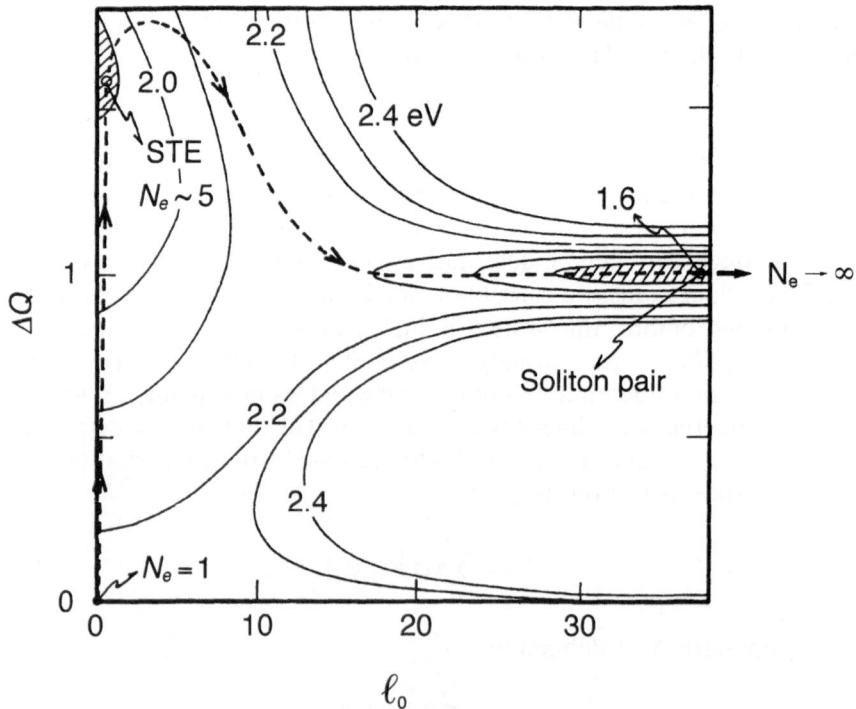

FIG. 2.3-10. Contour map of the adiabatic potential energy surface of an exciton for one-dimensional CDWs in the $\Delta Q - \ell_0$ plane. ΔQ is the amplitude of the Peierls distortion in the CDW ground state and ℓ_0 is the intersoliton distance. $U = 1.43\,\text{eV}$, $V = 0.79\,\text{eV}$, $S = 0.29\,\text{eV}$, and $T = 1\,\text{eV}$. N_e is explained in the text. *Shaded areas* represent energy minima, and the *dashed line* is the decay path of the exciton. (From [17], with permission)

FIG. 2.3-11. Charge- and spin-density profiles of **a** STE, **b** first excited state of the soliton pair; spin solitons, **c** ground state of the soliton pair; charge solitons in one-dimensional CDWs. The *solid line* denotes the charge density $<n_{\ell\alpha} + n_{\ell\beta} - 1>$ and the *dashed line* is the spin density $<n_{\ell\alpha} - n_{\ell\beta}>$. (From [17], with permission)

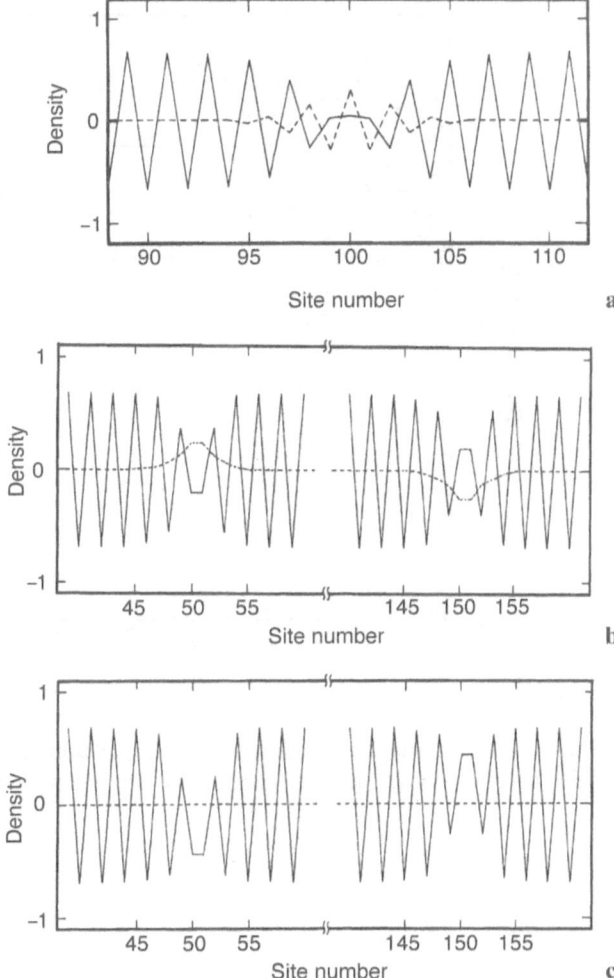

$$Q_\ell = (-1)^\ell Q\left\{1 + \Delta Q\left[\tanh\theta\left(|\ell| - \ell_0/2\right) - 1\right]\right\}, \qquad (2.3\text{-}21)$$

where $(-1)^\ell Q$ denotes the Peierls distortion in the CDW ground state, and Q should be determined beforehand within the mean-field theory. The braces $\{\ldots\}$ denote the local lattice displacement from this ground state. ΔQ is its amplitude and θ corresponds to the reciprocal width of a soliton. ℓ_0 denotes the intersoliton distance. When $\ell_0 = 0$, this pattern corresponds to the STE-type local lattice distortion. On the other hand, when $\ell_0 \gg 1$ and $\Delta Q = 1$, the phase of the Peierls distortion is completely inverted in the region $-\ell_0/2 < \ell < \ell_0/2$. This situation corresponds to a soliton–antisoliton pair.

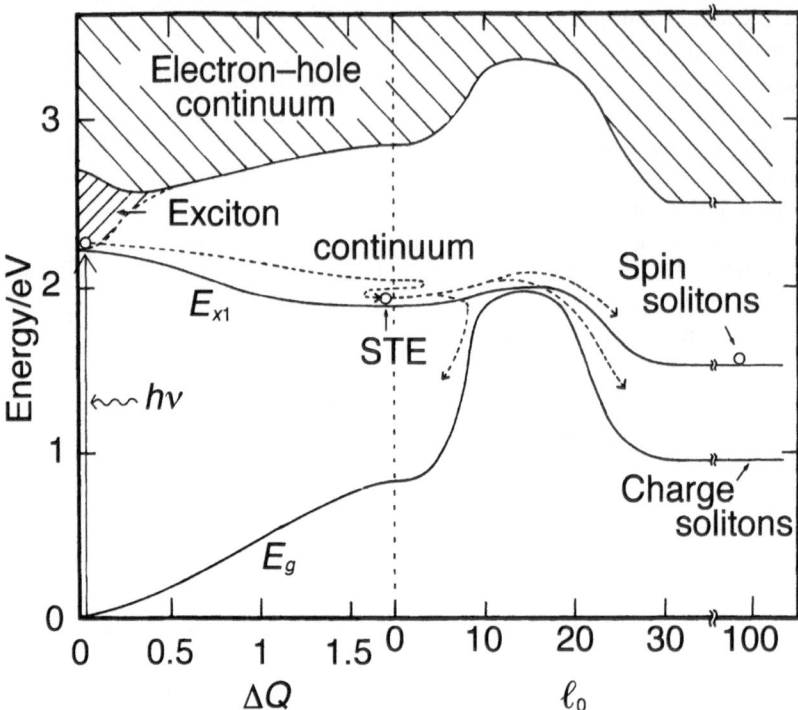

FIG. 2.3-12. Adiabatic potential energy surface along the decay path indicated in Fig. 2.3-10 as function of ΔQ and ℓ_0. E_{x1} and E_g are defined in the text. (From [17], with permission)

We have thus calculated the contour map of the adiabatic potential-energy surface of the exciton. It is shown in Fig. 2.3-10 using the $\ell_0 - \Delta Q$ plane, wherein θ is determined to minimize the energy at each point of the surface.

The point $\Delta Q = 0$ and $\ell_0 = 0$ corresponds to the Franck–Condon state or the CT exciton, which is just produced by the photon from the CDW ground state. In this figure, there are two energy minima. The minimum at $\ell_0 = 0$ and $\Delta Q = 1.6$ corresponds to the STE, and the minimum at $\Delta Q = 1$ corresponds to the pair of solitons, which is the lowest relaxed excited state of this system. The charge and spin density profiles around the STE and the pair of solitons are shown in Figs. 2.3-11a and b, respectively. Figure 2.3-11a shows that the spin density oscillation has appeared in the STE over about five metallic sites. On the other hand, we can see from Fig. 2.2-11b that the soliton pair has the spins around the boundaries between two phases. This is a spin soliton [17]. Incidentally, it can be seen from Fig. 2.3-11c that the ground state of the soliton has a charge [17].

Since the energy barrier between the STE and the soliton pair is only about 0.1 eV, the STE is expected to decay into the soliton pair, either through a thermal activation process over the barrier or through a tunneling process under the barrier. When ℓ_0 is more than 30, the energy of the soliton pair remains unchanged, being independent of ℓ_0. The decay path of the exciton on the adiabatic potential surface is indicated by the dashed line in Fig. 2.3-10. This figure also shows that N_e increases from unity to about five as the lattice relaxation proceeds from the Franck–Condon state to the STE. From this STE, relaxation proceeds further with an increase in N_e, and, finally, the spin soliton is created as the relaxed excited state. To visualize the decay path, Fig. 2.3-12 shows the adiabatic potential energy along the decay path, which is indicated by the dashed line in Fig. 2.3-10. In this figure, E_{x1} denotes the energy of the first excited state, namely, the energy of the exciton, and E_g denotes that of the ground state.

The formation energy of the soliton pair depends on the strain energy to create boundaries between the different phases. In the case of one-dimensional CDW, the size of the boundary region is always microscopic. Hence, as seen from Figs. 2.3-10 and 2.3-12, once the soliton has been created, its energy remains unchanged even if ℓ_0 increases. Therefore, the two spin solitons can completely separate from each other, and N_e will become infinite without any resistance. Therefore, in the case of a one-dimensional CDW, the other ground state can be created from the starting ground state by one photon only. This is one of the distinctive features of the nonlinear lattice relaxation in one dimension.

Thus we have seen that the HMMC is a typical example of a system in which the Peierls transition and the soliton are realized.

References

1. Knox R (1963) Theory of excitons. In: Seitz F (ed.) Solid state physics, vol 5: Academic, New York
2. Kroto H, Heath J, O'Brien S, Curl R, Smalley R (1985) C_{60}, Buckminsterfullerene. Nature 31:162
3. Yannoni C, Bernier P, Bethune D, Meijer G, Salem J (1991) NMR determination of the bond lengths in C_{60}. J Am Chem Soc 113:3190
4. Liu S, Lu Y, Kappes M, Ibers J (1991) The structure of the C_{60} molecule: X-ray crystal structure determination of a twin at 110k. Science 254:408
5. Weaver J, Martins J, Komeda T, Chen Y, Ohno T, Kroll G, Troullier N, Haufler R, Smalley R (1991) Electronic structure of solid C_{60}. Phys Rev Lett 66:1741
6. Kelly M, Etchegoin P, Fuchs D, Kratschmer W, Fostiropulos K (1992) Optical transitions of C_{60} films in the visible and ultraviolet from spectroscopic ellipsometry. Phys Rev B46:4963
7. Sohmen E, Fink J, Kratschmer W (1992) Electron energy-loss spectroscopy studies on C_{60} and C_{70} fullerite. Z Phys B86:87
8. Tsubo T, Nasu K (1994) Theory for exciton effects on optical absorption spectra of C_{60} molecules and C_{60} crystals. J Phys Soc Jpn 63:2401; Tsubo T, Nasu K (1994) Theory for exciton effects on light absorption spectra of fcc-type crystals. Solid State Comm 91:907
9. David W, Ibberson R, Matthewman J, Prassides K, Dennis T, Hare J, Kroto H, Taylor R, walton D (1991) Crystal structure and bonding of ordered C_{60}. Nature 353:147

10. Harrison W (1980) Electronic structure and the properties of solids. Freeman, San Francisco
11. Heeger A, Kivelson S, Schricffer J, Su W(1988) Solitons in conducting polymers. Rev Mod Phys 60:781
12. Yamashita M, Nonaka Y, Kida S, Hamaue Y, Aoki R (1981) Characterization and classification of tervalent nickel complexes. Inorg Chem Acta 52:43
13. Peierls R (1955) Quantum theory of solids. Oxford University Press, Oxford, pp 108
14. Ueta M, Kanzaki H, Kobayashi K, Toyozawa Y, Hanamura E (1984) Excitonic processes in solids. Springer-Verlag, Berlin
15. Wada Y, Mitani T, Yamashita M, Koda T (1985) Charge transfer excitons in halogen bridged mixed-valent Pt and Pd complexes. J Phys Soc Jpn 54:3143
16. Mishima A, Nasu K (1989) Nonlinear lattice relaxation of photo-generated charge trasnfer excitation in halogen-bridged mixed-valence metal complexes I, II. Phys Rev b39:5758
17. Suzuki M, Nasu K (1994) Nonlinear lattice relaxations and proliferations of excitons in one- and two-dimensonal charge density waves. Synth Metal. 64:247

2.4
Electrons in Specific Molecular Systems

Kizashi Yamaguchi

Organic molecular systems are diamagnetic [1] except for specific molecular systems composed of organic radicals and related species [2]. Spin alignments and spin-mediated properties of these unusual systems are examined here theoretically. First, orbital symmetry rules for effective exchange interactions between organic radicals are derived on the basis of intermolecular interaction theories. Second, the reliability and utility of first-principle calculations are discussed in relation to the theoretical prediction of the sign and magnitude of effective exchange integrals (J_{ab}) between nitroxide molecules, which have been used as stable building blocks for organic ferromagnets. Third, the ferromagnetic phase transition in the case of the β-phase of *para*-nitrophenyl nitronyl nitroxide (*p*-NPNN) [3] is examined as a typical example that illuminates the crystal structure–ferromagnetism relationship. Last, possible molecular systems with both conduction and localized electrons are discussed from the standpoint of new specific molecular functional materials such as organic ferromagnetic metals and spin-mediated organic superconductors [4]. The formal relationship between molecular magnetism and molecular recognition with hydrogen bonding is also pointed out based on the effective spin Hamiltonian models.

2.4.1 Molecular Interactions Between Open-Shell Species

The magnetic measurements for pure organic ferromagnets [2, 3] demonstrate that the Heisenberg model [1] is applicable to the description of effective exchange interactions between spins of organic radicals,

$$H(\text{HB}) = -\sum J_{ab} S_a \cdot S_b \tag{2.4-1}$$

where S_c ($c = a, b$) denotes the spin localized at site C. The singlet (S)–triplet (T) energy difference for diradicals is given by this model as

$$\Delta E(\text{ST}) = {}^1E(\text{S}{:}{\uparrow}{\downarrow}) - {}^3E(\text{T}{:}{\uparrow}{\uparrow}) = 2J_{ab} \tag{2.4-2}$$

Therefore, the parallel (ferromagnetic) spin alignment ($\uparrow\uparrow$) is more favorable than the antiparallel (antiferromagnetic) spin alignment ($\uparrow\downarrow$) if J_{ab} is positive in sign [1].

2.4.1.1 Orbital Symmetry Rules

The orbital symmetry rules for effective exchange interactions have been presented for transition metal complexes and related crystals [5, 6]. Similarly, orbital symmetries of singly occupied molecular orbitals (SOMOs) [7, 8] should play important roles in the theoretical prediction of the sign of effective exchange interactions between organic radicals. The orbital symmetry rules for the effective exchange interactions are indeed derived on the basis of the intermolecular perturbation and configuration interaction (CI) schemes [9]. The primary SOMO–SOMO interaction is described by the intermolecular CI wavefunctions for the S and T diradical (DR) configurations [10]

$$^{1,3}\Phi(\mathrm{DR}) = \frac{1}{\sqrt{2(1 \pm S_{ab}^2)}} \left(\left| \phi_a \bar{\phi}_b \right| \pm \left| \phi_b \bar{\phi}_a \right| \right) \tag{2.4-3}$$

where 1 and 3 denote the S and T states, respectively and S_{ab} is the intermolecular orbital overlap, $S_{ab} = \int \phi_a \phi_b d\tau$. Since S_{ab} is small at the intermolecular distances determined for crystals of organic radicals, the total energies can be expanded as a Taylor series, and the ST energy difference for the DR configuration is truncated at the second-order level [1] as

$$\Delta E(\mathrm{ST:DR}) = {}^1E(\mathrm{DR}) - {}^3E(\mathrm{DR})$$
$$= 2K_{ab} - \left\{ S_{ab}^2 \left(H_{aa} + H_{bb} \right) - 2S_{ab}H_{ab} \right\} \tag{2.4-4}$$

where H_{cc} ($c = a,b$) and H_{ab} denote the Coulomb and resonance integrals, respectively, and K_{ab} is the Coulombic exchange integral defined by $K_{ab} = \int \phi_a(1)\phi_b(2)r_{12}^{-1}\phi_a(2)\phi_b(1)d\tau_1 d\tau_2$.

The lower-lying charge transfer (CT) excitations are feasible for a singlet radical pair, giving rise to a unique contribution to the spin state [11]:

$$^1\Phi(\mathrm{CT}:\ b \to a) = {}^1\Phi(\mathrm{CT\ I}) = \left| \phi_a \bar{\phi}_a \right| \tag{2.4-5}$$

$$^1\Phi(\mathrm{CT}:\ a \to b) = {}^1\Phi(\mathrm{CT\ II}) = \left| \phi_b \bar{\phi}_b \right| \tag{2.4-6}$$

where CT I and II denote the CT configurations $a^- - b^+$ and $a^+ - b^-$, respectively. The stabilization energies of the singlet state by the configuration mixings between singlet DR, CT I, and CT II are given by the second-order perturbation method. Thus, the total ST energy difference, which is given by $2J_{ab}$, is approximately given by

$$\Delta E(\text{ST}) = 2J_{ab} = \Delta E(\text{ST} : \text{DR}) + \Delta E(\text{ST} : \text{CT})$$
$$\cong 2K_{ab} - 2CS_{ab}^2 \qquad (2.4\text{-}7)$$

where C is a positive (nearly constant) value given by the intermolecular CI.

The sign of J_{ab} for organic radical pairs is usually negative (antiferromagnetic) since the latter orbital-overlap term ($-CS_{ab}^2$) dominates the former (K_{ab}). Stacking modes of radical pairs for guaranteeing zero orbital-overlaps are therefore particularly important in order to accomplish the ferromagnetic exchange interactions. *We thus obtain the no-overlap and orientation principle for molecular ferromagnetism* [12, 13].

2.4.1.2 Shapes of SOMOs

In conformity with the above principle, the symmetry and shape of SOMOs play important roles in the application of *symmetry rules* for exchange interactions. Figure 2.4-1 illustrates typical SOMOs **1–6** of organic radicals used as building blocks for organic ferromagnets. The SOMOs are easily characterized as symmetric (S) or antisymmetric (A) by using the symmetry plane σ_1. Alkyl-substituted carbon radicals such as the *t*-butyl radical have a *p*-type SOMO **1**. The cation radicals of vinyl compounds and isoelectronic species have a π-type SOMO **2** with a π-bonding nature. On the other hand, anion radicals of vinyl groups and isoelectronic compounds such as the ketyl radical and iminonitroxide have a π^*-type SOMO **3** with a π-antibonding nature, whose importance will be disclosed in relation to recent experimental results showing ferromagnetic intermolecular interactions. The nonbonding MO (NBMO)-type SOMO **4** of the allyl

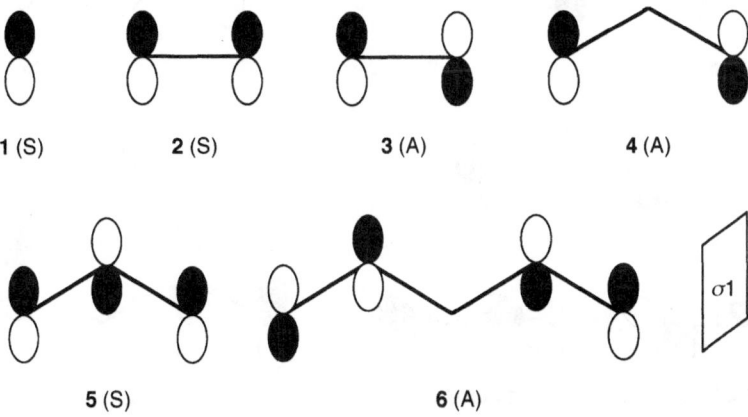

FIG. 2.4-1. Shape and symmetry of singly occupied molecular orbitals (SOMOs) **1–6** for typical organic radicals (see text). σ_1 denotes the first symmetry plane; *S*, symmetric; *A*, antisymmetric

radical and related species is also important for ferromagnetic interactions. The hyperconjugation of the alkyl group with a π^*-type SOMO **3** provides a π^*-type SOMO **5** which is essential for ferromagnetic interactions between alkyl nitroxides. Five-centered radicals have a π^*-type SOMO **6**, which consists of the p-type (**1**) and two π^*-type (**3**) SOMOs. Nitronyl nitroxide derivatives with **6** have been found to be stable spin sources for organic ferromagnets.

2.4.1.3 Intermolecular Hund Rules

The intermolecular Hund rule is one of the most useful guiding principles for molecular design of an organic ferromagnet [9]. For illustrative purposes, consider the SOMO–SOMO interactions **7–12** in Fig. 2.4-2, which are regarded as several combinations of the radical species in Fig. 2.4-1. For example, **7** and **8** represent the orbital interactions between **3** and between **5** respectively. These SOMO–SOMO interactions are orbital symmetry-allowed in the orientations since the interacting SOMOs have the same orbital symmetry concerning the symmetry plane σ_1. Therefore, the SOMO–SOMO overlap term in Eq. (2.4-7) plays a predominant role, giving rise to negative J_{ab} values. This means a larger

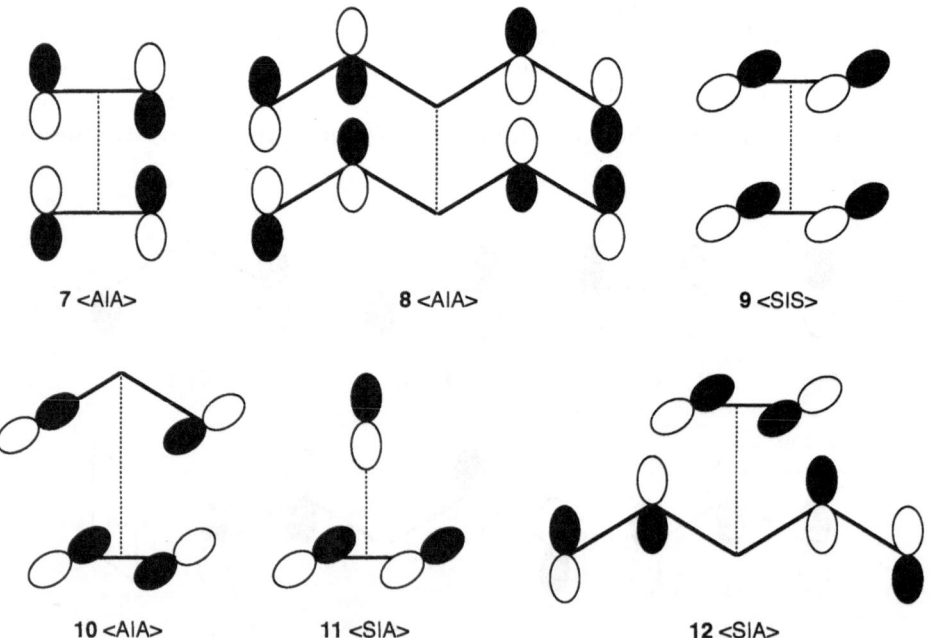

FIG. 2.4-2. The SOMO–SOMO interactions for typical radical pairs: **7** and **8** for the σ–σ conformation; **9** and **10** for the π–π conformation; **11** and **12** for the σ–π conformation. Their symmetry properties are given by the orbital overlap. The *dotted lines* represent the mode of contact of the two fragments

stabilization of the singlet DR state than the triplet DR state. SOMO–SOMO interactions are also orbital-symmetry allowed in the orientations **9** and **10** given by the **2–2** and **3–4** orbital combinations, respectively. They are accompanied by π-type bondings, leading to the formation of antiparallel spin alignments. In the past two decades, many experiments have shown that organic radical crystals exhibit antiferromagnetic interactions because of these SOMO–SOMO overlaps [2, 12, 13]. This in turn indicates that the ferromagnetic intermolecular interactions are orbital-symmetry forbidden because of the nonzero orbital overlaps. *Therefore, no SOMO–SOMO overlap should be emphasized as a guiding principle for achieving organic ferromagnetism* [9, 12, 13].

The SOMO–SOMO overlaps are zero in the $\sigma\pi$-type orientations **11** and **12**, which are composed of the **1–2** and **3–6** orbital pairs, respectively. On the other hand, the Coulombic exchange term (K_{ab}) given by the exchange integral ($<\phi_S\phi_A|$ $\phi_S\phi_A>$) is nonzero because of orbital symmetry, together with the close contact between SOMOs. The ferromagnetic interactions are symmetry allowed in these orientations **11–12**.

2.4.1.4 Orbital Symmetry Rules and Molecular Orientations

Following the above intermolecular Hund rule, effective exchange interactions are believed to be antiferromagnetic in the case of parallel orientations of organic radicals because of the nonzero overlaps between SOMOs. However, it should be noted that several specific SOMO–SOMO overlaps become zero even in these orientations [9, 11, 12]. For example, they become zero even at parallel conformations **13** and **14** in Fig. 2.4-3, for which the Coulombic exchange term K_{ab} is still nonzero, and the resulting net effective exchange interaction is ferromagnetic [11]. As shown in **13** (**2–3** pair), the π^*-nature of SOMOs plays an important role for ferromagnetic interaction since the orbital overlap disappears even in the rectangular form. The NBMO is also crucial for ferromagnetic interactions in parallel orientation **14** (**4–5** pair). The SOMO–SOMO overlap becomes zero even for more complex face-to-face stacking **15** since the phase relationship or orbital topology for interacting SOMOs is different.

There are several important sliding but parallel conformations with zero SOMO–SOMO overlaps even for *homocombinations* of radical species. For example, the orbital overlap between SOMOs becomes almost zero at the rhombus conformations **16** and **18** (**3–3** and **6–6** pairs) and the zigzag conformation **17** (**4–4** pair) in Fig. 2.4-3, for which the Coulombic exchange term K_{ab} is still nonzero, leading to ferromagnetic interactions [9]. In the sliding conformations, two symmetry planes (σ_1 and σ_2) are necessary to characterize the orbital interactions as shown in Fig. 2.4-3. As is apparent from **16–18**, the π^*-nature and NBMO property of SOMOs are essential for ferromagnetic interactions since the orbital overlaps disappear even in the parallel but sliding modes. Thus, there are many radical pairs characterized by symmetry-imposed zero SOMO–SOMO overlaps. Orbital symmetry consideration is therefore a primary step for the molecular design of organic ferromagnets [11].

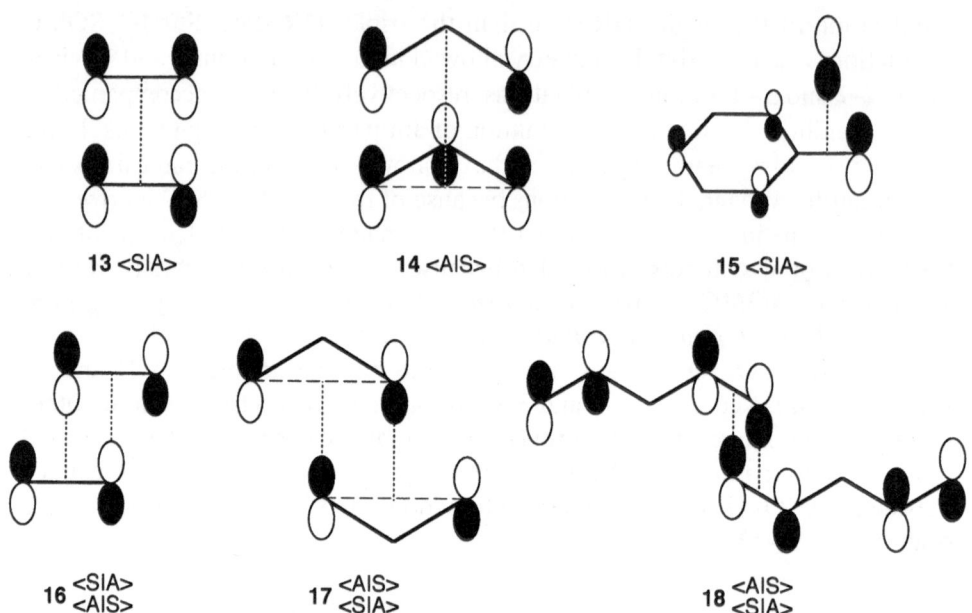

FIG. 2.4-3. The SOMO–SOMO interactions with zero orbital overlaps for the hetero (**13–15**) and homo (**16–18**) combinations of organic radicals. Two symmetry planes, σ_1 and σ_2, are used. The dashed line represents the symmetry plane

2.4.2 Organic Ferromagnetic Clusters

2.4.2.1 Ab Initio Calculations

The next step in the molecular design of organic magnets is to carry out first-principle calculations of J_{ab} for clusters of radical species. One of the most convenient and practical methods is the unrestricted Hartree–Fock (UHF) approach [13, 14], which is applicable to relatively large magnetic systems and magnetically ordered broken-symmetry states. The UHF method provides SOMOs for up and down spins of clusters as follows:

$$\psi_i^+ = \phi_a + \sum C_b \phi_b \qquad (2.4\text{-}8)$$

$$\psi_i^- = \phi_b + \sum C_a \phi_a \qquad (2.4\text{-}9)$$

where C_x $(x = a, b)$ denote the fractions of spin delocalization (SD) or charge transfer (CT) within clusters. As an extension of the preceding perturbation theory, the singlet and triplet projected UHF (PUHF) solutions for the radical pair are expressed by the intermolecular CI wavefunctions as [10, 11]

$$^{1,3}\Phi\left(\text{PUHF}\right)=\left\{\left|\psi_i^+\overline{\psi_i^-}\right|\pm\left|\psi_i^-\overline{\psi_i^+}\right|\right\}\Big/\sqrt{2\left(1\pm T_i^2\right)} \qquad (2.4\text{-}10)$$

$$=^1\Phi\left(\text{DR}\right)+2\sum C_a\Phi\left(\text{CT I}\right)+2\sum C_b\Phi\left(\text{CT II}\right) \quad \left(\text{for singlets}\right) \qquad (2.4\text{-}11)$$

$$=^3\Phi\left(\text{DR}\right) \quad \left(\text{for triplets}\right) \qquad (2.4\text{-}12)$$

where T_i is the SOMO–SOMO overlap: $T_i = \int\psi_i^+\psi_i^-dt$. The singlet PUHF wavefunction is expressed by mixing the DR–CT configurations, whereas the triplet UHF is given by DR. The UHF calculations provide both potential (PE) and kinetic (KE) exchange terms in a variational manner: they are the first and second terms in Eq. 2.4-7, respectively.

In addition to these two terms, higher-order intermolecular interactions may contribute to effective exchange interactions. They can be included by the Møller–Plesset (MP) and coupled-cluster (CC) SD(T) methods. The effective exchange integrals (J_{ab}) can be calculated by the energy difference between the total energies for the lowest (LS) and highest (HS) spin states by the post-UHF methods (X = MP, CC SD (T)) [14]:

$$J_{ab}\left(\text{APUHF}-\text{X}\right)=\frac{^{\text{LS}}E\left(\text{UHF}-\text{X}\right)-^{\text{HS}}E\left(\text{UHF}-\text{X}\right)}{^{\text{HS}}\left\langle S^2\right\rangle\left(\text{UHF}-\text{X}\right)-^{\text{LS}}\left\langle S^2\right\rangle\left(\text{UHF}-\text{X}\right)} \qquad (2.4\text{-}13)$$

where the effect of spin projection is included in an approximate manner in the denominator given by the total spin augular momentum. Here, AP means an approximate but *size-consistent spin projection* in contrast to other projection methods.

To confirm the selection rules presented in Sect. 2.4.1, *ab initio* APUHF MP4 (APUMP4) and APUHF CC (APUCC) SD(T) calculations have been carried out for clusters of nitroxide radicals for which the experimental results are available. The conclusions of these UHF-based methods were the same as those using the spin-restricted CASSCF and CASPT2(D) treatments, although the latter results are not discussed here [15].

2.4.2.2 Applications to Some Important Conformation Models

Figure 2.4-4 illustrates three important conformation models for nitroxide dimer: T-shape conformation (I), parallel syn-stacking mode (II) and parallel anti-stacking mode (III). Model I was first examined to clarify the importance of the π^*-nature of SOMO for ferromagnetic interaction. Since the π^*-type SOMO has a node near the center of the N–O bond, the $\pi^*\pi^*$ orbital overlap should disappear for a bridge (T-shape) structure as illustrated in Fig. 2.4-5A. On the other hand, the PE term remains positive ($K_{ab} = <\pi^*(1)\pi^*(2)|\pi^*(2)\pi^*(1)> = S_{13}^2U/2 > 0$) under the Mulliken approximation for the atomic orbital overlap and under the assumption for on-site Coulombic repulsion; $U = U(N) = U(O)$ [9]. The

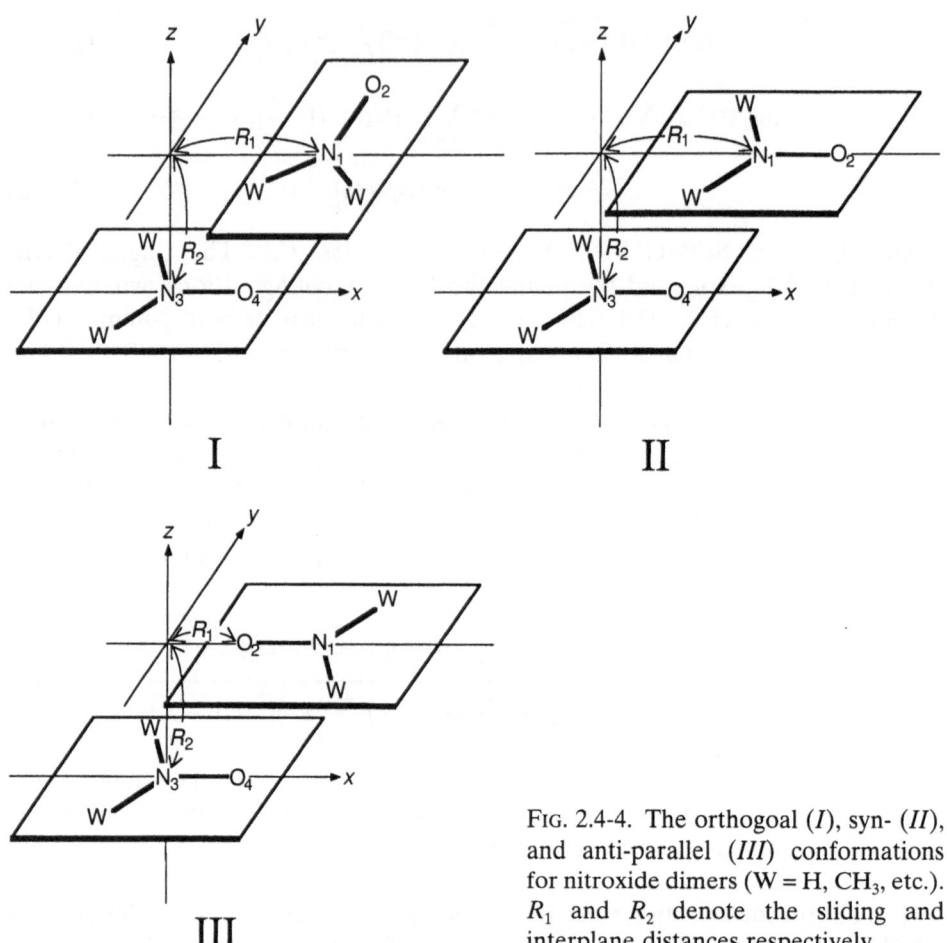

FIG. 2.4-4. The orthogoal (*I*), syn- (*II*), and anti-parallel (*III*) conformations for nitroxide dimers (W = H, CH_3, etc.). R_1 and R_2 denote the sliding and interplane distances respectively

parallel orientation is important for obtaining the nonzero PE, and the net effective exchange integral J_{ab} in Eq. 2.4-1 should become positive (ferromagnetic).

To confirm the above no-overlap and orientation principle, the T-shape stacking mode (I) of nitroxides was examined by changing the sliding distance (R_1) at a fixed interplane distance ($R_2 = 3.4\,\text{Å}$). The J_{ab} values were calculated for the parallel interplane stacking mode by APUMP2(4) and APUCC SD(T)/4-31G. Figure 2.4-6a shows variations of the calculated effective exchange integrals with R_1.

The J_{ab} values are positive in the bridge structure (Fig. 2.4-5A), whereas they are negative for no-bridge conformations (Fig. 2.4-5B). The maximum positive J_{ab} values are about 4.5 and 5.5 cm^{-1} by APUCC SD(T)/4-31G and 6-31G* respectively for a bridge structure with $R_1 = 0.6\,\text{Å}$. The magnitude of J_{ab} therefore increases by about $1\,\text{cm}^{-1}$ by addition of the *d*-polarization function.

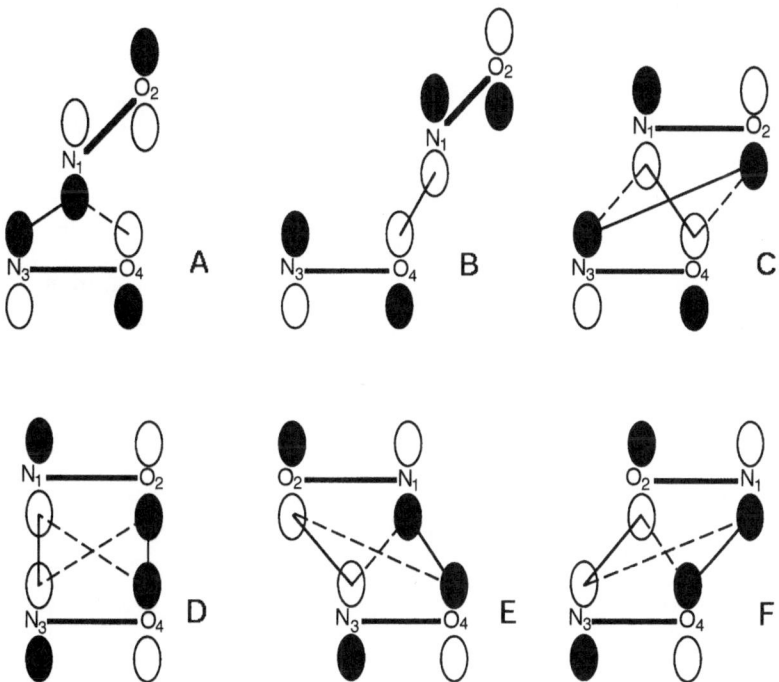

FIG. 2.4-5. The SOMO–SOMO interactions at several specific geometries of the nitroxide dimers in Fig. 2.4-4. The *solid* and *dashed lines* represent in-phase and out-of-phase contact between orbital lobes

As a second example, the syn-stacking mode II in Fig. 2.4-4 was examined by changing R_1 for $R_2 = 3.4$ Å. Because of the π^*-nature of SOMO for nitroxide, the SOMO–SOMO overlap should disappear for the rhombus conformation:

$$S_{\pi^*\pi^*} = \left(S_{14} + S_{23}\right) - \left(S_{13} + S_{24}\right) = 0 \tag{2.4-14}$$

$$K_{ab} = \left\langle \pi_a^*(1)\pi_b^*(2) \middle| \pi_a^*(2)\pi_b^*(1) \right\rangle = \left(S_{14}^2 + S_{23}^2 - S_{13}S_{24}\right)U/2 > 0. \tag{2.4-15}$$

as illustrated in Fig. 2.4-5C. Therefore the KE term is almost zero, but the PE term remains nonzero because of the close SOMO–SOMO contact in the parallel orientation.

To confirm this prediction, the APUMP4(2) and APUCC SD(T)/4-31G calculations were carried out. Figure 2.4-6b shows how the calculated J_{ab} values vary with R_1. The J_{ab} values are certainly positive (ferromagnetic) near the rhombus conformation in Fig. 2.4-5C, whereas they are negative near the rectangular conformation in Fig. 2.4-5D. The maximum positive J_{ab} values by APUCC SD(T)/4-31G and 6-31G* are 9.5 and 11.5 cm^{-1}, respectively, at $R_1 = 1.4$ Å.

The anti-parallel stacking mode III in Fig. 2.4-4 was examined by changing R_1 for $R_2 = 3.4$ Å. Figure 2.4-6c shows variations of the J_{ab} values with R_1 calculated by APUMP and APUCC SD(T)/4-31G. The J_{ab} values become positive (ferro-

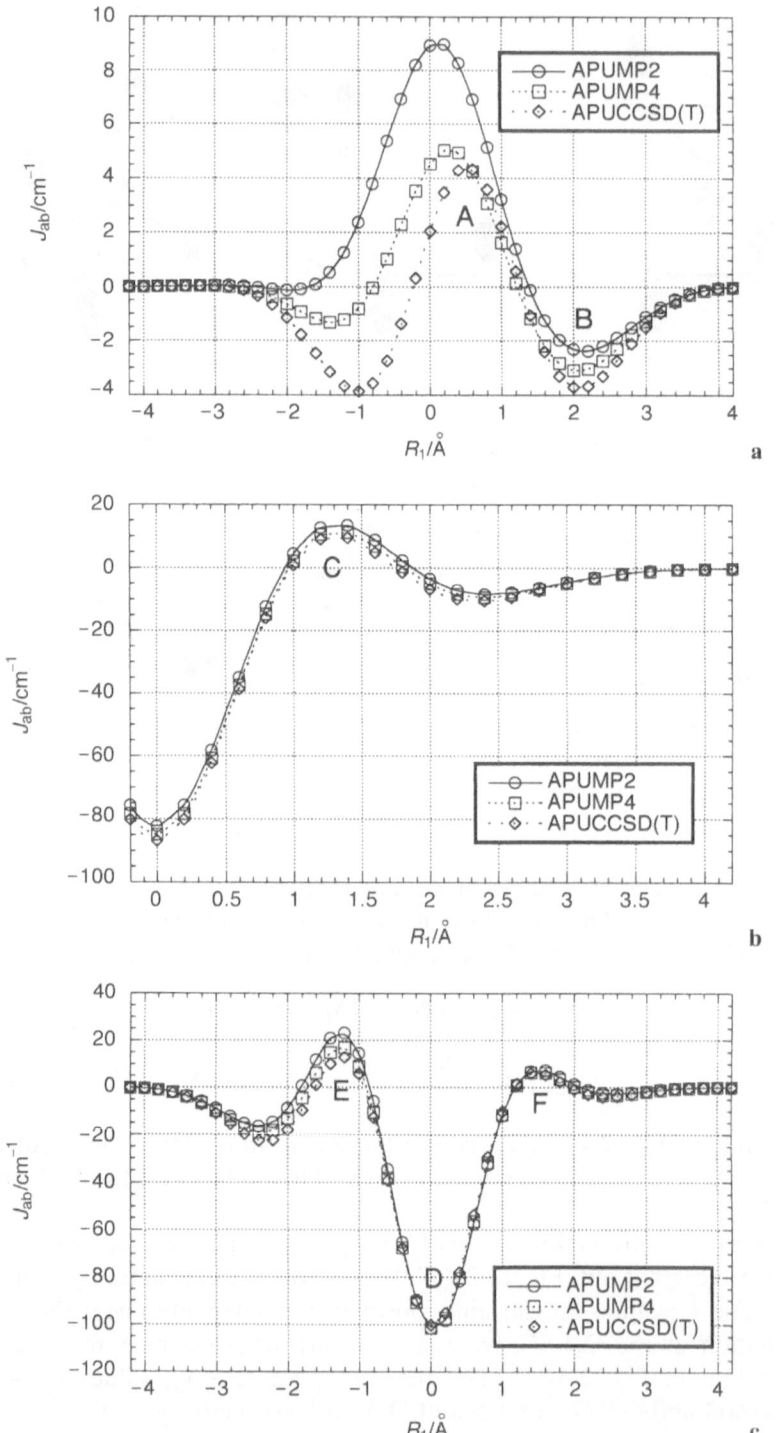

FIG. 2.4-6. Variations of J_{ab} values with R_1 by the approximate projected unrestricted Hartree–Fock Møller–Plesset perturbation theory, 2nd (or 4th) order [APUHF MP2(4)] and couple–cluster single and double (with triple) [CC SD(T)/4-31G] computations for the three parallel conformations in Fig. 2.4-4. A–F denote the key intermediates of Fig. 2.4-5

magnetic) at the rhombus conformations in Figs. 2.4-5E and 2.4-5F, whereas they are negative near the rectangular conformation. The maximal positive J_{ab} values by APUCC SD(T)/4-31G (6-31G*) are 9.5 (15.5) and 5.8 (5.4) cm^{-1} at E ($R_1 = -1.2$ Å) and F ($R_1 = 1.4$ Å), respectively. Thus, qualitative conclusions are independent of the level of the basis sets employed.

Ab initio computations have supported the no SOMO–SOMO overlap principle and the orientation principle for molecular ferromagnetism. The positive J_{ab} values calculated for the T-shape (I), syn-parallel (II) and anti-parallel (III) conformations are compatible with the observed ferromagnetic interactions in the crystals of TEMPO derivatives [16], MOTMP [17], adamantane bisnitroxide [18] and HQNN [19]. Chemical modifications of these nitroxides are essential for the increase of T_c. The crystal engineering approach is also necessary since the sign and magnitude of J_{ab} is largely dependent on stacking mode and crystal structure as shown in Fig. 2.4-6. As a typical example of ferromagnetic organic crystals, the β-phase crystal of *para*-nitrophenyl nitronylnitroxide (*p*-NPNN) will be discussed in the next section.

2.4.3 Organic Ferromagnetic Crystals

As is well known, molecular magnetism is a quantum cooperative phenomenon [20] on the macroscopic scale. Therefore, three-dimensional networks of spins should be formed. There are three chemical and synthetic approaches [12] to accomplish ferromagnetic long-range order: (1) crystalization of radical species, (2) through-bond coupling via polymer networks, and (3) constitutions of hybridized systems. Although several approaches have been initiated, only the *p*-NPNN crystal will be discussed here as an example of the first through-space approach. In contrast to TEMPO and MOTMP, the preceding direct KE and PE interactions between SOMOs were calculated to be negligible in the β-phase crystal of *p*-NPNN because of too large an intermolecular separation, showing the necessity of theoretical analysis of the indirect mechanisms. Among them, the spin polarization (SP) effect of the phenyl ring was found to be crucial for remote ferromagnetic interaction [21]. The effective exchange interaction via the SP mechanism of the phenyl ring (see, for example, Fig. 2.4-7b) is approximately given by the spin density product (SDP) term [11]:

$$J_{ab}(\text{SP}) = -\sum_{ab} B_{ab}\rho_a\rho_b \qquad (2.4\text{-}16)$$

where ρ_a is the spin density induced on a site and B_{ab} is the parameter determined by first-principle calculation.

2.4.3.1 Spin Polarization and Spin Densities

According to Eq. 2.4-16, spin population is a key factor for the theoretical prediction of the sign of J_{ab}. To elucidate spin populations, the APUHF and

unrestricted Kohn-Sham (UKS) Becke-Lee-Yang-Parr (B-LYP) 6-31G* calculations [22, 23] were carried out for phenyl nitronyl nitroxide (PNNO) and p-
NPNN, assuming the geometries taken from the X-ray structures as illustrated in
Fig. 2.4-7a. The neutron diffraction experiment revealed negative spin densities
(−0.121) on the C5 atom of PNNO, indicating a significant SP effect. The spin
densities are −0.08 and −0.146 by APUHF/6-31G* and APUHF/INDO, respectively, whereas it is −0.07 by UKS B-LYP/6-31G*. These methods therefore
reproduced the experimental tendency for PNNO. The situation was similar for
p-NPNN.

The spin densities on the benzene ring calculated for PNNO by APUHF/6-
31G* were alternating, as illustrated in Fig. 2.4-7b, and compatible with experimental results. Their signs and magnitudes by the UKS B-LYP/6-31G* method
without spin projection were qualitatively compatible with the observations. The
APUHF/6-31G* and intermediate neglect of differential overlap (INDO) values
were also similar to experimental findings. This is why the APUHF/INDO calculations have provided qualitatively reasonable effective exchange integrals (J_{ab})
between p-NPNN molecules as shown in the next section.

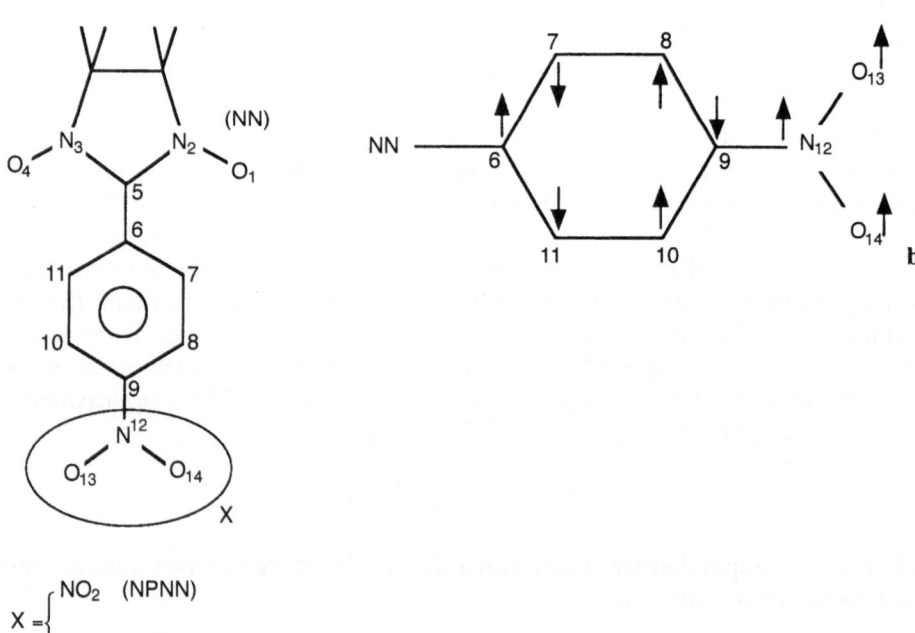

FIG. 2.4-7. The molecular structure of p-NPNN (**a**), its spin population on the benzene ring
(**b**) and the intermolecular interactions responsible for J_{12} (**c**) and J_{13} (**d**) in the β-phase
crystal (see Fig. 2.4-8)

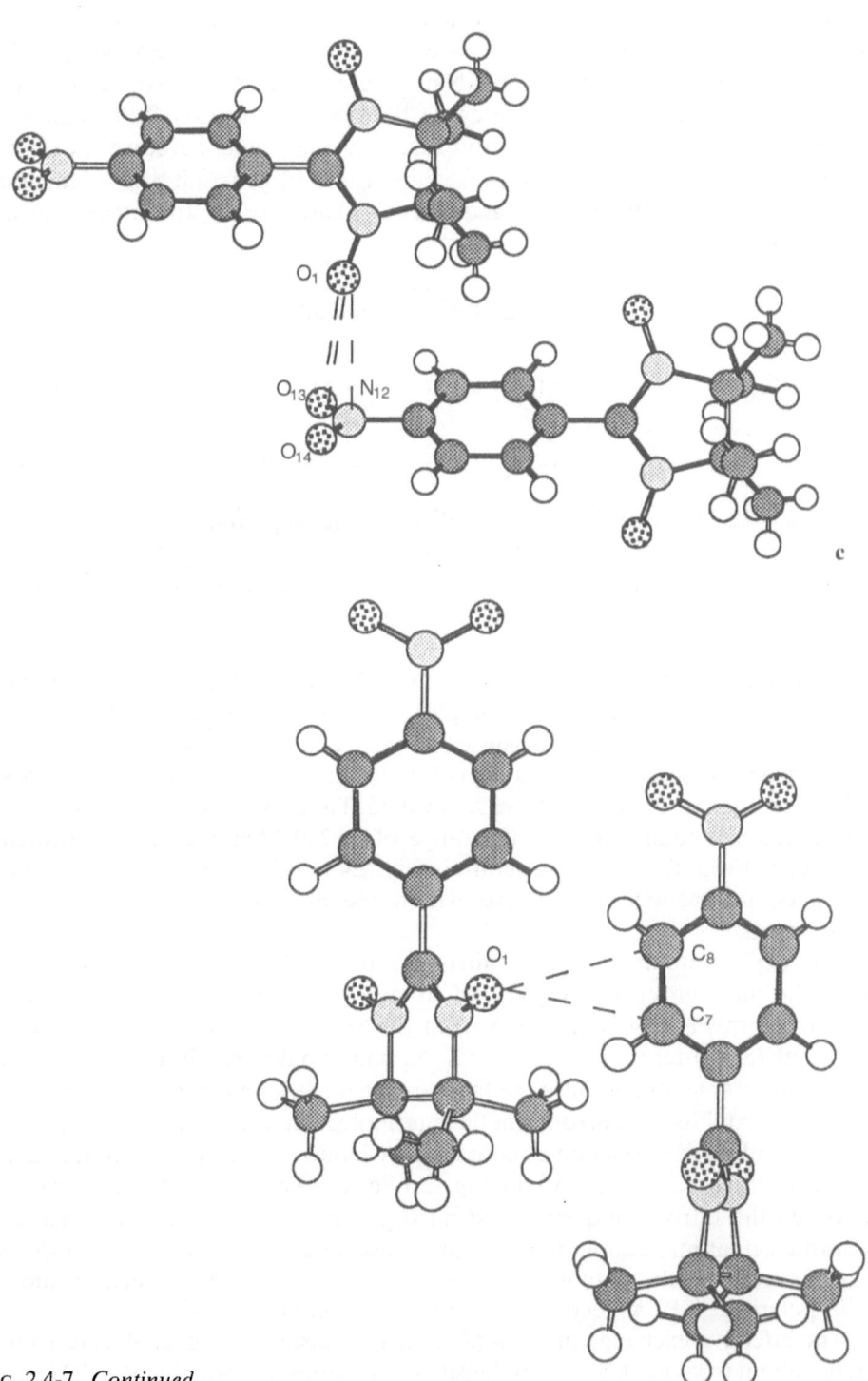

FIG. 2.4-7. *Continued*

2.4.3.2 Calculations of J_{ab} values for the β-phase of p-NPNN

Figure 2.4-8 illustrates the crystal structure of the β-phase of p-NPNN with Fdd2 space symmetry. As seen in Fig. 2.4-8, the twelve nearest neighbor p-NPNN molecules around the central p-NPNN molecule are divided into three equivalent groups. The effective exchange integrals between the central p-NPNN molecule and its nearest neighbors in each group are equivalent, and therefore one of them is explicitly defined as J_{1n} (n = 2–4) without loss of generality. The lattice distances r_{1n} between lattice point 1 and n ($n = 2$–4) are given by the lattice constants (a,b,c) as follows

$$r_{12} = \left(a^2 + c^2\right)^{1/2}\Big/2 = 11.49\,\text{Å} \tag{2.4-17}$$

$$r_{13} = \left(a^2 + b^2 + c^2\right)^{1/2}\Big/4 = 6.37\,\text{Å} \tag{2.4-18}$$

$$r_{14} = \left(a^2 + b^2 + 9c^2\right)^{1/2}\Big/4 = 10.04\,\text{Å} \tag{2.4-19}$$

where a = 12.36, b = 19.36 and c = 10.97 (Å). The magnitude of J_{1n} is sensitive to remote intermolecular interactions rather than r_{1n}, since p-NPNN is a long organic molecule (see Fig. 2.4-7). This is a unique characteristic of organic ferromagnets, in sharp contrast to that of inorganic magnetic solids or magnetic alloys.

To determine both intra- and interplane effective exchange integrals (J_{ab}), APUHF/INDO[24, 25] and complete active space self-consistent field (CASSCF)/4-31G[26] calculations were carried out for p-NPNN clusters and isoelectronic species m-N-methylpyrid inium nitronyl nitroxide (m-MPYNN$^+$), whose geometries are known from X-ray data. The J_{12} values for the clusters were calculated and found to be in the range of 0.17–0.20 cm^{-1} at the experimental geometry. From the crystal structure in Fig. 2.4-8, it is concluded that the ferromagnetic interaction is operative within the a–c plane in accordance with experiment.

The oxygen atom (O$_1$) of the nitronyl nitroxide group of p-NPNN interacts with the nitro group (N$_1$–O$_2$–O$_3$) of its nearest neighbor in the a–c plane as illustrated in Fig. 2.4-7c. Judging from the spin populations, the spin density products (SDP) term between the O$_1$–N$_1$ pair should exhibit a small antiferromagnetic interaction, because the face-to-face dimer of a planar model system of p-NPNN exhibits antiferromagnetic intermolecular interaction via the O$_1$–N$_1$ contact and the introduction of nonplanarity induces the inversion from negative J_{ab} to positive J_{ab}. As shown in Fig. 2.4-7c, the nonorthogonal conformation between the nitronyl nitroxide and nitro groups is essential for ferromagnetic intermolecular interaction. Ferromagnetic exchange in the a–c plane of p-NPNN thus arises from the potential exchange between the SOMO electron and the spin-polarized (SP) π-electron of its nearest neighbor [21, 24].

The effective exchange integral (J_{13}) for the clusters in Fig. 2.4-8 is ferromagnetic (about 0.08 cm^{-1}). The spin density populations indicate that the SDP terms

Fig. 2.4-8. The crystal structure (Fdd2) of the β-phase of p-NPNN. The twelve nearest neighbors of the central molecule are divided into three equivalent groups of four molecules. The groups are denoted by *solid, shaded,* and *open circles*. J_{12}, J_{13}, and J_{14} denote the independent effective exchange integrals for each group of equivalent p-NPNN molecules

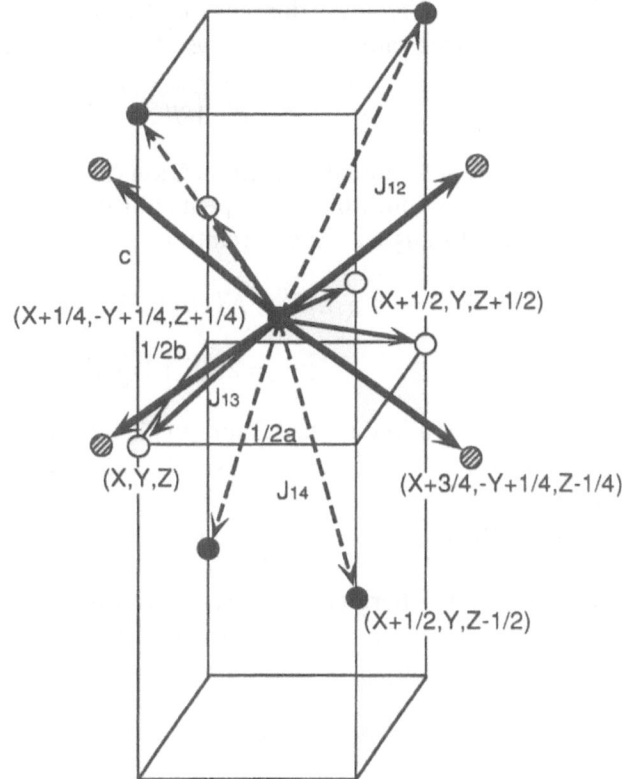

(see Eq. 2.4-16) between the O_1–C_1 and O_1–C_2 atomic pairs are positive and negative, respectively, as illustrated in Fig. 2.4-7d. The net interaction after mutual cancellation remains ferromagnetic in this stacking mode. Therefore, J_{13} should be sensitive to the pressure effect and others as has already been confirmed by the experiment. The J_{14} values for clusters in Fig. 2.4-8 are slightly antiferromagnetic ($-0.014\,\mathrm{cm}^{-1}$) because of the weak van der Waals interaction between methyl and nitro groups. However, this weak interaction is overcome by J_{13}. The net interplane J value remains positive (ferromagnetic).

2.4.3.3 Langevin–Weiss Model for the Ferromagnetic Phase Transition

To estimate the ferromagnetic phase transition temperature (T_c), the Langevin–Weiss mean-field theory is applied to molecular crystals of p-NPNN in Fig. 2.4-8. The spin S_b is replaced by the expectation value $\langle S_b \rangle = \langle S \rangle$ under the mean-field approximation [21, 25]. Then Eq. 2.4-1 can be rewritten as

$$E_{ex} = -2\sum J_{ab}\langle S \rangle \cdot S_a \tag{2.4-20}$$

$$= \mathbf{A}\mathbf{M} \cdot g\mu_B \mathbf{B}_a \tag{2.4-21}$$

where AM ($M = -Ng\mu_B<S>$) denotes the Weiss molecular magnetic field and N is the number of free radicals in the crystal. The A factor is defined by

$$A = N^{-1}\left(g\mu_B\right)^{-2} 2\sum J_{ab}. \tag{2.4-22}$$

With the above situation, the magnetization of a molecular crystal is described by the Brillouin function, which disappears at the critical temperature (T_c), leading to the so-called Langevin–Weiss relation

$$k_B T_c = N\left(g\mu_B\right)^2 S\left(S+1\right) A/3 \tag{2.4-23}$$

$$= 2S\left(S+1\right)\sum J_{ab}/3 \tag{2.4-24}$$

where S is 1/2 for a free radical. Equation 2.4-24 is applied to theoretical estimations of T_c for the β-phase of p-NPNN and molecular crystals of other free radicals.

The magnitude of $(J_{13} + J_{14})$ is about one-third of J_{12} in the a–c plane in the case of the β-phase of p-NPNN. This indicates the adequacy of the quasi-three-dimensional Heisenberg-type model. Therefore, the A factor of the Weiss effective molecular magnetic field is approximately given by taking the twelve nearest neighbor p-NPNN as

$$A = 2N^{-1}\left(g\mu_B\right)^{-2}\left(4J_{12} + 4J_{13} + 4J_{14}\right) \tag{2.4-25}$$

Then, T_c is calculated to be 0.64 K, in accordance with the experimental value of 0.60 K [3, 21].

The present theoretical calculations for the β-phase of p-NPNN have revealed that the Langevin–Weiss (LW) model utilizing calculated J_{ab} values works well for a theoretical explanation of organic ferromagnetism. The LW model was successfully used to estimate T_c for the γ-phase of p-NPNN [25], although its refinement by inclusion of low-dimensional effects is essential for quantitative purposes. It was also used for theoretical predictions of possible high-T_c organic ferromagnets based on J_{ab} values calculated by APUHF/INDO, UKS B-LYP/4-31G, and CASSCF/4-31G [26]. Thus, the first-principle cluster calculations of J_{ab} are applicable to the design of molecular magnetic insulators for which the Heisenberg model is reliable for theoretical description.

2.4.4 Possibilities of Organic Magnetic Metals and Other Exotic Molecular Systems

2.4.4.1 CASSCF and CASSCF-Based Second-Order Perturbation Theory (CASPT2) Calculations of Hole-Doped Systems

It is well known that the so-called high-T_c super-conductivity of copper oxides is induced by the hole (electron) doping of two-dimensional antiferromagnetic sheets composed of CuO_2 units [27]. In the past decade, there has been much interest shown in the hole (electron) doping of antiferromagnetic or ferromagnetic insulators in relation to magnetic conductivity, spin crossover, spin-mediated high-T_c super-conductivity, and so on. After this discovery, we immediately initiated theoretical studies of the hole doping of antiferromagnetic or ferromagnetic polycarbenes and other organic magnetic systems, and explored the theoretical possibility of organic analogs of copper oxides and organic ferromagnetic conductors, as illustrated in Fig. 2.4-9 [4, 28].

Electronic structures of hole-doped polymers and organic crystals in Fig. 2.4-9 are described by a common π-R (radical) model Hamiltonian for electron carriers plus localized electron systems. This Hamiltonian is an analog to the so-called s–d, d–f, and Anderson Hamiltonians in condensed matter physics [4, 29]:

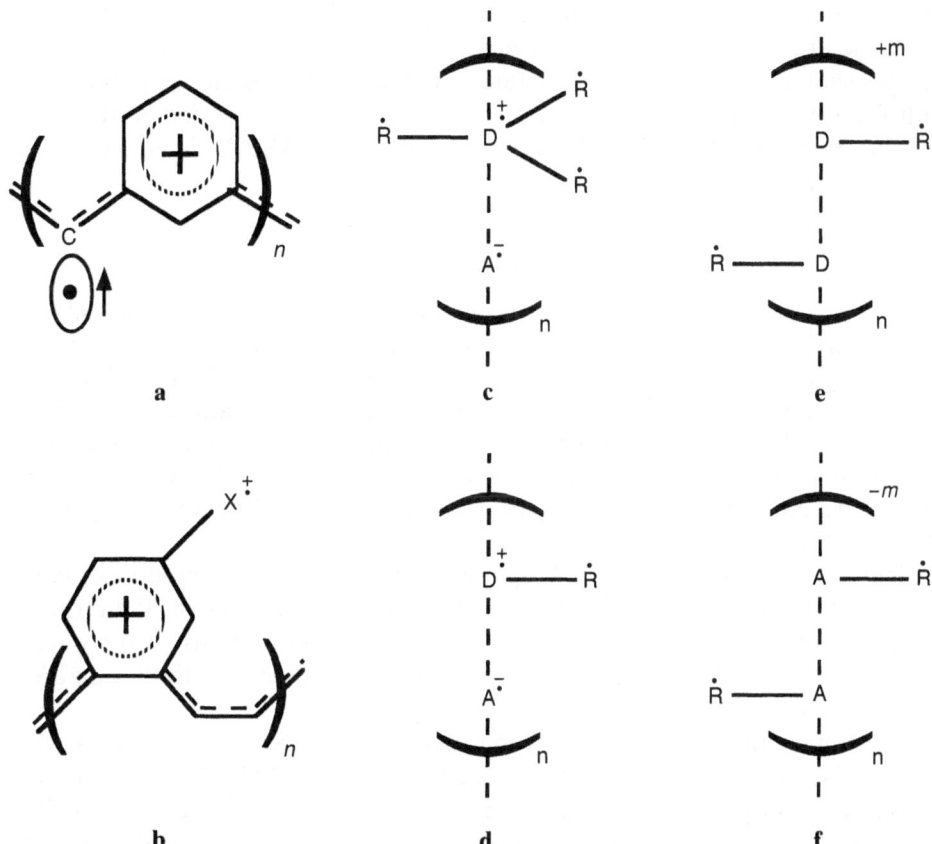

FIG. 2.4-9. Possible organic magnetic conductors with remarkable characteristics **a–f**: (1) low-dimensionality, (2) electron correlation and (3) electron-lattice interaction. D and A represent electron donor and acceptor molecules and R denotes radical groups as spin sources

$$H = \sum \varepsilon_K \sigma_{ks}^* \sigma_{ks} + \sum \left(V_K \sigma_{ks}^* R_{ks} + \sum V_K R_{ks}^* \sigma_{ks} \right) + \sum \varepsilon_d R_{ks}^* R_{ks}$$
$$+ U R_{ks}^* R_{ks} R_{ks}^* R_{ks} \qquad\qquad (2.4\text{-}26)$$

where σ_{ks}^* and R_{ks}^* denote the creation operators of conduction and localized electrons, and V is the coupling constant between them. The specific molecular systems in Fig. 2.4-9 are therefore possible models for spin-mediated superconductors and organic dense Kondo models [30]. Several groups have already initiated experimental approaches to hole- or electron-doped polyradicals [31].

The preceding UHF-based MP and CC approaches are often less reliable in the case of hole (electron)-doped systems, since hopping (delocalization) effects become particularly important. On the other hand, the CASSCF method can be used to determine exchange coupling parameters involved in the π-R Hamiltonian since the method includes all the configurations feasible within complete active space (CAS) {m-orbital, n-electron} for the multiconfiguration (MC) SCF procedure. The CASSCF wavefunction involves all the valence-bond the (VB) configurations responsible for the resonating VB (RVB) effects. To obtain the RVB picture, hole densities are calculated by the difference between total charges shared by the ground spin state of a neutral species and its monocation with the spin multiplicity under consideration. The second-order perturbation method based on CASSCF, which is referred to as CASPT2, is employed for refinement of CASSCF results.

2.4.4.2 Hole Doping of Ferromagnetic Insulators

As an example of the hole doping of ferromagnetic insulators [32], hole-doped *meta*-phenylene bis(methylene) is examined. There are several important VB structures of the species, for example VB I–IV as illustrated in Figs. 2.4-10c–f. To elucidate the RVB effect for I–IV, the CASSCF 4-31G calculations were carried out varying the size of CAS space. The HS quartet state is more stable by $118\,\mathrm{cm}^{-1}$ than the LS doublet state at the CASSCF {4, 3} level. Moreover, the HS-LS gaps become larger (490 and $1261\,\mathrm{cm}^{-1}$ by {6, 5} and {8, 7}) with the expansion of the CAS space.

Figure 2.4-10c shows the hole population for quartet hole-doped *meta*-phenylene bis(methylene). The hole densities are delocalized over the terminal carbon atoms, together with the *ortho*- and *para*-carbon atoms of the phenyl group, showing that VB I and II are resonating, together with three other VB structures (for example III and IV).

To estimate the dynamic correlation corretions, CASPT2 calculations were also conducted. The ground state is a quartet at the CASPT2 {6, 5} level: the quartet–doublet gap, $\Delta E(\mathrm{QD})$, is $716\,\mathrm{cm}^{-1}$. The present CASSCF and CASPT2 calculations are consistent with the experiment for hole-doped polycarbenes with m-phenylene bridges [31]. Thus, the spin polarization (SP) effect for hole-doped polycarbenes with m-phenylene skeletons overcomes the spin delocalization (SD) effect introduced by hole doping. Therefore, an organic ferromagnetic metal state is feasible for the polymers in Fig. 2.4-9a.

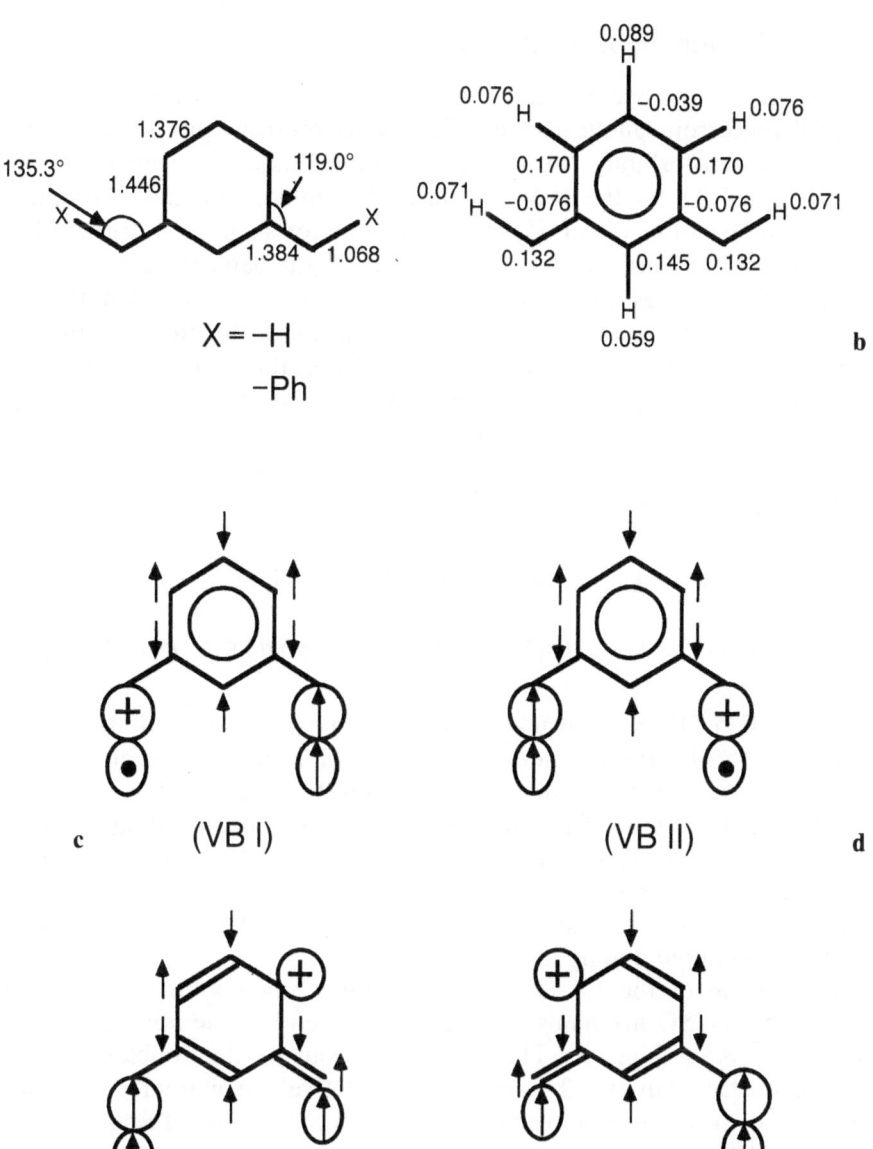

FIG. 2.4-10. **a** The molecular structure of *m*-phenylene bis (methylene), **b** its hole population by CASSCF, and **c–f** the valence bond (*VB*) structures (*I–IV*) showing the hole (+)

2.4.4.3 Hole Doping of Antiferromagnetic Insulators

Hole-doped bis(methylene) amine was examined as an example of the hole doping of antiferromagnetic insulators [33]. Since there are three active π-orbitals and two active σ-orbitals, the five active orbitals and six electrons {5, 6} are regarded as a CAS for the neutral state of the amine. Assuming the optimized singlet geometry of CASSCF {5, 6}, calculations for singlet (S), triplet (T), and quintet (Q) states were carried out to elucidate adiabatic HS–LS energy gaps. The ground state was calculated to be a singlet, in accord with the SD or superexchange mechanism via the hetero atom. Since the energy gaps for S–T and S–Q excitations are given by $2J_{ab}$ and $6J_{ab}$, on the basis of the Heisenberg model, the J_{ab} values have been calculated as –871 and –1233 cm^{-1}. The values are similar to that of copper oxides [34].

The hole-doped amine with radical groups is isoelectronic to the hole-doped copper and nickel oxides [34].

$$(\uparrow)Cu - O(\downarrow) - Cu(\uparrow) \leftrightarrow (\uparrow)CH_2 - NH(\downarrow)^+ - CH_2(\uparrow) \quad (2.4\text{-}27)$$

$$(\uparrow\uparrow)Ni - O(\downarrow) - Ni(\uparrow\uparrow) \leftrightarrow (\uparrow\uparrow)CH - NH(\downarrow)^+ - CH_2(\uparrow\uparrow) \quad (2.4\text{-}28)$$

The CASSCF {5, 5}/4-31G calculations have been performed for doublet, quartet, and sextet states for the bis(methylene) amine monocation. The ground state is HS (quartet) for the π-hole-doped ion; $\Delta E(QD) = 0.517$ eV. Since the ground state of the neutral species is a singlet, the LS–HS relative stability is reversed upon π-hole doping. Thus, the CASSCF and CASSPT2 results are consistent with the theoretical prediction that the spin correlation for the terminal methylene pair should be converted from the anti-parallel alignment to the parallel alignment upon hole doping [34, 35].

The inversion of the LS–HS relative stability upon π-hole doping can be explained by the SD mechanism, which is followed by hole populations on the basis of the CASSCF results. The hole populations for the doublet and quartet states are illustrated in Figs. 2.4-11c and d, respectively. Apparently, the holes are delocalized over the whole molecular skeleton in both HS and LS states, showing resonating VB states.

Tris(methylene) amine was examined as a model compound of tris(*para*-phenymethylene phenyl) amine. The CASSCF {7, 7} calculations were carried out for the π-hole state. The ground state for the monocation was calculated to be a sextet. Thus, the LS–HS crossover occurs upon hole doping as in the case of bis(methylene) amine.

A logical extension of the calculated results for bis(methylene) amine and tris(methylene) amine provides an expectation that the SD rule gives rise to the HS ground state for monocation of *bis*- and tris(para-phenylmethylene phenyl) amines, as illustrated in Fig. 2.4-11. Stable high-spin ion radicals are also possible if carbene groups are replaced by stable nitroxide and nitronyl nitroxides as

FIG. 2.4-11. The molecular
structures of secondary (**a**)
and tertiary (**b**) amines
with (phenyl)methylene
groups, and the hole
populations in the doublet
(**c**) and quartet (**d**) states

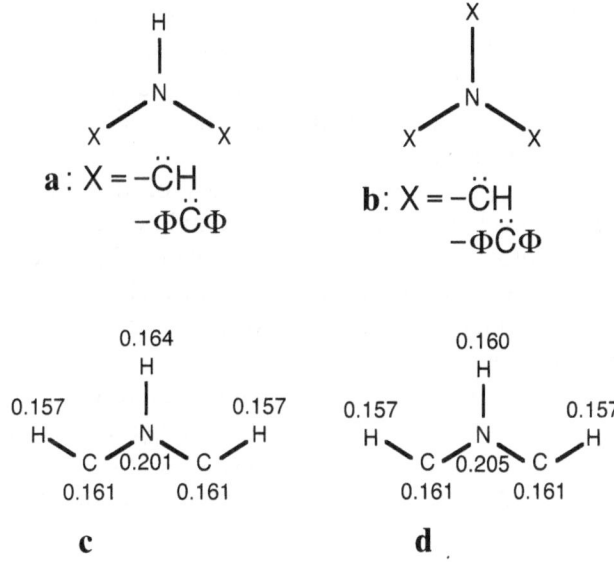

a : X = –ĊH
–ΦC̈Φ

b: X = –ĊH
–ΦC̈Φ

shown in Fig. 2.4-11. They are potential building blocks for organic ferrimagnets
and ferrimagnetic metals as shown in Fig. 2.4-9.

2.4.4.4 Conclusion

Organometallic conjugated systems with the conductive π-electron and localized
d-electron can be described by the π-d Hamiltonian, which is isoelectronic to the
π-R Hamiltonian for pure organic systems [4, 28]. Efforts to synthesize such
organometallic systems have been extensively carried out and several new systems are now under investigation [22, 26]. The CASSCF calculations of these
systems provide useful information on the couplings between delocalized π- and
localized d-electrons, and they are to be investigated in the future.

In this review, attention was focused on specific molecular systems with organic
radical groups or organometallic groups as spin sources. The systems under
consideration are therefore organic radical crystals, conducting polymers with
radical groups, organic metals (for example, the TTF derivatives of Fig. 2.4-9)
with radical groups, and other π-electron systems coupled with localized d-
electrons. These specific molecular systems, with and without electron carriers,
are electronically characterized with strong or intermediate electron correlations.
Therefore they are useful targets for first-principle calculations using the recently
developed accurate post-Hartree–Fock methods and density functional theory
(DFT) methods involving self-energy corrections.

The CASSCF calculations are hardly applicable to larger organic systems with
conduction electrons and spins, although they are useful for construction of
reliable model Hamiltonians such as the t-J, π-R, Anderson, and Hubbard Hamiltonians [4]. To discuss various macroscopic properties, there model Hamiltonians

should be solved using quantum simulation techniques developed in solid-state physics, although these techniques are not touched upon in this review. First-principle (CI, DFT, etc.) calculations plus quantum simulations are promising approaches toward designing specific molecular systems which are isolobal to doped copper oxides [4].

Here, the theoretical possibilities of dynamic (switching-type) molecular magnetisms [36] controlled by external perturbations (e.g., electric or magnetic field, photoexcitation, electron or hole doping) are not discussed although they are fascinating from the standpoint of molecular functionality. Both theoretical and experimental efforts toward these functional materials will be reviewed in volume II of this book.

Molecular magnetism is formally related to molecular recognition on the assumption that spin alignments in π-conjugated nitroxides are regarded as spatial alignments of hydroxyl groups in glucose and related species as illustrated in Fig. 2.4-12a and b; the upward- and downward-hydroxyl groups for molecular planes

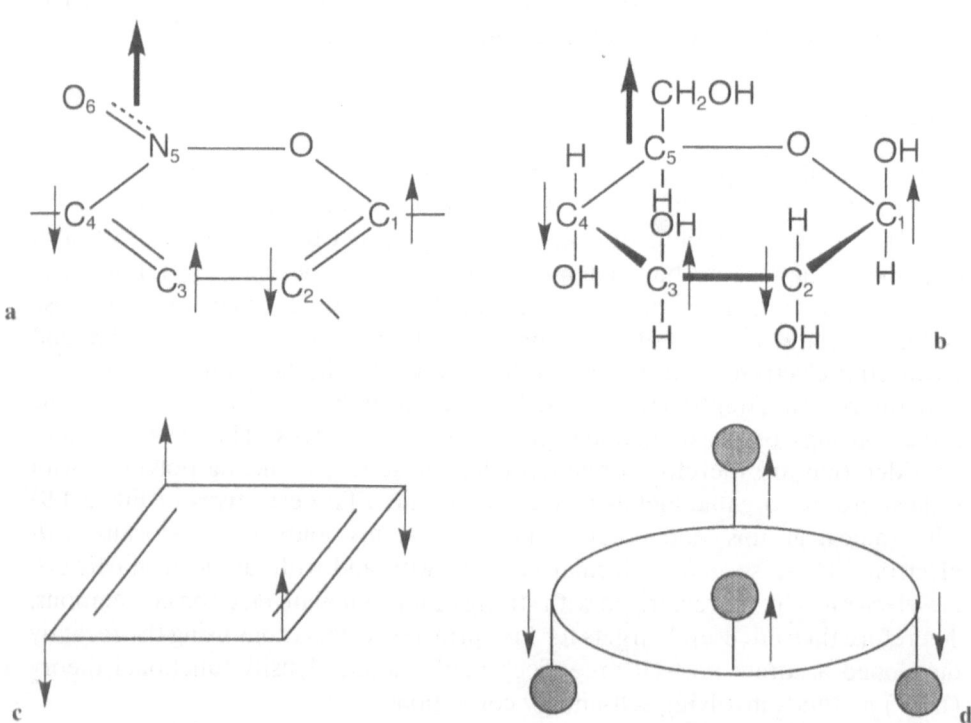

FIG. 2.4-12. The formal analogy of spin alignments in nitroxide (**a**) with spatial orientations of hydroxyl groups in glucose (**b**). A similar analogy is conceivable from the classical spin vector model in the case of spins in cyclobutadiene (**c**) and hydroxyl groups in artificial porphyrin (**d**)

correspond to the up- and down-spins in the Ising model [20]. The intermolecular effective exchange interactions (J_{ab}) between spins are replaced by the intermolecular effective interactions (X_{ab}) between hydroxyl groups ($X_{ab} = 1$ for the on-state of the hydrogen-bonding interaction and $X_{ab} = 0$ for its off-state, and $J_{ab} = 2X_{ab} - 1$). The spin Hamiltonian model [1, 7] is also applicable to molecular recognition of artificial hydrogen-bonding systems as illustrated in Fig. 2.4-12c and d. Aperiodic oligomers and polymers composed of saccharides and artificial model compounds [37] could form molecular memories described by spin Hamiltonians. Spin glass theories [20] used in alloys and complex inorganic solids, together with radical solids [12], have already been applied to protein folding problems [38].

References

1. Salem L (1982) The role of spin: the various manifestations of electron exchange. In: Electrons in chemical reactions: first principles. John Wiley New York, pp 183–213
2. Iwamura H, Miller JS (1993) Chemistry and physics of molecular-based magnetic materials. Mol Cryst Liq Cryst Vols 232 and 233
3. Nakazawa Y, Tamura M, Shirakawa N, Shiomi D, Takahashi M, Kinoshita M, Ishikawa M (1992) Low-temperature magnetic properties of the ferromagnetic organic radical, p-nitrophenyl nitronyl nitroxide. Phys Rev B46:8906–8914
4. Yamaguchi K (1990) N-band Hubbard models for copper oxides and isoelectronic systems. New models for organic and organometallic magnetic conductors and superconductors. Int J Quant Chem 37:167–196
5. Goodenough JB (1958) An interpretation of the magnetic properties of the perovskite-type mixed crystals $La_{1-x}Sr_xCoO_{3-\lambda}$. Phys Chem Solid 6:287–297
6. Kanamori J (1959) Superexchange interaction and symmetry properties of electron orbitals. Phys Chem Solid 10:87–98
7. Yamaguchi K, Yoshioka Y, Fueno T (1977) Heisenberg models for radical reactions I:Local spin(magnetic) symmetry conservations of biradical species. Chem Phys 20:171–181
8. Yoshioka Y, Yamaguchi K, Fueno T (1978) Heisenberg models of radical reactions II: Conservation of the local spin-permutation symmetry in reactions of biradical species. Theor Chim Acta 45:1–20
9. Kawakami K, Yamanaka S, Mori W, Yamaguchi K, Kajiwara A, Kamachi M (1995) No-overlap and orientation principle for ferromagnetic interactions between nitroxide groups. Chem Phys Lett 235:414–421
10. Yamaguchi K, Fueno T (1977) Diradical and zwitterionic intermediates in the excited state. Chem Phys 23:375–386
11. Yamaguchi K, Fueno T, Nakasuji K, Murata I (1986) Semiempirical molecular orbital (MO) calculations of the effective exchange integrals for sandwich dimers of free radical species. Chem Lett 629–632
12. Yamaguchi K, Namimoto H, Fueno T (1989) Ab initio calculations of effective exchange integrals. Possibilities of superparamagnetic, mictomagnetic and amorphous ferromagnetic states for aggregates of aromatic free radicals and polymer radicals. Mol Cryst Liq Cryst 176:151–161

13. Yamaguchi K (1975) The electronic structures of biradicals in the unrestricted Hartree–Fock approximation. Chem Phys Lett 33:330–335
14. Yamaguchi K, Okumura M, Mori W, Maki J, Takada K, Noro T, Tanaka K (1993) Comparison between spin-restricted and unrestricted post-Hartree–Fock calculations of effective exchange integrals in Ising and Heisenberg models: Chem Phys Lett 210:201–210
15. Yamanaka S, Okumura M, Yamaguchi K, Hirao K (1994) CASPT2 and MR MP2 calculations of potential curves and effective exchange integrals for the dimer of triplet methylene. Chem Phys Lett 225:213–220
16. Nogami T, Tomioka K, Ishida K, Yoshikawa H, Yasui M, Iwasaki F, Iwamura H, Takeda N, Ishikawa M (1994) A new organic ferromagnet: 4-benzylideneamino-2,2,6,6-tetramethylpiperidin-1-oxyl. Chem Lett 29–32
17. Kajiwara A, Sugimoto H, Kamachi M (1994) Crystal structure of organic metamagnet: 4-methacryloyloxy-2,2,6,6-tetramethyl-1-piperidinyloxyl (MOTMP). Bull Chem Soc Jpn 67:2373–2377
18. Chiarelli R, Novak MA, Rassat A, Tholence JL (1993) A ferromagetic transition at 1.48 K in an organic nitroxide. Nature 363:147–149
19. Sugawara T, Matsushita MM, Izuoka A, Wada N, Takeda N, Ishikawa M (1994) An organic ferromagnet: α-phase crystal of 2-(2′, 5′-dihydroxyphenyl)-4,4,5,5-tetramethy-4,5-dihydro-1H-imidazoyl-1-oxy-3-oxide (α-HQNN). J Chem Soc Chem Commun 1723–1724
20. White RM (1970) Quantum theory of magnetism. McGraw Hill, New York
21. Okumura M, Yamaguchi K, Nakano M, Mori W (1993) A theoretical explanation of the organic ferromagnetism in the β-phase of para-nitrophenyl nitronyl nitroxide. Chem Phys Lett 207:1–8
22. Becke AD (1988) Density-functional exchange-energy approximation with correct asymptotic behavior. Phys Rev A38:3098–3100
23. Lee C, Yang W, Parr RG (1988) Develpoment of the Colle–Salvetti correlation-energy formula into a functional of the electron density. Phys Rev B37:785–789
24. Yamaguchi K, Okumura M, Maki J, Noro T, Namimoto H, Nakano M, Fueno T, Nakasuji K (1992) MO theoretical studies of magnetic interactions in clusters of nitronyl nitroxide and related species. Chem Phys Lett 190:353–360
25. Okumura M, Mori W, Yamaguchi K (1994) A MO-theoretical calculation of the antiferromagnetism in the γ-phase of p-nitrophenyl nitronyl nitroxide. Chem Phys Lett 219:36–44
26. Okumura M, Yamaguchi K, Awaga K (1994) Ferromagnetic intermolecular interaction of the cation radical of m-N-methylpyridinium nitronyl nitroxide. Chem Phys Lett 228:575–582
27. Bednorz JG, Muller KA (1986) Possible high-Tc superconductivity in the Ba–La–Cu–O system. Z Phys B64:189–193
28. Yamaguchi K, Toyoda Y, Fueno T (1987) Ab initio GMO calculations of intermolecular effective exchange integrals and designing of organic magnetic polymers. Synth Metal 18:81–86
29. Yamaguchi K, Takahara Y, Fueno T, Nasu K (1987) Ab initio MO calculation of effective exchange integrals between transition-metal ions via oxygen dianions: Nature of the copper-oxygen bonds and superconductivity. Jpn J Appl Phys 26:L1362–L1364
30. Yamaguchi K, Okumura M, Fueno T, Nakasuji K (1991) New models for organic magnetic conductors or organic Kondo and dense Kondo systems. Synth Metals 41–43:3631–3634

31. Matsushita M, Nakamura T, Momose T, Shida T, Teki, Y, Takui T, Kinoshita T, Itoh K (1993) Novel organic ions of high-spin species III. ESR and H ENDOR studies of a monocation of m-phenylenebis(phenylmethylene). Bull Chen Soc Jpn 66:1333–1342

32. Yamanaka S, Kawakami T, Okumura M, Yamaguchi K (1995) CASSCF and CASPT2 calculations of hole-doped polycarbenes. Possibilities of organic ferromagnetic conductors and metals. Chem Phys Lett 233:257–265

33. Yamanaka S, Okumura M, Nagao H, Yamaguchi K (1995) CASSCF and CASPT2 calculations of hole-doped amines with triplet carbene groups. Possibilities of high-Tc organic ferrimagnets. Chem Phys Lett 233:88–97

34. Yamaguchi K, Takahara Y, Fueno T, Nogami T, Shirota Y (1988) Possible organic analogues to copper oxides: applications of a J-model. Jpn J Appl Phys 27:L766–L769

35. Yamaguchi K, Namimoto H, Fueno T, Nogami T, Shirota Y (1990) Possibilities of organic ferromagnets and ferrimagnets by the use of charge-transfer (CT) complexes with radical substituents. Ab initio MO studies. Chem Phys Lett 166:408–414

36. Okumura M, Mori W, Yamaguchi K (1993) Theoretical study of organic magnetisms: nitronyl nitroxide and related species, In: Doyama M, Kihara J, Tanaka M, Yamamoto R (eds) Computer aided innovation of new materials II. Elsevier, Tokyo, pp 1785–1788

37. Hayashi T, Asai T, Hokazono H, Ogoshi H (1993) Dynamic molecular recognition in a multifunctional porphyrin and ubiquinone analogs. J Am Chem Soc 115:12210–12211

38. Peliti E (1991) Biologically inspired physics. Plenum, New York

Part III
Fundamental Processes in Molecules and Molecular Systems

3.1
Molecular Recognition and Self-Regulation

Kazuo Kitaura

The phenomena associated with molecular recognition and self-regulation are found in both biological and chemical molecular systems. At an early stage in the catalytic reaction of an enzyme in vivo, the enzyme recognizes a substrate and forms a substrate–enzyme complex. Self-regulation of the enzyme occurs at the substrate-binding site to identify the shape and positions of functional groups of substrate. Chemical molecular systems called host–guest complexes also have molecular recognition and self-regulation properties.

In order to understand the phenomena associated with molecular recognition and self-regulation, it is necessary to understand molecular interactions. Molecular interactions are of prime importance in the association of molecules. The concepts of molecular interaction are useful in understanding not only intermolecular interactions but also interactions between subunits within a molecule.

Theoretical methods of studying molecular interactions are described in Sect. 3.1.1. Section 3.1.2 explains physical interactions such as electrostatic, polarization, exchange, and charge transfer interactions based on the orbital interaction picture of molecular interactions. In Sect. 3.1.3 the nature of molecular interactions in some hydrogen-bonding and electron donor–acceptor complexes is discussed from a unified perspective.

3.1.1 Theoretical Descriptions of Molecular Interactions

Molecular interactions determine the structures and properties of molecular assemblies. Quantum mechanics provides a theoretical basis for understanding intermolecular interactions, and the formalism of perturbation theory can be used to understand molecular interactions [1, 2]. The exchange perturbation method describes molecular interactions in terms of component interactions that have physical meaning, such as electrostatic (ES) interaction, exchange repulsion (EX), polarization (PL), charge transfer (CT), and dispersion (DISP), and plays an important role in our understanding of these phenomena. However, in practice, typical applications of perturbation theory have been limited to small molecules because of the amount of computer time needed. For theoretical studies of molecular interactions, variational approaches have been employed as well as perturbation theory approaches. The variational, or "supermolecule," approach is applied to many molecular systems, and component interactions corresponding to those defined by perturbation theory are also obtained by energy decomposition analysis [3]. We therefore describe molecular interactions using the supermolecule approach.

3.1.1.1 The Supermolecule Approach

The supermolecule method treats all interacting molecules as a single molecule (supermolecule) and carries out electronic structure calculations in the same manner as for a normal molecule. The *ab initio* method has been successfully applied to predict the geometries, bonding energies, and properties of molecular complexes and has played an important role in the study of molecular interactions. General computer programs such as GAUSSIAN 94 are available for *ab initio* calculations.

For simplicity, let us consider two interacting closed-shell molecules A and B, or a molecular complex AB. (One can easily extend the descriptions introduced in this section to the interaction of three or more molecules.) The interaction energy of the complex, ΔE, is simply obtained as the difference between the total energy of the complex and the sum of the monomer energies:

$$\Delta E = E_{AB} - E_0$$
$$E_0 = E_A + E_B \tag{3.1-1}$$

where E_{AB}, E_A, and E_B are the total energies of the complex, monomer A, and monomer B, respectively. The total energy is obtained at either the self-consistent field (SCF) or the correlated wave function level of calculation. In practice it is convenient to divide the interaction energy into two parts: the SCF energy, ΔE_{SCF}, and the electron correlation energy, ΔE_{corr}.

$$\Delta E = \Delta E_{SCF} + \Delta E_{corr} \tag{3.1-2}$$

The electron correlation energy is obtained as the difference between the correlation energy of the complex and the sum of the correlation energies of the monomers. ΔE_{corr} corresponds closely to the dispersion energy (intermolecular correlation energy), but contains not only the dispersion energy but also changes in the intramolecular correlation energies of monomers in the complex. To obtain "pure" dispersion energy, it is necessary to perform a separate calculation using the perturbation formula for dispersion energy.

3.1.1.2 Ab Initio Calculations

The quantities calculated by the *ab initio* method depend upon the basis set used in the calculations. For quantitative results a sufficiently large basis set must be used. Interaction energies for the water dimer from calculations using large basis sets are given in Table 3.1-1. The experimental energy is $5.4 \pm 0.7 \, \text{kcal} \, \text{mol}^{-1}$ while the value at the Hartree–Fock limit is estimated to be $3.67\,\text{kcal mol}^{-1}$. This indicates the significance of the effect of electron correlation. To obtain the quantitative or semi-quantitative interaction energy, therefore, it is necessary to use a basis set which is larger than the triple-zeta plus polarization set, augmented with diffuse functions and a correlated wavefunction. A diffuse function is especially important for molecular interactions.

It is difficult to study molecular interactions using the supermolecule approach because of a problem connected with the so-called "basis set superposition error" (BSSE). BSSE becomes particularly important when a lower level of basis set is used. The origin of BSSE may be explained as follows. When a monomer is brought into a complex, it improves its own wavefunction using the basis functions of the partner molecule. Thus a "monomer in a complex" may be variationally more stabilized than one which is isolated. Therefore, the interaction energy measured from the energy of an isolated monomer is overestimated because of BSSE.

In practice, BSSE is corrected using the counterpoise (CP) correction method [5]. Using this method, the interaction energy is calculated as

$$\Delta E^{CP} = E_{AB}\left(\chi_{AB}; R_{AB}\right) - \left\{E_A^{CP}\left(\chi_{AB}; R_{AB}\right) + E_B^{CP}\left(\chi_{AB}; R_{AB}\right)\right\} \qquad (3.1\text{-}3)$$

TABLE 3.1-1. Intermolecular interaction energies (kcal mol^{-1}) for the water dimer at different levels of basis set and different theories (from [4])

Theory	Basis set[a]			
	A	B	C	D
SCF[b]	−4.79	−4.21	−3.73	−3.72
MP2[c]	−5.32	−6.21	−5.16	−5.19
MP3[c]	−5.19	−6.21	−5.03	−5.09
MP4(SDQ)[c]	−5.31	−6.02	−4.89	−4.94

[a] A, the 6-311G basis set augmented with two sets of p-functions on the oxygen atom and one set of p-functions on the hydrogen atoms; B, triple-zeta set augmented with two sets of p-functions and diffuse s- and p-functions on the oxygen; C, the same as B with an additional set of polarization functions (one set of d-functions and two sets of f-functions on the oxygen atom and one set of p-functions and one set of d-functions on the hydrogen atoms); D, the same as C with an additional diffuse s-function and an additional set of d-functions on the hydrogen atoms.
[b] SCF, the self-consistent-field theory.
[c] MP2, second-order Møller–Plesset (MP) perturbation theory; MP3, third-order MP theory; MP4 (SDQ), fourth-order MP theory with single (S), double (D), and quadruple (Q) substitutions.

where χ_{AB} is the basis set for the complex AB. This combination of the basis sets χ_A and χ_B for the isolated monomers A and B, respectively, is given by

$$\chi_{AB} = \chi_A \oplus \chi_B \tag{3.1-4}$$

Thus the CP correction uses the same basis set for both the monomers and the complex.

A systematic examination of the CP correction at the SCF level has been carried out for various hydrogen-bonding dimers [6]. It is clear that basis-set dependency is attenuated by the CP correction; corrected results using small basis sets agree reasonably well with those from the larger basis set 6-31G** (Figs. 3.1-1 and 3.1-2). It should be noted, however, that corrected results from the STO-3G basis set are much less satisfactory. CP correction at the SCF level works well for interacting molecules and is frequently used, but at the correlated wavefunction level some problems arise [7].

3.1.2 Orbital Interaction View of Molecular Interactions

Energy decomposition (ED) analysis [8], the procedures for which are described in Sects. 3.1.2.1. to 3.1.2.4, provides component interactions which correspond to those defined by perturbation theory within the framework of the supermolecule

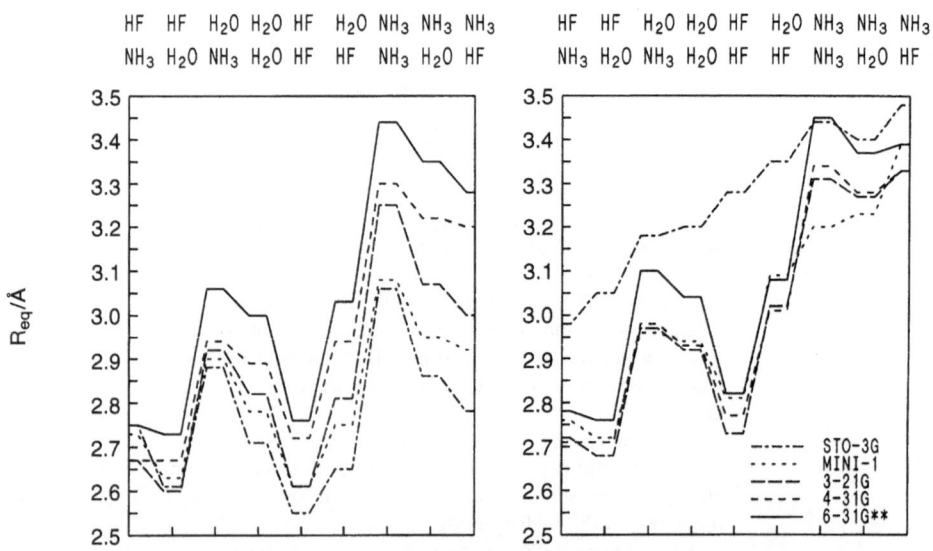

FIG. 3.1-1. Equilibrium distances (R_{eq}) for hydrogen-bonding dimers (electron acceptor monomers are shown in the first line and electron donor monomers in the second line) with various basis sets at the SCF level. **a** Without counter-poise (CP) corrections. **b** With CP corrections. STO-3G and MINI-1 belong to a minimal basis set and 3-21G and 4-31G belong to a valence–double-ζ basis set. 6-31G** is a valence–double-ζ plus polarization basis set. (Modified from [6])

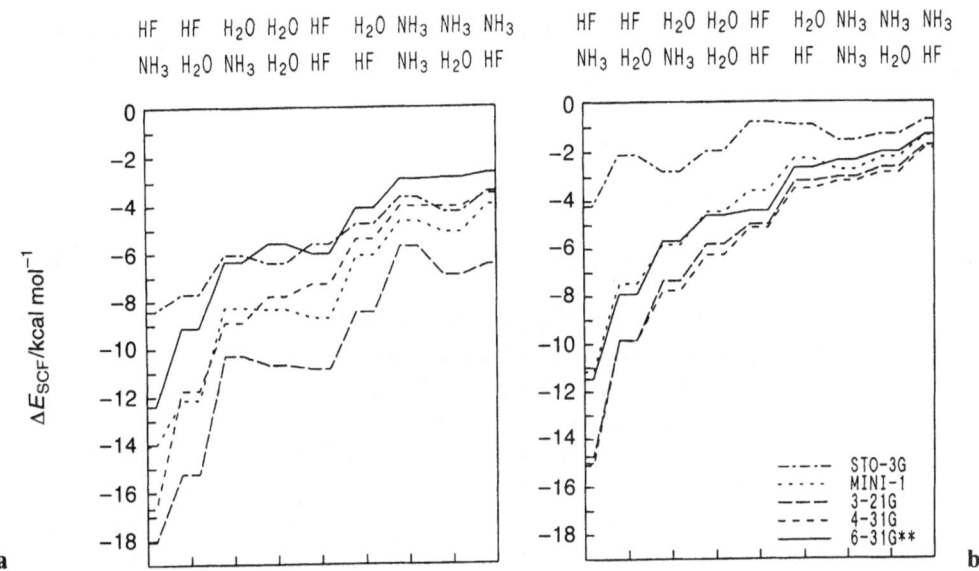

FIG. 3.1-2. Interaction energies ΔE_{SCF} at R_{eq} for hydrogen bonding dimers with various basis sets at the SCF level. **a** Without CP corrections. **b** With CP corrections. (Modified from [6])

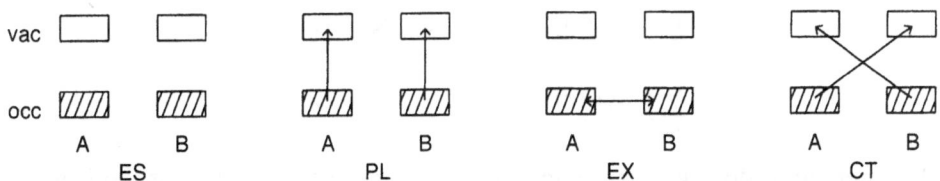

FIG. 3.1-3. The modes of orbital interactions among occupied (*occ*) and vacant (*vac*) orbitals of monomers *A* and *B* involved in component interactions. *ES*, electrostatic interaction; *PL*, polarization interaction; *EX*, exchange repulsion interaction; *CT*, charge transfer interaction

approach. ED analysis is the practical analysis not only of the interaction energy but also of other properties based on physically meaningful component interactions. For example, it is possible to separate the change in electron density distribution (difference density distribution) which accompanies molecular interactions into its components. ED analysis also gives an orbital interaction perspective of molecular interactions that is useful to chemists. The theoretical basis for this approach to chemical reactions was formulated by Fukui [9] and Woodward and Hoffmann [10]. Orbital interaction diagrams help in understanding and predicting the structures and energies of interacting molecular systems by using the shapes of the orbital of isolated molecules. A diagrammatic representation of the modes of orbital interaction for various component interactions is given in Fig. 3.1-3.

ΔE_{SCF} and the difference density distribution $\Delta\rho_{\text{SCF}}$ at the SCF level can be divided into its component parts.

$$\Delta E_{\text{SCF}} = E_{\text{ES}} + E_{\text{PL}} + E_{\text{EX}} + E_{\text{CT}} + E_{\text{MIX}} \tag{3.1-5}$$

$$\Delta\rho_{\text{SCF}}(r) \equiv \rho_{\text{AB}}(r) - \left\{\rho_{\text{A}}(r) + \rho_{\text{B}}(r)\right\}$$
$$= \rho_{\text{PL}}(r) + \rho_{\text{EX}}(r) + \rho_{\text{CT}}(r) + \rho_{\text{MIX}}(r) \tag{3.1-6}$$

E_{MIX} is the highest-order term in the perturbation expansion and is the energy of the coupling interaction between the other four well-defined components. E_{MIX} is usually small in magnitude (see Tables 3.1-3 and 3.1-4) and does not play an important role in weak molecular interactions. ρ_{AB}, ρ_{A}, and ρ_{B} are the electron density distributions associated with the wavefunctions for the complex and isolated monomers A and B, respectively. There is no ES contribution to the difference density, since by definition the change in electron density distribution is not accompanied by an ES interaction, as described in the following section.

3.1.2.1 Electrostatic Interaction

An electrostatic interaction is defined as a classical Coulombic interaction between the charge distributions of isolated monomers A and B.

$$E_{\text{ES}} = \int dr\, dr'\, \rho_{\text{A}}(r)\rho_{\text{B}}(r')/|r - r'| - \sum_{t \in B} \int dr \rho_{\text{A}}(r) Z_t /|r - r_t|$$
$$- \sum_{s \in B} \int dr \rho_{\text{B}}(r) Z_s /|r - r_s| + \sum_{s \in A} \sum_{t \in B} Z_s Z_t /|r_s - r_t| \tag{3.1-7}$$

The interaction does not cause any mixing between the orbitals of the isolated monomers (no deformation of the electron distribution). According to ED analysis procedure, the ES energy is calculated as follows:

$$\Psi_1 = \Phi_{\text{A}}^0 \cdot \Phi_{\text{B}}^0$$
$$E_1 = \left\langle \Psi_1 \middle| H_{\text{AB}} \middle| \Psi_1 \right\rangle$$
$$E_{\text{ES}} = E_1 - E_0 \tag{3.1-8}$$

Ψ_1 is a hypothetical or model wavefunction which is the Hartree product of the total wavefunctions (Slater determinants) of isolated monomers Φ_{A}^0 and Φ_{B}^0. H_{AB} is the total Hamiltonian and E_1 is the total energy of complex AB associated with the model wavefunction. The electrostatic interaction energy E_{ES} is obtained by subtracting the total energy of isolated monomers, E_0, from E_1. The model wavefunction does not allow an exchange of electrons between molecules. Therefore E_1 does not include intermolecular exchange interactions, and the ES energy calculated in this manner is exactly the same as the energy calculated using Eq. (3.1-7).

At large intermolecular distances, E_{ES} asymptotically becomes the sum of the Coulombic interactions between the net atomic point charges on constituent atoms Q:

$$E_{ES} \cong \sum_{s \in A} \sum_{t \in B} Q_s Q_t / |r_s - r_t| \qquad (3.1\text{-}9)$$

This expression for ES is familiar to chemists. To estimate the ES interaction energy, it is more appropriate to use the net atomic charges derived from the electrostatic potential, which are determined in order to find the electrostatic potential of the molecule, rather than using those from Mulliken population analysis.

The electrostatic potential of molecule A, defined as

$$V_A(r) = \int \rho_A(r_i) / |r - r_i| dr_i - \sum_{s \in A} Z_s / |r - r_s| \qquad (3.1\text{-}10)$$

is obtained from molecular orbital calculations. Using the net atomic charge Q_s, then

$$V_A(r) \cong \sum_{s \in A} Q_s / |r - r_s| \qquad (3.1\text{-}11)$$

Thus $QV_A(r)$ represents the electrostatic interaction energy between a point charge Q at r and the molecule. The electrostatic potential may be used to estimate E_{ES} and to predict electrostatically favorable configurations of a partner molecule without performing calculations on dimers. This is useful, since the ES interaction plays an important role in many molecular systems. Electrostatic potential has been used in the study of large molecules, especially biological molecules [11].

3.1.2.2 Polarization Interaction

The polarization interaction is described by the orbital interactions within each monomer. Owing to the electrostatic potential of the partner molecule, orbital mixing occurs between the occupied orbitals of a monomer and the vacant orbitals of the same monomer (Fig. 3.1-3). The polarization energy, E_{PL}, is calculated as follows:

$$\Psi_2 = \Phi_A^2 \cdot \Phi_B^2$$
$$E_2 = \langle \Psi_2 | H_{AB} | \Psi_2 \rangle$$
$$E_{PL} = E_2 - E_1 \qquad (3.1\text{-}12)$$

where Ψ_2 is a Hartree product of the total wavefunctions of monomers Φ_A^2 and Φ_B^2, which are Slater determinants derived from the deformed occupied orbitals of the monomers. The polarization energy is the Coulombic interaction energy between the induced charge distribution (difference charge distribution) of a monomer and the charge distribution of the partner molecule. E_{PL} corresponds to the classical polarization energy

$$E_{PL}^{classical} = -\frac{1}{2}\alpha_A E_A^2 - \frac{1}{2}\alpha_B E_B^2 \qquad (3.1\text{-}13)$$

where α_A is the polarizability of molecule A, and E_A is the external electric field on A.

As an example, the difference density distribution, $\Delta\rho_{PL}$, of the water dimer is shown in Fig. 3.1-4. As expected, the charge distribution of the proton donor molecule has been pushed away from the proton acceptor molecule to some extent. Polarization appears to propagate through a local polarization of bonds. This polarization interaction is the key to account for the deformation of the density distribution in this case.

3.1.2.3 Exchange Repulsion Interaction

Exchange repulsion interaction is the interaction of the total wavefunction which satisfies the Pauli principle. The exchange repulsion energy E_{EX} is calculated as

$$\Psi_3 = \mathcal{A}\left(\Phi_A^0 \cdot \Phi_B^0\right)$$
$$E_3 = \left\langle \Psi_3 \middle| H_{AB} \middle| \Psi_3 \right\rangle$$
$$E_{EX} = E_3 - E_1 \qquad (3.1\text{-}14)$$

where $\mathcal{A}$ is an antisymmetrizer. Ψ_3 is an antisymmetrized product of the total wavefunctions of the isolated monomers, or a Slater determinant formed from the orthonormalized occupied orbitals of those monomers. The orbital interaction between occupied orbitals of monomers (see Fig. 3.1-3) is the result of orthonormalization among those monomers. The EX interaction results in a decrease in the electron distribution in the interaction region and an increase in the electron distribution in the outer region (Fig. 3.1-4).

3.1.2.4 Charge Transfer Interaction

Charge transfer interaction is the delocalization interaction of electrons. The electrons of a monomer delocalize to its partner. The CT interaction is caused by orbital mixing between the occupied orbitals of monomer A and the vacant orbitals of monomer B, resulting in a charge transfer from A to B, as well as orbital mixing between the occupied orbitals of B and the vacant orbitals of A, resulting in a charge transfer from B to A. The charge transfer interaction energy E_{CT} is calculated as follows:

FIG. 3.1-4. Difference density maps for the water dimer from the 6-31G** basis set. The *solid lines* show the increases in electron density and the *dotted lines* show the decreases. The water molecule on the *left* is an electron donor or a proton acceptor, and the water molecule on the *right* is an electron acceptor or a proton donor. $\Delta\rho_{SCF}$, total difference density at the SCF level; ρ_{EX}, difference density of exchange repulsion; ρ_{PL}, difference density of polarization; ρ_{CT}, difference density of charge transfer

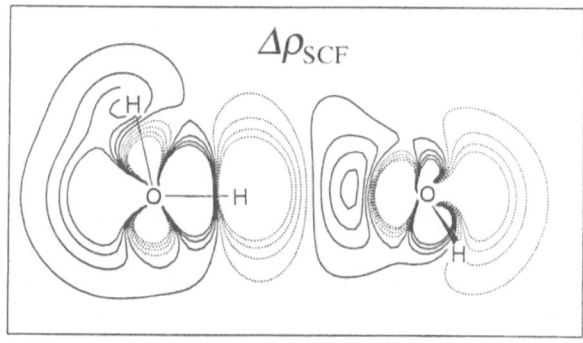

$\Delta\rho_{SCF}$

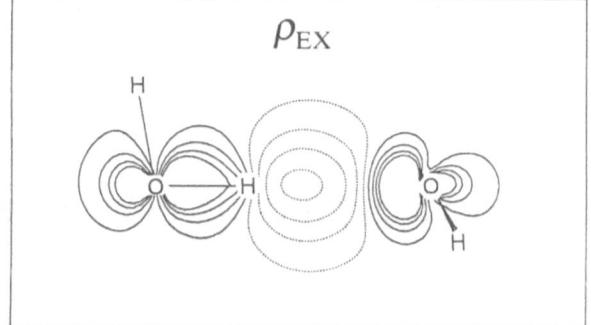

ρ_{EX}

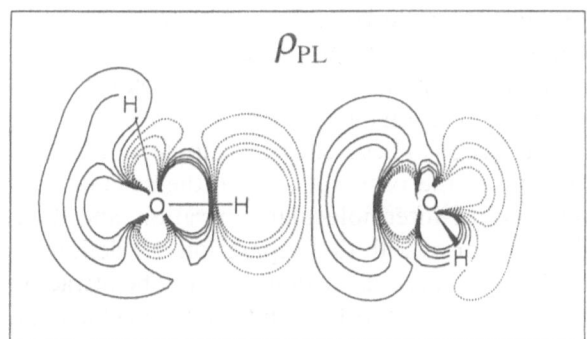

ρ_{PL}

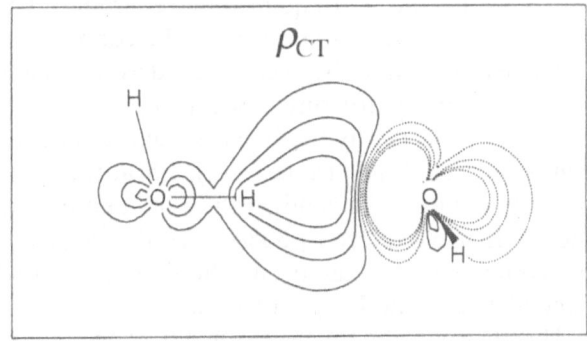

ρ_{CT}

$$\Psi_4 = \mathscr{A}\left(\Phi_A^4 \cdot \Phi_B^4\right)$$

$$E_4 = \left\langle \Psi_4 \middle| H_{AB} \middle| \Psi_4 \right\rangle$$

$$E_{CT} = E_4 - E_1 \tag{3.1-15}$$

Ψ_4 is an antisymmetrized product of the total wavefunctions of monomers Φ_A^4 and Φ_B^4, which involve the deformed occupied orbitals of monomer A and the deformed occupied orbitals of monomer B, respectively. Ψ_4 corresponds to the ground state wavefunction, Eq. (3.1-16) introduced by Mulliken in his theory of charge transfer interactions.

The difference density of the CT interaction is shown in Fig. 3.1-4. A substantial charge transfer from the electron donor to the $O \cdots H{-}O$ bonding region is observed. In the water dimer, the charge transfer in the opposite direction is too small to illustrate.

3.1.3 Bonding of Molecular Complexes

3.1.3.1 Hydrogen Bonding

A hydrogen bond is usually formed between a polar X–H bond and an electronegative atom $Y: X{-}H \cdots Y$. The bond is generally weak compared with chemical bonds, and the bonding energy is several kcal mol^{-1}. The water dimer is a typical example of a hydrogen bonding complex.

The interaction energy of the water dimer, as well as its components, is shown in Fig. 3.1-5. (These energies should not be taken as being precise since they depend on the basis set used in the calculations.) The component energies depend on the intermolecular separation, and the relative importance of the components varies with that separation. By definition, ES interaction is a long-range force and EX is a short-range force. The attractive forces overcome the EX force at intermediate and large intermolecular distances, although EX becomes dramatically large at short intermolecular distances. The attractive and repulsive forces are balanced at equilibrium separation. ES energy and CT are the main contributors to the stabilization of the complex.

Three models have been considered as candidates for the structure of the water dimer. The linear structure model is supported both experimentally and theoretically. The reason why the linear structure model is the most stable is shown by component analysis (Table 3.1-2). ES attractions are reduced by 25%–40% in bifurcated and cyclic configurations compared with the linear model. Although the EX is also reduced by 30%–60%, this does not balance the ES reduction and the complex is not stabilized. The linearity of $O{-}H \cdots O$ is due to the angular dependence of the ES interaction.

The energy components of some hydrogen-bonding dimers are given in Table 3.1-3. From the viewpoint of component interactions, they have many characteristics in common. ES energy is the largest attractive force and determines the

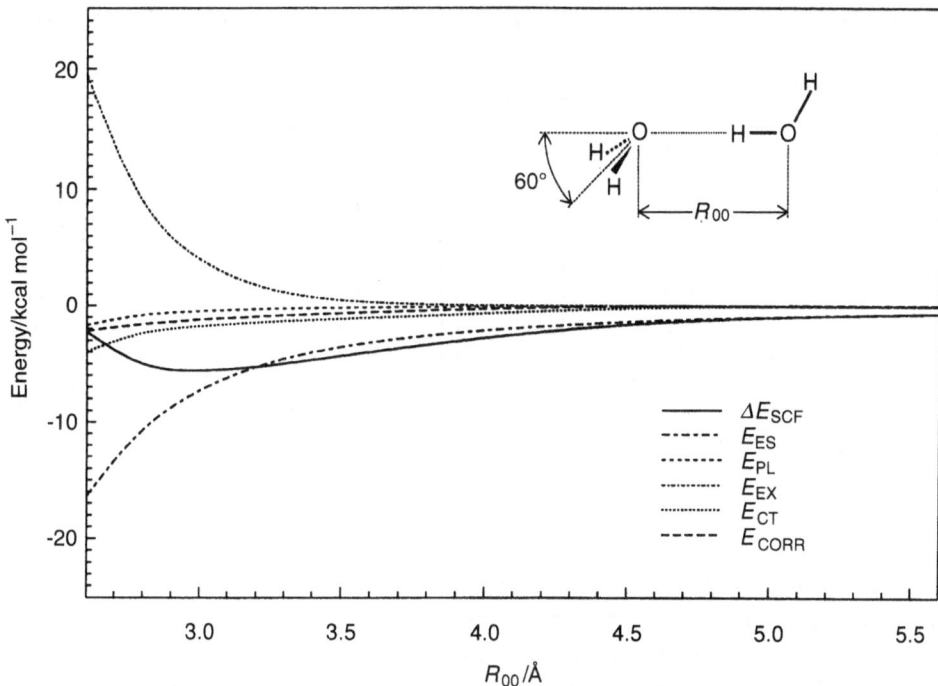

FIG. 3.1-5. The interaction energy ΔE_{SCF} and its components obtained with the 6-31G** basis set. The correlation contribution ΔE_{CORR} is calculated using the the second-order Moller–Plesset perturbation theory (MP2). E_{ES}, electrostatic energy; E_{PL}, polarization interaction energy; E_{EX}, exchange repulsion energy; E_{CT}, charge transfer energy

TABLE 3.1-2. Comparison of energy components (kcal mol^{-1}) for three models of the water dimer (from [3])

	Linear	Bifurcated	Cyclic
R_e(Å)	2.88	2.90	2.85
ΔE_{SCF}	−7.8	−6.4	−6.1
E_{ES}	−10.5	−7.4	−7.4
E_{EX}	6.2	2.3	4.7
E_{PL}	−0.6	−0.3	−0.3
E_{CT}	−2.4	−1.0	−2.8
E_{MIX}	−0.5	−0.1	−0.3

ΔE_{SCF}, interaction energy from the self-consistent field (SCF) calculation; E_{ES}, electrostatic energy; E_{EX}, exchange repulsion energy; E_{PL}, polarization energy; E_{CT}, charge transfer energy; E_{MIX}, higher-order interaction energy.

TABLE 3.1-3. Energy components (kcal mol^{-1}) of hydrogen-bonding dimers[a] (from [3])

Proton acceptor	Proton donor	R_e (Å)	θ (deg)	ΔE_{SCF}	E_{ES}	E_{EX}	E_{PL}	E_{CT}	E_{MIX}
H_3N	HF	2.68	0	−16.3	−25.6	16.0	−2.0	−4.1	−0.7
H_2O	HF	2.62	6	−13.4	−18.9	10.5	−1.6	−3.1	−0.4
HF	HF	2.71	60	−7.6	−8.2	4.5	−0.4	−3.2	−0.3
H_3N	HOH	2.93	0[b]	−9.0	−14.0	9.0	−1.1	−2.4	−0.4
H_2O	HOH	2.88	60[b]	−7.8	−10.5	6.2	−0.6	−2.4	−0.5
H_3N	HNH_2	3.30	0[b]	−4.1	−5.7	3.6	−0.6	−1.3	−0.2
H_2O	HNH_2	3.22	60[b]	−4.1	−4.6	2.5	−0.3	−1.5	−0.2
H_3N	HCH_3	4.02	0[b]	−1.1	−0.6	0.5	−0.3	−0.7	−0.0
H_2O	HCH_3	3.80	60[b]	−1.1	−0.5	0.5	−0.1	−0.9	−0.0

[a] All the complexes are linear, having a linear X $\cdots$ H–Y bond. θ is the angle between this bond and the molecular axis of the proton acceptor.
[b] Assumed.

strength of the bond. The results show that these hydrogen bonds are strongly electrostatic in nature with small but significant contributions from charge transfer. Charge transfer is usually the second strongest attraction term in hydrogen bonding. The relative importance of these contributions varies from one complex to another.

3.1.3.2 Bonding of Electron Donor–Acceptor Complexes

The theory of charge transfer interactions was proposed by Mulliken to explain an absorption band in a benzene–iodine complex that did not appear in the spectrum of either benzene or iodine alone. A charge transfer band has been observed in many molecular complexes, and these are called charge transfer or electron donor–acceptor (EDA) complexes.

According to the Mulliken theory, the wavefunction of the electronic ground state, Ψ_N, is written as

$$\Psi_N \cong a\mathcal{A}\Psi\left(\Phi_A \cdot \Phi_D\right) + b\mathcal{A}\Psi\left(\Phi_{A^-} \cdot \Phi_{D^+}\right), \quad |a| >> |b| \tag{3.1-16}$$

where $\mathcal{A}\Psi(\Phi_A \cdot \Phi_D)$ is an antisymmetrized product of the wavefunctions of the electron acceptor (Φ_A) and donor (Φ_D), and $\mathcal{A}\Psi(\Phi_{A^-} \cdot \Phi_{D^+})$ represents a wavefunction corresponding to the transfer of an electron from the donor to the acceptor [Ψ_N corresponds to Ψ_4 in Eq. (3.1-15)]. The wavefunction of the electronic excited state, Ψ_E, is written as

$$\Psi_E \cong a^*\mathcal{A}\Psi\left(\Phi_A \cdot \Phi_D\right) + b^*\mathcal{A}\Psi\left(\Phi_{A^-} \cdot \Phi_{D^+}\right), \quad |a^*| << |b^*| \tag{3.1-17}$$

The electronic transition from ground state Ψ_N to excited state Ψ_E gives a charge transfer absorption that does not correspond to the excited state of the donor or the acceptor.

TABLE 3.1-4. Energy components (kcal mol⁻¹) and qualitative classifications of EDA complexes (from [12])

Donor–acceptor	Type	Symmetry	R_e(Å)	E_{ES}	E_{EX}	E_{PL}	E_{CT}	E_{MIX}	ΔE_{SCF}	E_{DISP}	Classification
H_3N–BF_3	n–ν	C_{3v}	1.60	−142.3	136.3	−42.7	−52.7	29.9	−71.5		Strong ES
H_3N–BH_3	n–ν	C_{3v}	1.70	−92.9	86.9	−17.2	−27.1	5.6	−44.7		Strong ES
OC–BH_3	σ–ν	C_{3v}	1.63	−60.9	98.9	−61.8	−68.3	63.6	−28.5		Strong CT–PL–ES
H_3N–ClF	n–σ^*	C_{3v}	2.72	−11.2	7.4	−1.1	−3.6	0.2	−8.2		Intermediate ES
H_2O–OC(CN)$_2$	n–π^*	C_s	2.70	−9.7	4.4	−1.0	−1.8ᵇ		−8.0	−1.2	Intermediate ES
C_6H_6–OC(CN)$_2$	π–π^*	C_s	3.6ᵃ	−2.8	1.8	−1.7	−1.6	0.1	−4.2	−2.6	Intermediate ES–DISP–PL–CT
HF–ClF	n–σ^*	C_s	2.74	−3.6	1.8	−0.2	−1.4	0.1	−3.4		Weak ES
H_3N–Cl_2	n–σ^*	C_{3v}	2.93	−4.0	3.9	−0.8	−2.3	0.3	−2.9		Weak ES–CT
C_6H_6–ClF	π–σ^*	C_s	3.6	−1.6	0.6	−0.1	−0.8	0.0	−1.8	−0.5	Weak ES–CT
H_3N–F_2	n–σ^*	C_{3v}	3.00	−0.8	0.6	−0.3	−0.6	0.0	−1.1		Weak ES–CT
H_2CO–F_2	n–σ^*	C_s	2.91	−0.4	0.3	−0.1	−0.5	0.0	−0.7		Weak CT–ES–DISP
C_2H_4–H_2CO	π–π^*	C_s	3.75	−0.5	0.4	−0.1	−0.5ᵇ		−0.7	−0.4	Weak CT–ES–DISP
C_6H_6–Cl_2	π–σ^*	C_s	3.6	−0.5	0.7	−0.1	−0.8	0.0	−0.6	−0.7	Weak CT–DISP–ES
C_6H_6–F_2	π–σ^*	C_s	3.3	−0.2	0.3	−0.0	−0.4	0.0	−0.3	−0.4	Weak DISP–CT–ES
F_2–F_2	n–σ^*	C_s	2.7	−0.1	0.3	0.0	−0.4	0.0	−0.2	−0.2	Weak DISP–CT
	π–σ^*										

ᵃ Not optimized.
ᵇ E_{CT} + E_{MIX}.

The bonding nature of EDA complexes in the electronic ground state can be explained by examining the component interactions. Many EDA complexes have been analyzed and classified based on their component energies [12]. Table 3.1-4 shows the component energies for various EDA complexes. It is necessary to know the CT interaction in order to obtain the charge transfer spectra as elucidated by the Mulliken theory. However, the ES interaction is the most important bond in an EDA complex in the ground state except for complexes composed of non-polar molecules. The right-hand column in Table 3.1-4 shows the EDA complexes classified according to the relative importance of their component interactions to the total bonding energy.

3.1.4 A Theoretical Perspective on Molecular Recognition

We have seen how molecular interactions are described by the quantum mechanical theory. The concepts and physical pictures introduced in this section may be useful in understanding various types of molecular interactions.

Most molecular systems associated with molecular recognition are large with many intra- and intermolecular interaction sites, which may work simultaneously or sequentially. Cooperation between interaction sites may be one of the keys to understanding the mechanism of molecular recognition. Electronic structure theory has provided very little information about cooperation, because interesting systems such as host–guest complexes are usually too large for the necessary computations to be made. Rapid improvements in computers, as well as further development of the theory, should make it possible to investigate the theory of cooperation in more detail.

It is also important to study the statistical and dynamic aspects of the phenomena associated with molecular recognition and self-regulation. For this the technique of molecular dynamics simulation [13] may be a powerful tool, as this method has been successfully applied to liquids, solids, and proteins.

In molecular simulations the reliability of the results largely depends on the quality of the potential functions used. Classical potential functions such as Lennard–Jones-type potentials have generally been used until now, but quantum mechanical potential functions may be required in order to obtain reasonable descriptions of the various types of molecular interactions which contribute to a complex molecular system. In practice, however, this will be very difficult for some considerable time, since it requires a huge amount of computer time.

References

1. Kaplan I (1987) Theory of intermolecular interactions. Elsevier, Amsterdam
2. Jeziorski B, Moszynski R, Szalewicz K (1994) Perturbation theory approach to intermolecular potential energy surfaces of van der Waals complexes. Chem Rev 94:1887–1930

3. Morokuma K, Kitaura K (1980) Variational approach (SCF *ab initio* calculations) to the study of molecular interactions: The origin of molecular interactions. In: Ratajczak H, Orville-Thomas WJ (eds) Molecular interactions. Wiley, Chichester

4. Saebo S, Tong W, Pulay P (1993) Efficient elimination of basis set superposition errors by the local correlation method: Accurate *ab initio* studies of the water dimer. J Chem Phys 98:2170–2175

5. Boys SF, Bernardi F (1970) The calculation of small molecular interactions by the difference of separate total energies. Some procedures with reduced errors. Mol Phys 19:553–566

6. Alagona G, Ghio C, Cammi R, Tomasi J (1988) A reappraisal of the hydrogen-bonding interaction obtained by combining energy decomposition analyses and counterpoise corrections. In: Marina J (ed.) Molecules in physics, chemistry, and biology, vol 2. Kluwer, London

7. van Duijneveldt FB, van Duijneveldt-van de Rijdt GCM, van Lenthe JH (1994) State of the art in counterpoise theory. Chem Rev 94:1873–1885

8. Kitaura K, Morokuma K (1976) A new energy decomposition scheme for molecular interactions within the Hartree–Fock approximation. Int J Quantum Chem 10:325–340

9. Fukui K (1966) An MO-theoretical illustration for the principle of stereoselection. Bull Chem Soc JPN 39:498–503

10. Hoffmann R, Woodward RB (1968) The conservation of orbital symmetry. Acc Chem Res 1:17–22.

11. Politzer P, Truhlar DG (eds) (1981) Chemical applications of atomic and molecular electrostatic potentials. Plenum, New York

12. Morokuma K (1977) Why do molecules interact? The origin of electron donor-acceptor interaction, hydrogen bonding and proton affinity. Acc Chem Res 10:294–300

13. Allen MP, Tildesley DJ (1987) Computer simulation of liquids. Oxford University Press, New York

3.2
Energy Transfer

OKITSUGU KAJIMOTO

Energy transfer is the fundamental process for energy accumulation and dissipation in a molecular ensemble or assembly. Chemical reactions, for example, occur as a result of vibrational energy accumulation into a specific bond of the molecule. Vibrational energy can be accumulated via a series of collisions with other molecules or via internal conversion from an other electronic state without collisions.

Energy transfer can be divided into two categories, intermolecular and intramolecular energy tranfer. Intermolecular energy transfer occurs when two molecules collide. The electronic/vibrational energy of one molecule is lost to or gained from the electronic, vibrational, rotational, or translational degree of freedom of the other molecule. Even vibrational energy transfer in the liquid phase can be modeled as successively occurring collisional energy transfer. On the other hand, intramolecular vibrational energy transfer, or vibrational energy redistribution (IVR), in large molecules occurs without collision. An excited electronic state can also be converted without collision to another electronic state of the molecule via internal conversion or intersystem crossing. Intramolecular energy transfer is, of course, enhanced by bimolecular collisions.

3.2.1 Vibrational Energy Transfer in the Gas and Liquid Phases

3.2.1.1 Collisional Energy Transfer

Bimolecular Collisions. Energy transfer through bimolecular collisions can be treated quantum mechanically [1]. Let us consider a collision between atom A and diatomic molecule BC with a single vibrational mode. The stationary Schrödinger equation of the system is written as.

$$\left[-\frac{\hbar^2}{2M}\nabla_R^2 - \frac{\hbar^2}{2\mu}\nabla_r^2 + V(R,r) - E \right]\Psi = 0 \tag{3.2-1}$$

where R denotes the distance between A and BC (center of mass), r is the internal coordinate of BC, amd M and μ represent the reduced masses of the system A–BC and the oscillator BC, respectively. The potential V is a function of both R and r. E expresses the total energy of the system. When R is sufficiently large, V becomes a function of r alone and the above equation can be decomposed into two independent Schrödinger equations:

$$\left[-\frac{\hbar}{2M}\nabla_R^2 - E_n\right]f_n(R) = 0 \tag{3.2-2}$$

$$\left[-\frac{\hbar}{2\mu}\nabla_r^2 + V(r) - \varepsilon_n(r)\right]\phi_n(r) = 0 \tag{3.2-3}$$

where

$$E = E_n + \varepsilon_n, \qquad \Psi_n = f_n(R)\phi_n(r) \tag{3.2-4}$$

and n is the vibrational quantum number of the BC molecule. The asymptotic solution of Eq. 3.2-2 is

$$f_n(R) = A_n\exp(-ik_nR) + B_n\exp(ik_nR) \tag{3.2-5}$$

where $k_n = (2ME_n)^{1/2}/\hbar$. The first term expresses the incident wave whereas the second term represents the reflected wave. The solution of the second Schrödinger equation Eq. 3.2-3 may be the harmonic oscillator wavefunction, which depends on the potential function $V(r)$.

The general solution of Eq. 3.2-1 can be expressed as a linear combination of $f_n\phi_n$.

$$\Psi = \sum_n f_n(R)\phi_n(r) \tag{3.2-6}$$

Substitution of Eq. 3.2-6 into Eq. 3.2-1, and integration with respect to r after multiplying by $\phi_j(r)$ on both sides of the equation, produces n simultaneous differential equations:

$$\frac{\hbar^2}{2M}(\nabla_R^2 + k_j^2)f_j(R) = \sum_n U_{jn}(R)f_n(R) \qquad j = 1, 2, \dots \tag{3.2-7}$$

where $k_j^2 = 2ME_j/\hbar^2$ and

$$U_{jn}(R) = \int_{-\infty}^{\infty}\phi_j^*(r)V(R,r)\phi_n(r)dr \tag{3.2-8}$$

Let us consider the case where BC is originally in the ith state and as a result of collisional interaction it is excited to the jth state. When the probability of the $i \rightarrow j$ transition is very small, solving the following *two* simultaneous differential equations is sufficient to evaluate the transition probability P_{ij}.

$$\left[\frac{\hbar}{2M}\left(\nabla_R^2 + k_i^2\right) - U_{ii}\right]f_i = U_{ij}f_j \tag{3.2-9}$$

$$\left[\frac{\hbar}{2M}\left(\nabla_R^2 + k_j^2\right) - U_{jj}\right]f_j = U_{ij}f_i \tag{3.2-10}$$

The right-hand side of Eq. 3.2-9 is assumed to be zero because f_j is sufficiently small compared with f_i. The asymptotic solution of the equation as $R \to \infty$ can be expressed as

$$f_i(R) \sim \exp(-ik_iR) + a_i\exp(ik_iR) \tag{3.2-11}$$

$$f_j(R) \sim a_j\exp(ik_jR) \tag{3.2-12}$$

These equations imply that most of the reflected wave remains in the initial vibrational state and only a small portion is converted to the jth state. The ratio between these two states after the intraction is $|a_j^2/a_i^2|$, where $a_i^2 \sim 1$. Therefore the transition probability becomes

$$P_{ij} = \frac{k_j}{k_i}|a_j^2| \tag{3.2-13}$$

where the factor k_j/k_i is the correction for the velocity of the incident and outgoing waves.

The general form of the transition probability is found to be

$$P_{ij} = C\left|\int_{-\infty}^{\infty} \Psi_j^* U_{ij}\Psi_i dR\right|^2 \tag{3.2-14}$$

If the interaction $V(R,r)$ can be expressed as a product of $V_1(R)$ and $V_2(r)$, the probability becomes

$$P_{ij} = C\left|\int_{-\infty}^{\infty} f_j^*V_1(R)f_i dR\right|^2\left|\int_{-\infty}^{\infty} \phi_j^*V_2(r)\phi_i dr\right|^2 \tag{3.2-15}$$

The first integration term is called the *translational overlap* and expresses the energy matching between the initial and final states. Since the function $f_i(R)$ can be approximated as $\exp(ik_iR)$, the integrand is the product of $V_1(R)$ and $\exp(i(k_i-k_j)R)$. $V_1(R)$ is large only in the small range of R (i.e. ΔR) where the collision occurs. If the plane wave, $\exp(i(k_i-k_j)R)$, oscillates many times within ΔR, the integrand $V_1(R)\exp(i(k_i-k_j)R)$ averages zero and the integral also becomes zero. Therefore, the integral can be large only when the wavelength of the plane wave is of the order of ΔR. The wavelength is given by

$$\lambda = \frac{2\pi}{k_i - k_j} = \frac{2\pi(k_i + k_j)}{k_i^2 - k_j^2} = \frac{2\pi(2M\bar{v}/\hbar)}{(2ME_i - 2ME_j)/\hbar^2} = \frac{vh}{\Delta E} \qquad (3.2\text{-}16)$$

and equating this to ΔR leads to the so-called *Massay criterion* [2] as a condition for efficient vibrational energy transfer, i.e.

$$\frac{\Delta E}{h} \frac{\Delta R}{v} \sim 1 \qquad (3.2\text{-}17)$$

This condition also suggests that P_{ij} becomes large when the collision time $\Delta R/v$ is equal to the period of oscillation $1/v(= h/\Delta E)$. According to this criterion, P_{ij} must increase with an increase in the collision velocity and hence with increasing temperature, because at around room temperature $\Delta E\Delta R/h$ is much larger than the velocity of molecules.

The second integration term in Eq. 3.2-15 expresses the selection rule in vibrational energy transfer. If the interaction is linear for the internal coordinate r, this selection rule becomes $\Delta n = 1$ for a harmonic oscillator, just like the selection rule for the dipole transition. Usually, the $0 \rightarrow 2$ transition is much harder than the $0 \rightarrow 1$ transition.

Jackson and Mott [3], using an exponential repulsive interaction $V(R) = \exp(-R/\ell)$ together with the Harmonic oscillator approximation, obtained the probability of the $\Delta n = 1$ transition as

$$\langle P_{i \rightarrow i+1} \rangle = \left(i + \frac{1}{2} \pm \frac{1}{2} \right) \left(\frac{M}{\mu} \right) m'^2 \left(\frac{\varepsilon}{hv} \right) \left(\frac{2\pi\varepsilon}{3kT} \right)^{1/2} \exp\left[-\frac{3}{2} \left(\frac{\varepsilon}{kT} \right)^{1/3} \pm \frac{hv}{2kT} \right] \qquad (3.2\text{-}18)$$

where $\varepsilon = \mu(4\pi^2 v\ell)^2$ and $m' = m_C/(m_B + m_C)$. This equation predicts the $T^{-1/3}$ dependence of the transition probability in the logarithm scale, which agrees with the observed trend (as is demonstrated in Fig. 3.2-1) [4].

For three-dimensional systems, Schwartz et al. [5] and Herzfeld and Litovitz [6] have derived an expression for thermally averaged $P_{1\rightarrow0}$. Cottrell and McCoubrey [7] compared the predicted $P_{1\rightarrow0}$ with the observed probability for various molecules.

Vibrational Predissociation in van der Waals Clusters. Since the 1970s the field of cluster chemistry has developed rapidly and significantly. Predissociation of van der Waals (vdW) clusters is one of the most successful topics in cluster chemistry, which was originated by Levy [8] using I_2–rare-gas vdW clusters. When the vibrational mode of the I_2 component of the vdW cluster is excited, the vibrational energy of I_2 rapidly dissipates to the intermolecular vdW mode, which causes the dissociation of the cluster. This vibrational predissociation is a specific example of *V–T* energy transfer between molecules on collision, and provides

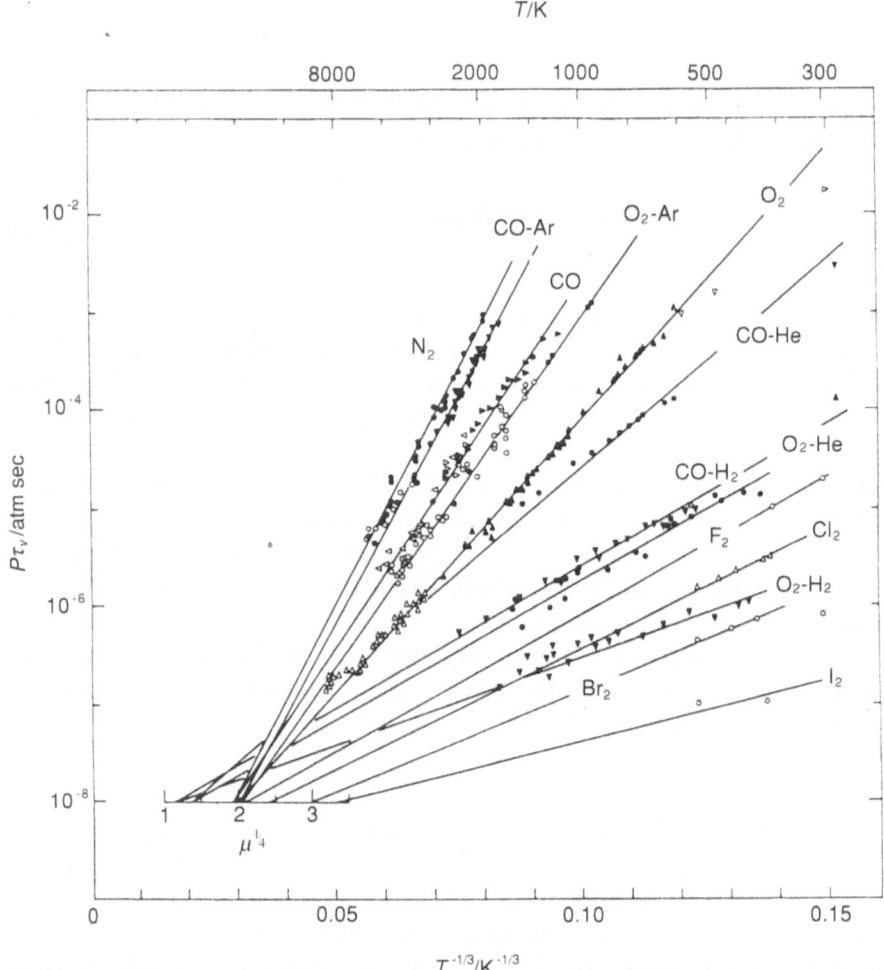

FIG. 3.2-1. The $T^{-1/3}$ dependence of vibrational relaxation times in various combinations of diatomic gases. (From [4], with permission)

much important information on the energy dependence, mode dependence, and even interaction potential dependence of the vibrational energy transfer between molecules. For example, Levy and co-workers also found that the $\Delta v = 1$ selection rule holds in $V–T$ energy transfer within vdW molecules. More recently, Gutmann and co-workers [9, 10], using real-time measurements, demonstrated that the rate of I_2–He vibrational predissociation increases with increasing vibrational quantum numbers of I_2; at higher vibrational levels, the energy gap between the excited energy level and the dissociation limit of the next lower vibrational level decreases and the situation becomes more favorable for $V–T$ energy transfer in terms of the Massay criteria.

Collisional Energy Transfer in the Liquid Phase. Vibrational and translational/ vibrational energy transfer in the liquid phase has been analyzed based on the gas-phase theory. Litovitz [11] and Herzfeld and Litovitz [12] put forward the independent binary collision (IBC) theory. They assumed that the rate of vibrational energy transfer causing the $i–j$ transition in the liquid phase, k_{ij}, can be expressed as

$$k_{ij}(\rho,T) = P_{ij}(T)Z(\rho,T)$$
(3.2-19)

where P_{ij} is the energy transfer probability per collision, and is assumed to be independent of the density of the liquid. Z denotes the collision number in the liquid phase, and is usually estimated from the gas-phase collision rate and the value of the radial distribution function $g(r^*)$ at a critical distance where the energy transfer mainly occurs, i.e.

$$Z_{liq} = Z_{gas} \frac{\rho_{liq} g_{liq}(r^*)}{\rho_{gas} g_{gas}(r^*)}$$
(3.2-20)

This simple treatment gives an accurate estimate of the energy transfer in the liquid phase in many cases. For example, Simpson and co-workers [13, 14] investigated the energy transfer between N_2 and NO in both the gas phase and in liquid xenon at the same temperature and found that the rate constant for nonresonant ($\Delta E > 150\,cm^{-1}$) $V–V$ energy transfer is the same in both phases. For near-resonant $V–V$ transfer, when $\Delta E/h$ comes closer to collision frequency, the IBC model seems to be inadequate [15]. Recently, Harris and co-workers [16, 17] studied the vibrational relaxation of a hot I_2 molecule in liquid Xe of various densities and found that the IBC model is correct even for a molecular system with large amplitude motions ($170–210\,cm^{-1}$).

3.2.1.2 Intramolecular Vibrational Energy Redistribution (IVR)

IVR in Free Molecules. In a relatively large molecule, intramolecular vibrational energy redistribution can occur without collisions. Figure 3.2-2 shows the fluorescence decay curve of anthracene excited to two vibronic levels with excess energies of 766 and 1792 cm^{-1} [18]. The decay rate increases about three orders of magnitude with increasing excess energy. This occurs as a result of intramolecular vibrational energy transfer from the original vibrational level to many nearby dark vibrational levels. The rate constant for the vibrational energy transfer from the ith vibrational level to the manifold of the jth state can be expressed using the Fermi golden rule as

$$k_{ij} = \frac{2\pi}{\hbar} V_{ij}\rho_j$$
(3.2-21)

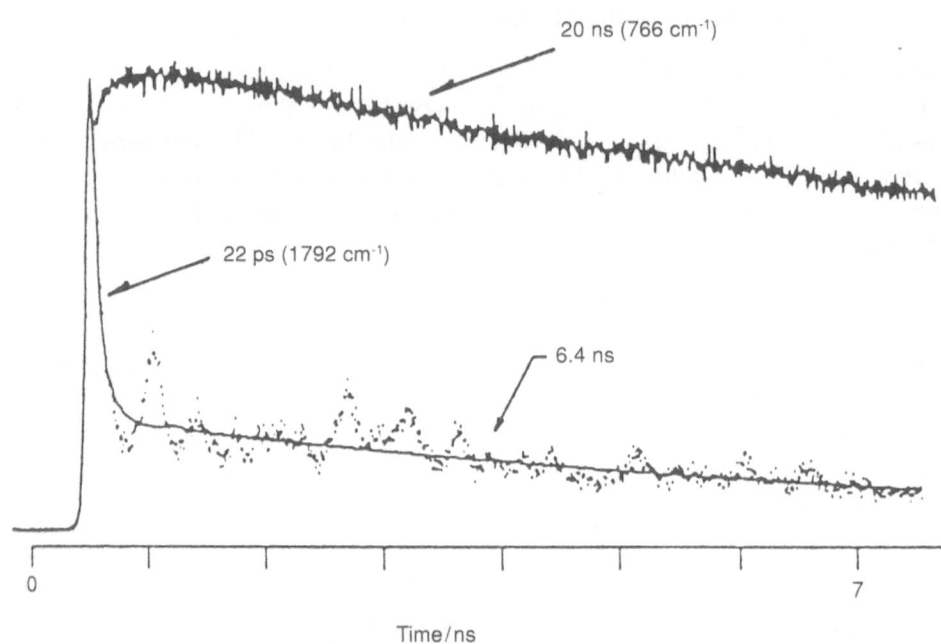

20 ns (766 cm⁻¹)

22 ps (1792 cm⁻¹)

6.4 ns

0

7

Time/ns

Fig. 3.2-2. Rate of vibrational energy redistribution from the two vibronic levels of anthracene. IVR from the 1792 cm⁻¹ level is three orders of magnitude greater than that from the 766 cm⁻¹ level. (From [18], with permission)

where V_{ij} depicts the off-diagonal coupling matrix element between the vibrational energy states i and j, and ρ_j corresponds to the level density of the jth state. Ordinarily, as the excess energy increases, the coupling between fundamental vibrational modes becomes more significant and the vibrational level density ρ_j also becomes larger. Figure 3.2-3 shows the vibrational density of state as a function of excess vibrational energy, together with the observed lifetimes of several vibronic levels of anthracene. Since the vibrational level density increases more rapidly for molecules with more vibrational modes and also those with low-frequency vibrational modes, methyl substitution (low-frequency internal rotation mode) to benzene, for example, enhances the intramolecular vibrational energy transfer.

Near the dissociation threshold, IVR must be very efficient and complete energy randomization is assumed to occur before the dissociation or other reactions. This rapid IVR is the basis for statistical treatments in reaction rate theories such as Rice-Ramsperger-Kassel-Marcus (RRKM) and transition state theories.

The time required for IVR has been measured for molecules of various sizes, as shown in Table 3.2-1. It is apparent that as the number of the vibrational mode and the excess energy increase, the IVR rate increases from microseconds to femtoseconds.

FIG. 3.2-3. The vibrational state density of anthracene as a function of the excess energy. The rates of IVR from the various excess energy levels are also given. (From [18], with permission)

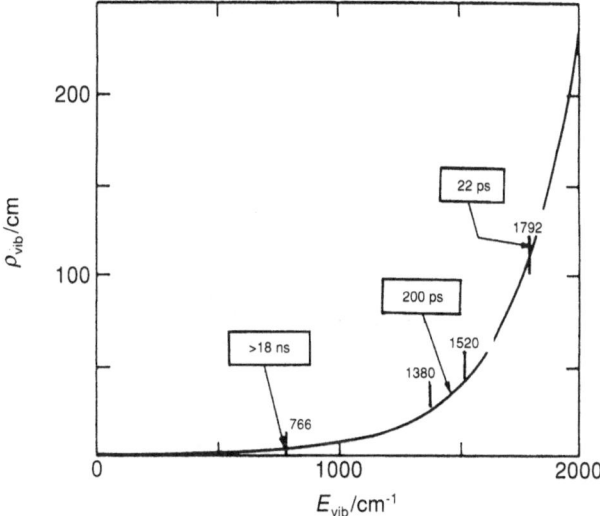

TABLE 3.2-1. IVR rates measured for molecules of various sizes

Molecules	Vibrational freedom	Excess energy (cm^{-1})	Relaxation time (ps)
Propylbenzene	56	$530(6b_0^1)$	>2 000 000
		$932(12_0^1)$	10 000
Butylbenzene	66	$530(6b_0^1)$	200 000
		$930(12_0^1)$	<1000
p-difluorobenzene	30	$2191(3_0^1 5_0^1 30_0^1)$	10
Naphthalene	48	$3069(6_0^1 9_0^1 8_0^1)$	11
		3763	5
Coumarine 6	126	$5950(2\nu_{CH})$	4
Benzene	30	$14072(5\nu_{CH})$	0.05
		$3090(\nu_{CH}, \text{liq.})$	1

IVR in Clusters and Condensed Phases. In the case of clusters, the presence of intermolecular vibrations must enhance IVR. The low-frequency vibrational mode in a free molecule is usually of the order of a few hundred cm^{-1}, while the frequency of the intermolecular mode of small clusters is less than 100 cm^{-1}, which significantly facilitates IVR by increasing the state density even at low excess vibrational energies.

Figure 3.2-4 shows the dispersed fluorescence spectra of free benzonitrile and its 1:1 CF$_3$H complex excited at the same vibronic band (31^1) [19]. In the case of free molecules, the fluorescence with sharp structure occurs only from the originally excited vibronic level, which indicates that IVR does not occur within the fluorescence lifetime. Conversely, in the CF$_3$H complex, IVR is completed very rapidly and the fluorescence with broad features occurs from the vibrationally relaxed level combined with low-frequency dark levels.

The above observation shows that even interaction with a single solvent molecule significantly enhances the IVR process. Molecules in the condensed phase

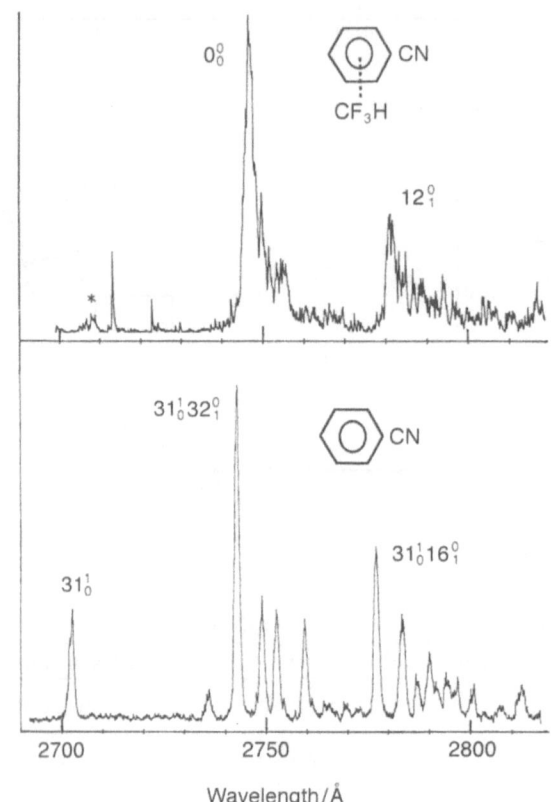

FIG. 3.2-4. The dispersed fluorescence spectra of benzonitrile and its van der Waals complex with CF_3H excited to vibronic level 31_0^1. The fluorescence from the van der Waals complex shows the completion of IVR. (From [19], with permission)

are surrounded by many solvent molecules and generally possess quite a wide spectrum of vibrational frequency, in the low-frequency region in particular, and therefore IVR must be occurring very rapidly. For larger molecules, this IVR or vibrational relaxation within solvated molecules is probably complete within a few picoseconds. However, for molecules of fewer than five atoms, the IVR rate could be measured using ordinary picosecond systems [20].

3.2.2 Electronic Energy Transfer in the Gas and Liquid Phases

3.2.2.1 Bimolecular Electronic Energy Transfer

A Simple Molecular Orbital Description. The electronic states of donor and acceptor molecules which are relevant to electronic energy transfer are shown in the simple diagram in Fig. 3.2-5. In the energy transfer process, an electron in the ϕ_D^* orbital of the donor molecule is deactivated into the ϕ_D orbital while an

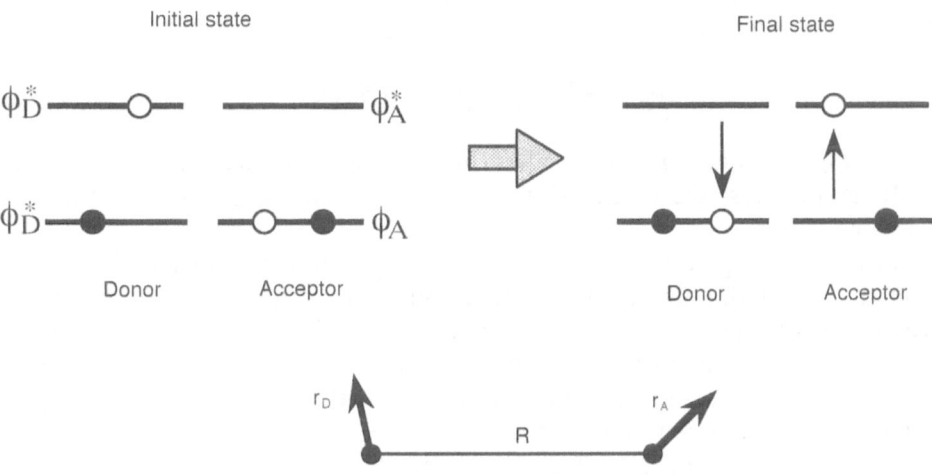

FIG. 3.2-5. A simple molecular orbital description of the donor and acceptor molecules relevant to electronic energy transfer

electron in the ϕ_A orbital of the acceptor molecule is excited into the ϕ_A^* orbital. Therefore, the initial and the final wavefunctions can be expressed as

$$\Psi_I = \frac{1}{\sqrt{2}}\left(\left|\phi_D\bar{\phi}_D^*\phi_A\bar{\phi}_A\right| - \left|\phi_D^*\bar{\phi}_D\phi_A\bar{\phi}_A\right|\right)$$

$$\Psi_F = \frac{1}{\sqrt{2}}\left(\left|\phi_D\bar{\phi}_D\phi_A\bar{\phi}_A^*\right| - \left|\phi_D\bar{\phi}_D\phi_A^*\bar{\phi}_A\right|\right) \tag{3.2-22}$$

The perturbation H_1 causing electronic energy transfer could be a dipole–dipole or a dipole–quadrupole interaction as

$$H_1(R) = \frac{e^2}{\varepsilon R^3}\left(r_D r_A - 3\left(r_D\frac{R}{R}\right)\left(r_A\frac{R}{R}\right)\right) + (\text{dipole–quadrupole term}) + \ldots \tag{3.2-23}$$

where r_D and r_A are the electron coordinates of the donor and acceptor, respectively, and R represents the direction vector between them with a distance of R. ε and e are the dielectric constants of the medium and the electron charge, respectively. Then, using the Fermi golden rule, the rate of electron transfer from D to A can be expressed as

$$k_{ET} = \frac{2\pi}{\hbar}\rho_F\left|\left\langle\Psi_I\left|H_1\right|\Psi_F\right\rangle\right|^2 \tag{3.2-24}$$

where ρ_F represents the state density of the final state. Using Eq. 3.2-22, the term in parentheses in Eq. 3.2-24 becomes

$$\langle \phi_D^* \phi_A | H_1 | \phi_D \phi_A^* \rangle - \langle \phi_D^* \phi_A | H_1 | \phi_A^* \phi_D \rangle \tag{3.2-25}$$

The first and second terms are called *Coulomb* and *exchange* integrals, respectively, and express the origin of two typical mechanisms of electronic energy transfer.

For a dipole–dipole interaction, the Coulomb term is approximately proportional to the multiple of the electronic transition moments of the donor and the acceptor molecules, i.e. $|\langle \phi_D^* | r_D | \phi_D \rangle \langle \phi_A^* | r_A | \phi_A \rangle|^2$. The rate of electronic energy transfer arising from this term therefore increases with increasing transition moments of the donor and acceptor molecules. In other words, the rate is proportional to the extinction coefficients of $S_0 \rightarrow S_1$ absorption of the respective molecules, i.e. ε_D and ε_A. This idea is the origin of the so-called *Förster mechanism* of electronic energy transfer. The rate of electronic energy transfer with this mechanism is expressed as

$$k_{ET} \propto \frac{\kappa^2}{R^6 \varepsilon^2} \int \varepsilon_D(\nu) d\nu \int I_D(\nu) \varepsilon_A(\nu) d\nu \tag{3.2-26}$$

where $\varepsilon(\nu)$ is the extinction coefficient of the molecule (A or D) as a function of the wavenumber ν of the light. $I_D(\nu)$ depicts the emission spectrum of the donor molecule and κ is the relevant constant according to the orientation of D and A. The Förster mechanism is usually applied to S–S electronic energy transfer between organic molecules, and is known to be effective for long-distance energy transfer on a short time-scale. For example, energy transfer from acetone to 9,10-diphenylanthracene occurs within 10 ps between molecules 25 Å apart.

On the other hand, the exchange integral is approximately proportional to $|\langle \phi_D^* | \phi_A^* \rangle \langle \phi_A | \phi_D \rangle|^2$, i.e. the overlap integral between the donor and acceptor molecules in both the ground and excited states. This means that the energy transfer arising from this mechanism becomes efficient when two molecules come close to each other. This type of short-range electronic energy transfer is called the *Dexter mechanism*, and is often applied to triplet–triplet energy transfer. The efficiency can be expressed as a function of the distance as

$$k_{ET} \propto C \exp\left(-\frac{2R}{L} \right) \tag{3.2-27}$$

where L is the characteristic distance determining the interaction range between the donor and acceptor, and in most cases is around 10 Å or less. Therefore, this mechanism is considered to be important for energy transfer within 10 Å.

Experimental Observations of Bimolecular Electronic Energy Transfer. The electronic energy transfer process has been extensively studied in both the gas and liquid phases. In the gas phase, collision between the donor and acceptor molecules induces energy transfer mainly as a result of the dipole–dipole interaction. Vast amounts of data have been accumulated on atom–atom collisions, and the importance of spin conservation and energy conservation have been demonstrated. The resonance condition $\Delta E = 0$ considerably enhances the rate of electronic energy transfer since the conversion of excess energy into translational motion is not necessary. The relationship $k_{ET} = C\exp(-\alpha\Delta E)$ has often been observed when a series of experiments is performed.

In the liquid phase, more complicated organic molecules have been studied. Both S–S and T–T electronic energy transfer have been extensively studied and the importance of the two mechanisms, the Förster and the Dexter, are well established [21].

3.2.2.2 Conversion of the Electronic State in an Isolated Molecule

In ordinary molecules, the electronic ground state is a singlet state. When the molecule absorbs a photon, the photo-induced dipole transition promotes the molecule from its singlet ground state to another singlet excited state. The excited state then decays to a lower electronic state via either a radiative or a nonradiative process. Among the nonradiative processes, the change of state to a different spin multiplet (singlet to triplet or vice versa) is called *intersystem crossing*, while conversion to a state of the same spin multiplicity is known as *internal conversion*. The energy levels relevant to these energy conversions are best represented by a Jabronski diagram, as shown in Fig. 3.2-6.

Internal Conversion. Conversion to a state with a different electron configuration can be regarded as the deformation of electron clouds from one stable form to another, since an electronic state can be represented by a specific electron distribution in a molecular frame. Such an interpretation naturally suggests that molecular vibrations must enhance the conversion of the electronic state. The interaction between an electron cloud and molecular vibrations is called a *vibronic interaction*, and this produces a correction term for the Born–

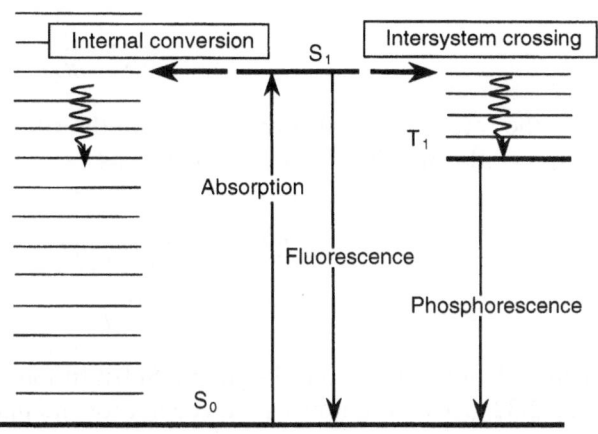

FIG. 3.2-6. A Jablonski diagram showing the energy levels relevant to internal conversion and intersystem crossing

Oppenheimer description of electron-nuclear interaction. A simple description of such an interaction in terms of electronic wavefunctions and normal mode vibrations is given by the *Herzberg–Teller expansion* of the electronic state.

First, the potential V for electron motion is expanded using the coordinates of normal vibrations Q_k as

$$V = V_0 + \sum_k \left(\frac{\partial V}{\partial Q_k} \right) Q_k + \dots \tag{3.2-28}$$

where V_0 is the potential of the structure at equilibrium (without vibrations). Then, the electronic wavefunction of the sth electronic state $\psi_s(q,Q)$ as a function of both electronic (q) and nuclear (Q) coordinates is also expanded, using the first-order perturbation theory, as

$$\psi_s(q,Q) = \psi_s(q,0) + \sum_{e \neq s} \sum_k \frac{\left\langle \psi_s(q,0) \left| \left(\frac{\partial V}{\partial Q_k} \right)_0 \right| \psi_e(q,0) \right\rangle}{E_s^0 - E_e^0} \psi_e(q,0) Q_k \tag{3.2-29}$$

where $\psi_s(q,0)$ and E_s^0 denote the wavefunction at structure equilibrium and the energy of the sth electronic state, respectively. Using this Herzberg–Teller expression for the electronic wavefunction, a vibronic state can be represented as

$$\Psi_s = \psi_s(q,Q) \chi_s(Q)$$
$$\chi_s(Q) = \prod_k \chi_k^{(s)} \tag{3.2-30}$$

where $\chi_k^{(s)}$ is the vibrational wavefunction along the Q_k normal coordinate in the sth electronic state.

Since the origin of internal conversion is the nuclear motion, the perturbation Hamiltonian can be expressed as

$$H' = -\sum_k \hbar^2 \frac{\partial^2}{\partial Q_k^2} \tag{3.2-31}$$

Then the rate of internal conversion from the sth to the lth electronic state, k_{sl}^{IC}, is expressed using the Fermi golden rule as

$$k_{sl}^{IC} = \frac{2\pi}{\hbar} \sum_l H_{sl}^2 \, \delta(E_s - E_l)$$
$$H_{sl} = \left\langle \Psi_l \left| H' \right| \Psi_s \right\rangle \tag{3.2-32}$$

where δ denotes the Kronecker delta. Substitution of Eqs. 3.2-30 and 3.2-31 into Eq. 3.2-32 gives the following expression for the matrix element H_{sl}:

$$H_{sl} = -\hbar^2 \sum_k \beta_e^{IC} \left\langle \chi_k^{(s)} \left| \frac{\partial}{\partial Q_k} \right| \chi_k^{(l)} \right\rangle \prod_{i \neq k} \left\langle \chi_i^{(s)} \middle| \chi_i^{(l)} \right\rangle$$

$$\beta_e^{IC} = \left\langle \psi_l(q,Q) \left| \frac{\partial}{\partial Q_k} \right| \psi_s(q,Q) \right\rangle \tag{3.2-33}$$

According to Eq. 3.2-33, the internal conversion occurs only when the electronic integral β_e^{IC} is nonvanishing. The vibrational mode, Q_k, which makes β_e^{IC} nonzero is called the promoting mode. In other words, only through vibration along the Q_k coordinates does the lth electronic state become accessible from the initially excited sth state. As Eqs. 3.2-32 and 3.2-33 show, the actual rate constant of internal conversion is a function of vibrational transition and overlap as well as of the electronic integral β_e^{IC}. Yeung and Moore [22] measured the detailed rate of internal conversion for formaldehyde in a free jet and compared the values obtained with those predicted by the above treatment.

Intersystem Crossing. For intersystem crossing to occur, the electron spin must be flipped from α to β, or vice versa. The most important cause of such spin flip is the orbiting motion of the electron, which provides a magnetic field to induce the spin flip. Such an interaction is called a *spin–orbit interaction*. The corresponding Hamiltonian operator is written as

$$H_{so} = \frac{1}{2m^2c^2} \sum_i \left(\nabla_{(i)} V \times P_i \right) \cdot S_i \tag{3.2-34}$$

where P and S denote the momentum and spin–flip operator for the ith electron, respectively. The first part of the summation ($\nabla V \times P$) rotates the spacial part of the electronic wavefunction, while S influences its spin.

As in the case of internal conversion, the rate of intersystem crossing can be evaluated using the Fermi golden rule (Eq. 3.2-32). Applying the Herzberg–Teller expansion to the initially prepared singlet state, the matrix element for the spin–orbit interaction, H_{st}, can be expressed as

$$\begin{aligned}
H_{st} &= \left\langle \psi_t(q,0)\chi_t(Q) \middle| H_{so} \middle| \psi_s(q,Q)\chi_s(Q) \right\rangle \\
&= \left\langle \psi_t(q,0) \middle| H_{so} \middle| \psi_s(q,0) \right\rangle \left\langle \chi_t(Q) \middle| \chi_s(Q) \right\rangle \\
&+ \sum_{e \neq s} \sum_k \frac{\left\langle \psi_s(q,0) \left| \frac{\partial V}{\partial Q_k} \right| \psi_e(q,0) \right\rangle}{E_s^0 - E_e^0} \left\langle \psi_t(q,0) \middle| H_{so} \middle| \psi_e(q,0) \right\rangle \left\langle \chi_t \middle| Q_k \middle| \chi_e \right\rangle
\end{aligned}$$

$$\tag{3.2-35}$$

where the running parameter e includes triplet electronic states. The first term of the matrix element corresponds to the direct singlet–triplet interaction between two nonvibrating electronic states, whereas the second term expresses the spin–orbit interaction through vibronic coupling. Usually, the first term is important for the interaction between (π, π^*) and (n, π^*) electronic states. For interaction between states of similar electron configuration, for example between two (π, π^*) or two (n, π^*) states, the first term is small and the second term becomes predominant. In the latter case, the vibrational mode Q_k, which gives a nonvanishing matrix element in the numerator of the second term, is called the promoting mode.

The recent development of jet spectroscopy enables us to evaluate the detailed rate of intersystem crossing for each vibronic level and quantitatively compare theoretical predictions with actual observations. Figure 3.2-7 shows the relative intensities of fluorescence and phosphorescence measured for benzonitrile in a free jet [23]. Roughly speaking, the ratio of the phosphorescence intensity to the fluorescence intensity expresses the efficiency of intersystem crossing from the excited vibronic level. Two features can be seen from the figure. First, the efficiency of intersystem crossing is enhanced at the 10^1_0 vibronic band, which indicates the presence of promoting modes. Second, intersystem crossing becomes significant when the benzonitrile dimer or trimer is excited at any vibronic bands. This enhancement is probably due to efficient IVR, which allows the participation of promoting modes (including some dark modes) just after

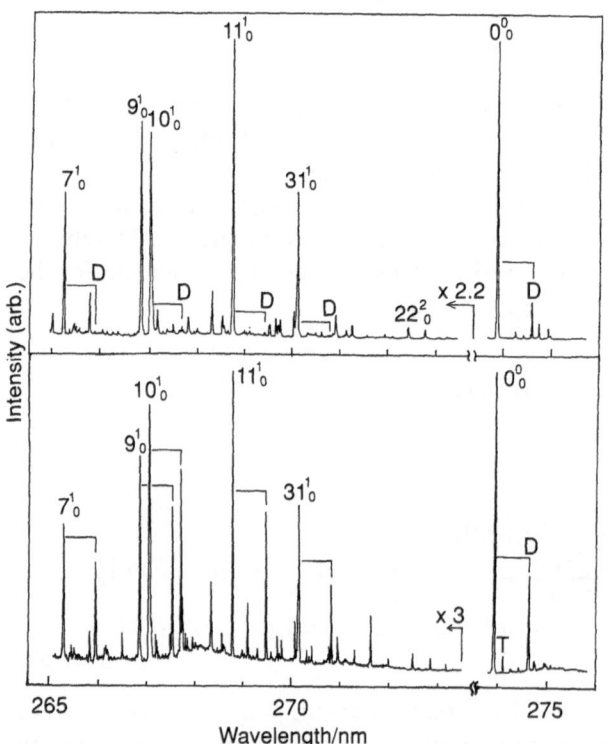

Fig. 3.2-7. The fluorescence and phosphorescence excitation spectra for benzonitrile in a free jet. The relative increase of phosphorescence intensity with increasing vibrational energy indicates that intersystem crossing is enhanced by the vibronic interaction. D indicates the peaks assignable to the benzonitrile dimer

photoexcitation. As mentioned in sect. 3.2.1.2, IVR occurs very rapidly for clusters because of the presence of low-frequency vibrations.

The conversion of electronic energy within a molecule, as mentioned above, also plays an important role in the process of energy flow within a molecular assembly. The initially excited molecule in the assembly may change its electronic state to triplet by intersystem crossing or to another singlet state by internal conversion. The electronic energy is then transferred to other molecules in the assembly by ordinary intermolecular energy transfer, and eventually reaches the reaction center.

References

1. Mott NF, Massay HSW (1965) The theory of molecular collisions. Clarendon, Oxford
2. Massay HSW (1949) Collisions between atoms and molecules at ordinary temperatures. Rep Progr Phys 12:248–267
3. Jackson JM, Mott NF (1932) Energy exchange between inert gas atoms and a solid surface. Proc R Soc Ser A 137:703–717
4. Millikan RC, White DR (1963) Systematics of vibrational relaxation. J Chem Phys 39:3209–3213
5. Schwartz RN, Slawsky ZI, Herzfeld KF (1952) Calculation of vibrational relaxation times in gases. J Chem Phys 20:1591–1599
6. Herzfeld KF, Litovitz TA (1959) Absorption and dispersion of ultrasonic waves. Academic, New York
7. Cottrell TL, McCoubrey JC (1961) Molecular energy transfer in gases. Butterworths, London
8. Levy DH (1981) van der Waals molecules. Adv Chem Phys 47(I):323–362
9. Willberg DM, Gutmann M, Breen JJ, Zewail AH (1992) Real-time dynamics of clusters. I. I_2X_n ($n = 1$). J Chem Phys 96:629–643
10. Gutmann M, Willberg DM, Zewail AH (1992) Real-time dynamics of clusters. II. I_2X_n (n = He, Ne, and H_2), picosecond fragmentation. J Chem Phys 97:8037–8047
11. Litovitz TA (1957) Theory of ultrasonic relaxation times in liquids. J Chem Phys 26:469–473
12. Herzfeld KF, Litovitz TA (1959) Absorption and dispersion of ultrasonic waves. Academic, New York
13. Williams HT, Gwynne V, Simpson CJSM (1987) Non- and near-resonant V–V energy transfer between the isotopes of N_2 and of CO in liquid Ar and in the gas phase at 85 K. Chem Phys Lett 136:95
14. Williams HT, Purvis MH, Simpson CJSM (1987) Non-resonant V–V transfer in liquid Ar and in the gas phase at 85 K. Chem Phys 115:7
15. Chesnoy J, Gale GM (1984) Vibrational energy relaxation in liquids. Ann Phys (Paris) 9:893–949
16. Paige ME, Harris CB (1990) Ultrafast studies of chemical reactions in liquids: Validity of gas phase vibrational relaxation model and density dependence of bound electronic state lifetime. Chem Phys 149:37
17. Harris CB, Smith DE, Russell DJ (1990) Vibrational relaxation of diatomic molecules in liquids. Chem Rev 90:481

18. Felker PM, Zewail AH (1985) Dynamics of intramolecular vibrational energy redistribution (IVR). II. Excess energy dependence. J Chem Phys 82:2975–2993
19. Kobayashi T, Kajimoto O (1987) Benzonitrile and its van der Waals complexes studied in a free jet. II. Dynamics in the excited state: The effect of changing the degree of freedom of partner molecules. J Chem Phys 86:1118–1124
20. Elsaesser T, Kaiser W (1991) Vibrational and vibronic relaxation of large polyatomic molecules in liquids. Annu Rev Phys Chem 42:83–107
21. Turro NJ (1978) Modern molecular photochemistry. Benjamin/Cummings, Menlo Park, CA
22. Yeung ES, Moore CB (1974) Predissociation model of formaldehyde. J Chem Phys 60:2139–2147
23. Kobayashi T (1988) Studies on the photophysical properties of van der Waals complexes in electronically excited states. Thesis, University of Tokyo

3.3
Electron Transfer

KEITARO YOSHIHARA

Electron transfer (ET) is one of the most important processes which give functionality to molecular systems. There are nationally occurring molecular systems which involve electron transfer, such as photosynthetic systems in plants. Animal metabolism is another example where ET plays an important role, and it also occurs in a wide range of phenomena such as organic and inorganic chemical reactions, molecular conductivity, photography, and electrochemical processes. Because of its universality, ET has been the subject of extensive research for many years [1–4].

Recent synthesis of a vast number of tailor-made donor–acceptor systems and donor–insulator–acceptor systems has brought new understanding of the mechanism of ET and new possibilities for controlling the free energy differences and the donor–acceptor distances and orientations. The recent development of spectroscopic methods for dynamic measurements with high temporal resolution has also led to new information about the dynamic aspects of ET. The basic assumptions in the theory of ET in equilibrium conditions are not valid for ET which occurs as fast as solvent relaxation times. ET has been found to occur even faster than solvent motions. Any theoretical description will have to take into account the relationship between the stochastic motion of complicated molecular systems and the multidimensional potential energy surface (PES) crossing between the reactant and the product.

The general concepts of nonadiabatic ET within transition state theory (TST) and related problems are discussed in Sect. 3.3.1, and the experimental observations of the energy-gap dependence of ET are given. The solvent dynamics and the dynamic aspects of adiabatic ET in the presence of solvent friction are described in Sect. 3.3.2. A different model of the solvent and nuclear two-dimensional PES model is given in Sect. 3.3.3 in order to explain ET which is faster than solvent dynamics. Some recent observations are given, with theoretical simulations.

3.3.1 Nonadiabatic Electron Transfer

3.3.1.1 Transition State Theory

ET occurs when a PES crosses from a reactant (electron donor) to a product (electron acceptor) as it does in any other reaction. Transition State Theory (TST) assumes that equilibration occurs much faster than the reaction. For a simple linear reaction (A → B), the reaction rate constant k_{TST} can be written as

$$k_{TST} = \frac{p\omega_0}{2\pi} \exp\left(-\frac{\Delta G^*}{k_B T} \right) \qquad (3.3\text{-}1)$$

where ω_0 is the frequency of motion in the reactant potential well, ΔG^* is the free energy of activation, p is the electronic transition probability in the transition region, k_B is the Boltzmann constant, and T is temperature. The solvent and reactant molecules are in equilibrium before the reaction. Then fluctuation of the solvent brings the reactant to the transition state. After reaching the transition state, p is the probability that the reaction will take place. Then the excess energy is transferred into the solvent heat bath and the system relaxes to a new equilibrium for the product. In some cases ET occurs in a polar solvent and not in the gas phase. Therefore, a solvent coordinate, which describes the solvent polarization, is used to describe the PES for ET. Polar solvent molecules organize around the solute molecule to minimize the dielectric interaction energy. Thus ET can be seen as a process for a reactant to pass through the transition state to form a product by fluctuation of the surrounding solvent molecules. The reorganization of the solvent molecules is often treated as being infinitely fast.

In the case of ET, the probability, p, can be determined by the electronic matrix element V_{el}. When V_{el} is small the PES becomes nonadiabatic, and when it is large the PES becomes adiabatic. In the nonadiabatic case, the rate constant of ET can be written as

$$k_{NA} = \frac{2\pi}{\hbar} \frac{V_{el}^2}{\sqrt{4\pi\lambda_s k_B T}} \exp\left(-\frac{\Delta G^*}{k_B T} \right) \qquad (3.3\text{-}2)$$

If a quadratic function with the same curvature is assumed for the PES of the reactant and product, the activation energy can be described by

$$\Delta G^* = \frac{(\lambda + \Delta G_0)^2}{4\lambda} \qquad (3.3\text{-}3)$$

with λ being the reorganization energy and ΔG_0 being the free energy gap of the reaction. λ is the deformation energy on the reactant PES at a position at the bottom of the product potential well, as shown in Fig. 3.3-1. There are two parts to the reorganization energy caused by changes in the bond lengths of the

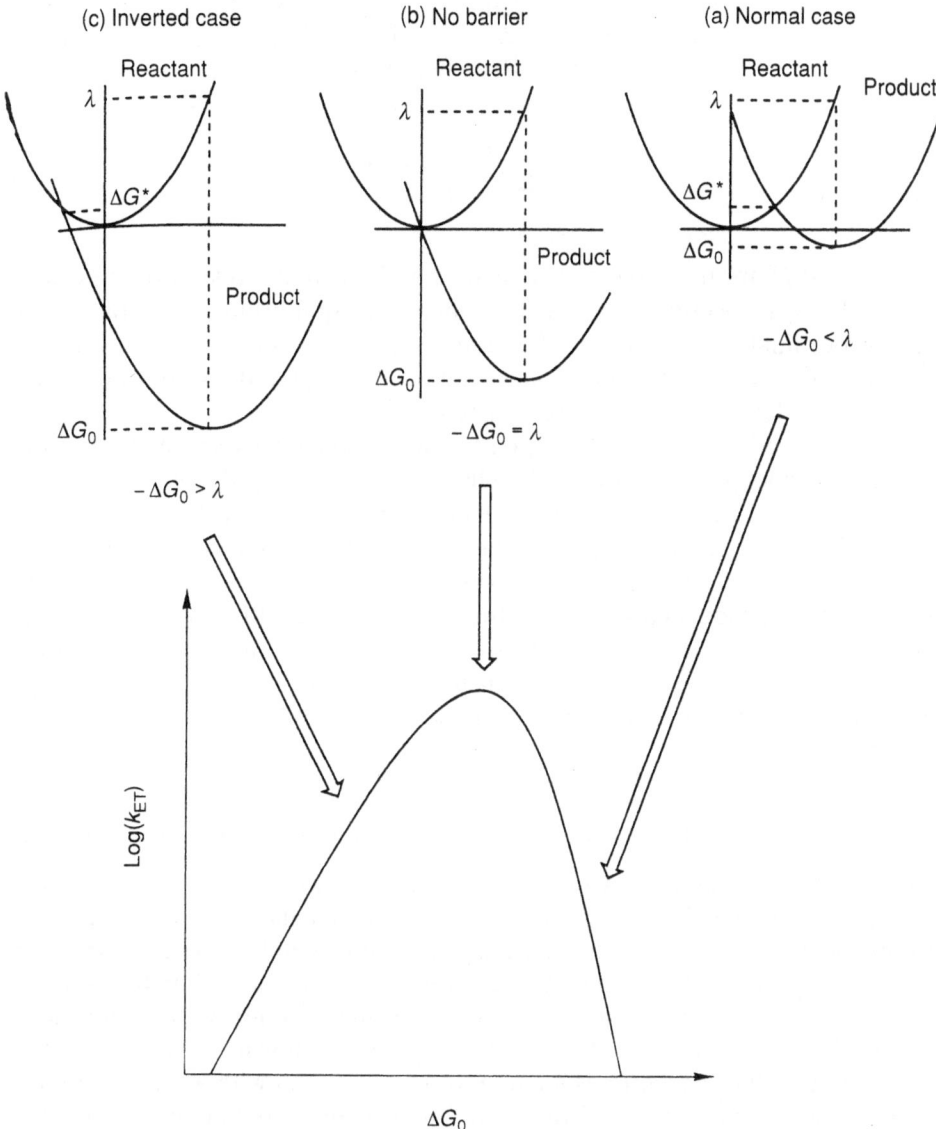

FIG. 3.3-1. The energy-gap dependence of the electron transfer rate constant k_{ET} and its relation to the relative position of the potential energy surfaces (free energy surface, ΔG_0) of the reactant and the product

reactants and changes in the solvent coordinates: λ_v, the inner sphere reorganization energy, and λ_s, the solvent reorganization energy. The following expressions are given for λ:

$$\lambda = \lambda_s + \lambda_v \qquad (3.3\text{-}4)$$

$$\lambda_{\mathrm{s}} = e^2 \left[\frac{1}{2r_{\mathrm{D}}} + \frac{1}{2r_{\mathrm{A}}} + \frac{1}{R_{\mathrm{DA}}} \right] \left(\frac{1}{\varepsilon_{\mathrm{s}}} + \frac{1}{\varepsilon_{\mathrm{op}}} \right) \tag{3.3-5}$$

$$\lambda_{\mathrm{v}} = \sum_i \frac{f_i^{\mathrm{r}} f_i^{\mathrm{p}}}{f_i^{\mathrm{r}} + f_i^{\mathrm{p}}} \left(\Delta q_i \right)^2 \tag{3.3-6}$$

where f_i^{r} and f_i^{p} are the force constants of the ith normal mode in the reactants and products, respectively, Δq_i is the change in equilibrium value of the ith normal coordinate, r_{D} and r_{A} are the radii of the two spherical reactants, R_{DA} is the center-to-center separation distance, and ε_{s} and $\varepsilon_{\mathrm{op}}$ are the static and optical dielectric constants of the solvents, respectively.

The free energy gap is used to represent the distance between the bottoms of the reactant and product potential wells. The energy gap dependence of the rate constant of ET can be separated into three regions. (a) $-\Delta G_0 < \lambda$, the region where the reaction becomes faster as $-\Delta G_0$ increases. (b) $-\Delta G_0 = \lambda$, the region where the reaction is fastest because the activation barrier vanishes. The PES of the product crosses the bottom of the reactant PES. (c) $-\Delta G_0 > \lambda$, the region where the reaction becomes slower again as $-\Delta G_0$ increases. Therefore, the rate constant of ET shows a bell-shape dependence on the free energy difference. Case (a) is called the "normal region" and case (c) is called the "inverted region."

3.3.1.2 Inclusion of Intramolecular High-Frequency Vibrational Modes

In the previous section we were mainly concerned with the effects of solvent motions on the reaction and ignored the intramolecular degree of freedom. Intramolecular motions include high-frequency vibrational modes and provide a channel to dispose of large amounts of energy. The inclusion of high-frequency modes brings in discrete quantum states. The high-frequency quantum mode is particularly important in the inverted region, as shown in Fig. 3.3-2. The ground state of the reactant has good vibrational overlap with a few vibrational states of the product. The ET rate constant becomes greater in the inverted region compared with that in the normal region, and makes the bell shape asymmetric.

First let us assume that the reaction takes place from the ground vibrational state of the reactant. The overall rate constant is the sum of the rate constants for the nth vibrationally excited state:

$$k_{\mathrm{NA}} = \sum_n k_{\mathrm{NA}}^{0 \to n} \tag{3.3-7}$$

and the free energy difference is quantized as

$$\Delta G_0^{0 \to n} = \Delta G_0 + nh\nu \tag{3.3-8}$$

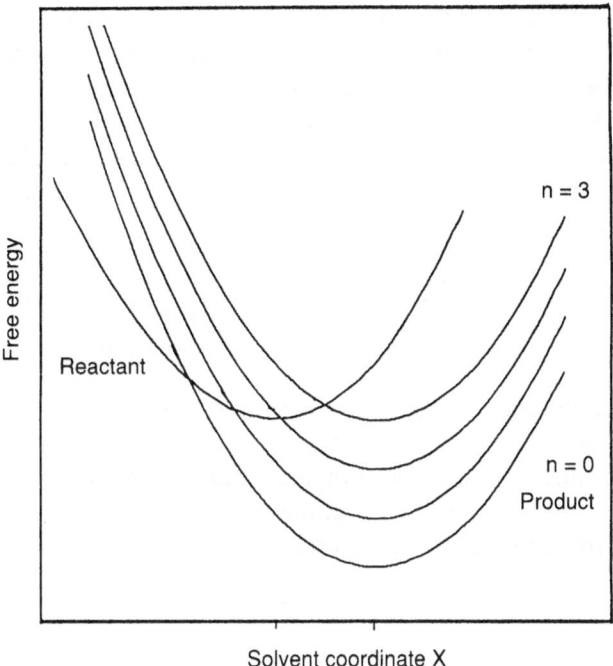

FIG. 3.3-2. The potential energy surface crossing at the "inverted region" showing the efficient contribution of the high-frequency vibrational modes of the product surface, which makes the "bell shape" of the free energy dependence of electron transfer asymmetric

The effective electronic matrix element, which takes into account the Franck–Condon overlap between the two vibronic states, is given as

$$\left(V_{\mathrm{el}}^{0\to n}\right)^2 = V_{\mathrm{el}}^2 \left\langle 0|n\right\rangle^2 \tag{3.3-9}$$

The Franck–Condon factor is expressed as

$$\left|\left\langle 0|n\right\rangle\right|^2 = \left(S^n/n!\right)\exp\left(-S\right) \tag{3.3-10}$$

where the electron vibrational coupling strength, S, is given by

$$S = \lambda_{\mathrm{hf.vib}}/h\nu_{\mathrm{hf.vib}} \tag{3.3-11}$$

Here $\lambda_{\mathrm{hf.vib}}$ and $\nu_{\mathrm{hf.vib}}$ are the reorganization energy and frequency of the quantized high-frequency mode, and we get

$$k_{\mathrm{NA}}^{0\to n} = \frac{2\pi}{\hbar}\frac{1}{\sqrt{4\pi\lambda_s k_{\mathrm{B}}T}}\left(V_{\mathrm{el}}^{0\to n}\right)^2 \exp\left(-\frac{\left(\Delta G_0^{0\to n} + \lambda_s\right)^2}{4\lambda_s k_{\mathrm{B}}T}\right) \tag{3.3-12}$$

Second, we will consider that the reactant molecule is also in its vibrational state with Boltzmann distribution, and the product molecule is in its higher vibrational excited state. This treatment is necessary for the temperature-dependence study of ET [5]. For simplicity, two typical vibrational modes are considered, low-frequency solvent vibrational mode (ω_s with a mean value of $<\omega_s>$) and high-frequency intramolecular vibrational mode (ω). At the low temperature limit, $k_B T \ll \hbar<\omega_s> \ll \hbar\omega$,

$$k_{ET} = \frac{2\pi V_{el}^2}{\hbar^2 \langle \omega_s \rangle} \exp(-S_s - S) \sum_{m=0}^{\infty} \frac{S_s^{p(n)} S^n}{[p(n)]! \, n!} \qquad (3.3\text{-}13)$$

with $S_s + \lambda_s / \hbar<\omega_s>$ and $S = \lambda_v / \hbar\omega$ being the coupling strengths for solvent low-frequency and intramolecular high-frequency modes, respectively, and $p(n)$ the Bessel function. λ_v is the reorganization energy for the intramolecular high frequency mode. The contributions of all product vibrational quanta are added. For the intermediate temperature region, $\hbar<\omega_s> \ll k_B T \ll \hbar\omega$, this is described as

$$k_{ET} = \frac{2\pi}{\hbar} \frac{V_{el}^2}{\sqrt{4\pi S_s \hbar \langle \omega_s \rangle k_B T}} \exp(-S)$$

$$\times \sum_{n=0}^{\infty} \exp\left[-\frac{\left(\Delta G - S_s \hbar \langle \omega_s \rangle - n\hbar\omega\right)^2}{4 S_s \hbar \langle \omega_s \rangle k_B T} \right] \frac{S^n}{n!} \qquad (3.3\text{-}14)$$

Equations (3.3-12) and (3.3-14) are different expressions of Eq. (3.3-2) introducing the effect of high-frequency quantum states, n, of the product state.

Expressions of ET theory which include both high- and low-frequency modes are powerful in analyzing ET over a wide temperature range. In photoinduced ET from cytochrome to chlorophyll in *Chromatium*, the rate constant of ET is only slightly temperature dependent at low temperatures, and strongly temperature dependent at higher temperatures [5]. Another example of diffusionless donor–acceptor systems is ET from a solid organic substrate to a dye adsorbate. This system seems to be in the inverted region of the free energy gap dependence. The temperature dependence of ET is very modest, and can be analyzed by including a high-frequency mode, but not without [6].

3.3.1.3 Energy Gap Dependence

Recent progress in synthesizing specific molecular systems has made it possible to study intramolecular ET while controlling the free energy difference, intermolecular distances, and molecular orientations. The functional molecules in which donor (D) and acceptor (A) are separated by an inert rigid spacer (S) of predetermined length and orientation have definite advantages over diffusive D and A

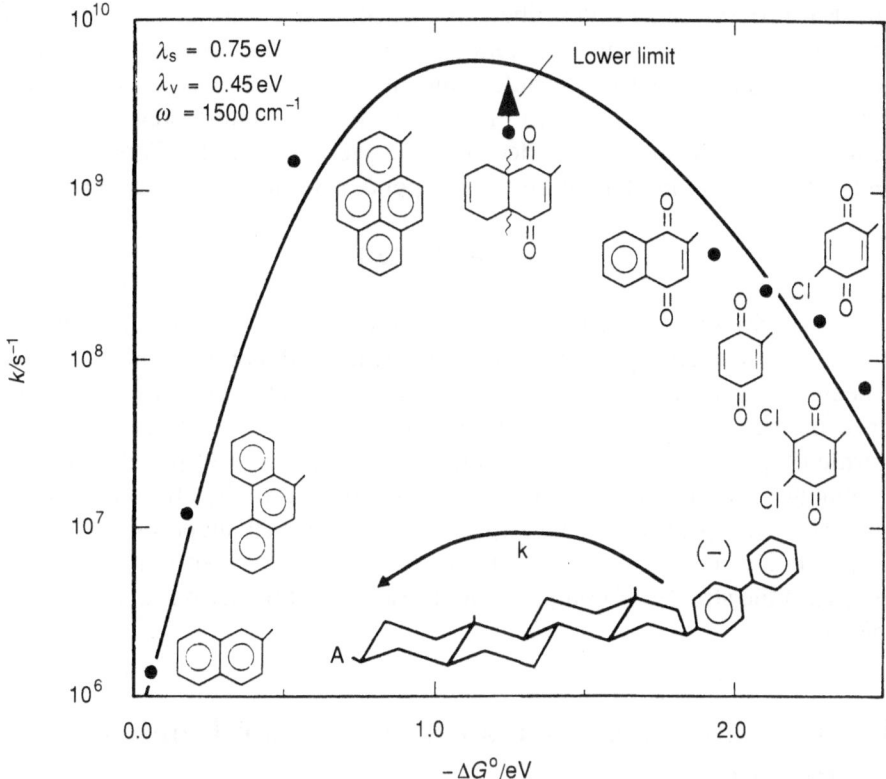

FIG. 3.3-3. The energy-gaps dependence of intramolecular electron transfer in a donor–spacer–acceptor system. (From [7], with permission)

in solution. In solvent systems in which D and A are dissolved independently, the rate of D and A encounters, i.e., the bimolecular reaction rate, limits the ET rate constants and the Marcus inverted region is not observed. D–S–A molecules, on the other hand, give a much more precise rate constant for ET [7]. The scheme in Fig. 3.3-3 shows the reaction and rate of intramolecular ET plotted against the free energy difference for the reaction. In this system the donor is a biphenyl negative ion and a charge is transferred to the neutral molecule, which is separated from it by a spacer molecule and is 17Å away. This clearly shows that the reaction rate constant diminishes with very negative ΔG_0 and demonstrates the existence of the inverted region, as the Marcus theory predicted. The reorganization energy is estimated to be about 0.75 eV for λ_s and about 0.45 eV for λ_v. The slightly asymmetric shape is due to a contribution from the vibrational states for PES crossing. Intermolecular ET is also studied in similar systems in solution at cryogenic temperature to avoid diffusional motion. A very clear bell-shaped free energy gap dependence was observed for the wide energy region [8]. Agreement with the theoretical description given by Eq. (3.3-12) is quite good.

The ET process can be described in the context of the general theory of radiationless transition. The exponential decrease of the nonradiative rate for large energy gaps between the initial and final states has been recognized in intramolecular processes, i.e., internal conversion and intersystem crossing in large molecules. This should be reflected in the averaged Franck–Condon density for ET, and the exponential energy gap dependencies are expected to be

$$k = \alpha \exp(-\gamma \Delta E) \qquad (3.3-15)$$

for large energy gaps (ΔE) between the initial and final states with relatively small changes in their equilibrium geometry (the weak coupling limit) [9]. This has been recognized in metal–pyridine complexes [10] and donor–acceptor bridged molecular systems. A theoretical formulation and analyses have been performed for a bridged donor–acceptor system [9]. It is also possible to avoid the influence of diffusive motion in the charge recombination reaction of photoexcited ion pairs [11]. The applicability of the bell shape or exponential energy gap law depends on the electronic structures of such geminate ion pairs, e.g., is the structure "compact" or "loose" [11]. Further research is needed in this area.

3.3.2 Adiabatic Electron Transfer (Solvation Limited Reaction)

3.3.2.1 Solvent Dynamics

When the electronic interaction between reactant and product PESs becomes large, the rate of ET could be of the same order as the solvent relaxation times. In such a case, ET in a solution is often a direct function of its solvent properties. In this subsection the nature of solvent dynamics and methods of obtaining solvent relaxation times are described. There are two main ways to estimate solvent relaxation times. One is to use dielectric dispersion and loss data. Dielectric spectra can be fitted by model functions of the frequency of an applied alternating electric field [12]. From this analysis, characteristic times of dielectric relaxation can be estimated. Information about solvation dynamics can be obtained by several theoretical methods that connect dielectric data with the dynamic behavior of solvent polarization. The simplest way is to obtain the longitudinal relaxation time of the solvent, τ_L, which is calculated from the Debye relaxation time τ_D,

$$\tau_L = \frac{\varepsilon_\infty}{\varepsilon_0} \tau_D \qquad (3.3-16)$$

where ε_∞ and ε_0 are the dielectric constants at the high-frequency limit and static limit, respectively. τ_D is obtained by dielectric measurements.

The second method is more direct: the energy relaxation of the excited electronic state of a proper probe molecule in a polar solvent is measured by monitoring the dynamic fluorescence Stokes shift of the probe molecule. When a probe dye molecule immersed in a liquid is photoexcited, the dipole moment of the molecule changes instantaneously. The polarization of the surrounding solvent molecules responds to this change and reorganizes. The energy relaxation of the excited probe is monitored by observing the shift in the fluorescence spectrum. The dynamic nature of solvent relaxation is defined by a normalized response function defined as

$$C(t) = \frac{v(t) - v(\infty)}{v(0) - v(\infty)}$$ (3.3-17)

where $v(t)$, $v(\infty)$, and $v(0)$ are the fluorescence peak frequencies at times t, ∞, and 0, respectively [13]. In this method, the probe molecule will change its dipole moment upon photoexcitation. The fluorescence decay of a probe molecule in the solvent of interest is recorded at different wavelengths. Because of the energy relaxation, the fluorescence peak shifts (dynamic Stokes shift) as a function of time. It is possible to observe fast decays at shorter wavelengths and correspondingly slower delays at longer wavelengths. By reconstructing time-resolved fluorescence spectra and using a static fluorescence spectrum, $C(t)$ can be determined experimentally. The $C(t)$ of aniline (AN) and N,N-dimethylaniline (DMA) shows a nonexponential feature and can be fitted by a biexponential function. These solvents are examined in more detail in Sect. 3.3.3.2. The solvent relaxation times obtained are 7.9ps (19%) and 18.7ps (81%) for DMA, and 1.2ps (28%) and 18ps (72%) for AN [14].

3.3.2.2 Adiabatic Electron Transfer

For ET, which occurs in the same time range as solvation dynamics, it is necessary to consider the effects of the finite response time of solvent polarization relaxation. The assumptions of TST, which ignores dynamic solvent processes, no longer hold. The frequency ω_0 in Eq. (3.3-1) is determined by solvent fluctuation. Fast ET is mainly the result of good electronic overlap between the reactant and the product. The reaction rate constant is expressed by Eq. (3.3-18), and the adiabaticity of the reaction is determined by Eq. (3.3-19) [15].

$$k_{ET} = k_{NA}/(1 + \kappa)$$ (3.3-18)

$$\kappa = \frac{4\pi V_{el}^2 \tau_s}{\hbar \lambda_s}$$ (3.3-19)

where κ is the adiabaticity parameter. Equation (3.3-18) gives ET from the nonadiabatic limit ($\kappa \ll 1$) to the adiabatic limit ($\kappa \gg 1$) in one expression.

The transition state is defined as the region of the activation maximum where the energy varies less than the thermal energy k_BT. For $\kappa \gg 1$, the adiabatic rate constant can be written as

$$k_A = \frac{1}{\tau_s}\sqrt{\frac{\lambda_s}{16\pi k_BT}}\exp\left(-\frac{\Delta G^*}{k_BT}\right) \qquad (3.3\text{-}20)$$

The reaction rate constant is inversely proportional to τ_s, and the reaction is called a solvent-controlled adiabatic reaction. At room temperature $16\pi k_BT = 1.2\,eV$, and the value of λ_s is usually smaller than this. Therefore, the maximum of k_A is about $1/\tau_s$. τ_s is usually in the order of picoseconds for ordinary simple solvents. Thus, for ultrafast ET occurring in the picosecond time range, solvent fluctuation should be the limiting step. This kind of effect has become observable only recently with the development of ultrashort-pulse lasers.

The correlation between solvent relaxation times and intramolecular ET rates has been investigated in various molecular systems. The excited state intramolecular ET of arylamino-naphthalene sulphonates in alcohol solutions was examined. Charge separation occurs from the arylamino moiety to the naphthalene moiety upon photoexcitation. Using many different alcohol solvents [16], a linear correlation was found between the ET rate constant and the inverse of the longitudinal relaxation time, τ_L. The charge separation of bianthryl from the locally excited state to the charge transfer state was also found to be a solvent-controlled adiabatic reaction [17]. Internal conversion from the S_2 to the S_1 state of 4-(9-anthryl)-N,N'-dimethylaniline (ADMA) was faster than the system response, and the succeeding process in the S_1 state was solvent controlled. The adiabaticity parameter was estimated to be about 80 in common polar solvents ($\lambda_s = 20\,kcal\,mol^{-1}$, $V_{el} = 1\,kcal\,mol^{-1}$, and $\tau_s = 1\,ps$).

3.3.3 Electron Transfer Faster than Solvation Times

3.3.3.1 Separation of Solvation and Nuclear Coordinates: Two-dimensional Theory

When ET occurs faster than the diffusive solvation process, distribution on the solvent coordinate may change during the reaction. Simple arguments which only take account of the solvent coordinate have to be revised. In this case the vibrational nuclear coordinate (q) and the solvent coordinate (X) are introduced as shown in Fig. 3.3-4 [18, 19]. Let us call this model the two-dimensional model (2D model).

$$G_r = \lambda_s X^2 + \lambda_v q^2 \qquad (3.3\text{-}21)$$

$$G_p = \lambda_s\left(X-1\right)^2 + \lambda_v\left(q-1\right)^2 + \Delta G \qquad (3.3\text{-}22)$$

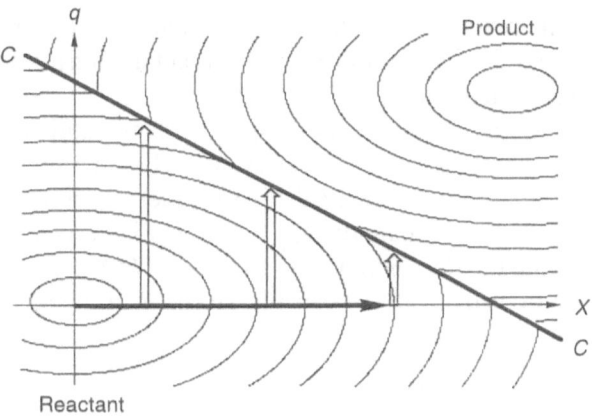

FIG. 3.3-4. A two-dimensional representation of the reactant and product potential energy surface spanned by solvent coordinate X and nuclear coordinate q. Curve C indicates the transition state. (From [14], with permission)

Here λ_v is the reorganization energy for the classical coordinate q. In this scheme, relaxation in the nuclear coordinate is assumed to be much faster than that in the solvent coordinate. The reaction takes place on the PES (X in Fig. 3.3-4) as the population moves along X by thermal fluctuation and transition from reactant to product. The reaction is most likely to occur at the region of X where the reaction barrier is low. The population can be expressed by the diffusion equation

$$\frac{\partial p(X,t)}{\partial t} = D\frac{\partial}{\partial X}\left[\frac{\partial}{\partial X}+\frac{1}{k_B T}\frac{dV(X)}{dX}\right]p(X,t)-k(X)p(X,t) \qquad (3.3\text{-}23)$$

where D is the diffusion coefficient, $V(X) = \lambda_s X^2$ is the PES on the solvent coordinate, and $p(X,t)$ is the population at X and t. The first term on the right-hand side represents the slow diffusion along the solvent coordinate X, and the last term represents the fast reaction along the nuclear coordinate q.

In the 2D or 2-mode treatment, the vibrational nuclear motion is assumed to be infinitely fast, but the solvent motion is assumed to be diffusive. Thus, although the distribution along X may evolve in time, the distribution along q is always at equilibrium. Therefore, one can define a rate constant $k(X)$ at each X with a suitable averaging over the population in the q coordinate:

$$k(X)=v_q \exp\left[-\frac{\Delta G^*(X)}{k_B T}\right] \qquad (3.3\text{-}24)$$

where $\Delta G^*(X)$ is the activation energy for the reaction observed at a point X, and v_q is the pre-exponential factor. The time-dependent full population on the reactant surface $P(t)$ is given by the sum of each population $p(X,t)$ at each X:

$$P(t)=\int p(X,0)\exp\left[-k(X)t\right]dX \qquad (3.3\text{-}25)$$

The main point of this theory is that the different reaction rate $k(X)$ at each X changes the distribution of the reactant during the reaction and gives rise to nonexponentiality.

The 2D model treats both the solvent and the intramolecular mode classically. However, the actual system should not only have classical modes but also quantum mechanical high-frequency modes. This effect can be examined in the same manner as for nonadiabatic ET (Sect. 3.3.1). The solvent coordinate dependent rate constant, $k(X)$, is expressed as the sum of contributions from all the vibrational states of the product in the form of Eq. (3.3-7), with

$$k_{NA}^{0 \to n}(X) = \frac{2\pi (V_{el}^{0 \to n})^2}{\hbar \sqrt{4\pi\lambda_{cl,vib}k_B T}} \exp\left(-\frac{(\Delta G_0^{0 \to n}(X) + \lambda_{cl,vib})^2}{4\pi\lambda_{cl,vib}k_B T}\right) \qquad (3.3\text{-}26)$$

Here $\lambda_{cl,vib}$ is the classical reorganization energy of the low-frequency vibration [20]. In this model the classical reorganization is divided into two parts, terms corresponding to either classical vibrations or solvent reorganization, and the values of these parameters can be obtained spectroscopically. $\Delta G_0^{0 \to n}(X)$, the effective energy gap between the ground vibrational state of the reactant and the nth vibrational state of the product, is solvent coordinate dependent as follows:

$$\Delta G_0^{0 \to n}(X) = \lambda_s + \Delta G_0 - 2\lambda_s X + nh\nu \qquad (3.3\text{-}27)$$

Here it is assumed that vibrational relaxation of the internal degrees of freedom in both the reactant and the product is fast on the time-scale of ET. We will later give examples of ET which can be rationalized by this extended model.

The dynamics of such a fast ET can be analyzed in the following way. To obtain D, we use experimental solvation data obtained by measurement of the dynamic fluorescence Stokes shift. The time-dependent diffusion coefficient $D(t)$ is described as

$$D(t) = -\frac{k_B T}{2\lambda_s} \frac{1}{\Delta(t)} \frac{d\Delta(t)}{dt} \qquad (3.3\text{-}28)$$

where the solvation dynamics is given by

$$D(t) = a_1 \exp(-t/\tau_1) + a_2 \exp(-t/\tau_2) \qquad (3.3\text{-}29)$$

where τ_1 and τ_2 are the solvation times for two exponential decays, as in the case of anilines in Sect. 3.3.2.1.

3.3.3.2 Observation and Analysis of Intermolecular Electron Transfer Faster than Solvation Dynamics

Ultrafast *intermolecular* ET has been found with various dyes in electron-donating solvents [20, 21]. The fluorescence decays of oxazine 1 (OX1) in donor solvents are shown in Fig. 3.3-5(a) and (b) as typical examples, and the results of the simulations Eq. (3.3-23) to (3.3-29) are shown in Fig. 3.3-5(a′) and (b′) [14]. The normalized population of the reactant, $P(t)$, is plotted against time (ps) on a logarithmic scale. The ET parameters are kept the same in these two cases, with changes only in the energy gaps $-\Delta G$: 0.248 eV in Fig. 3.3-5(a′) and 0.078 eV in

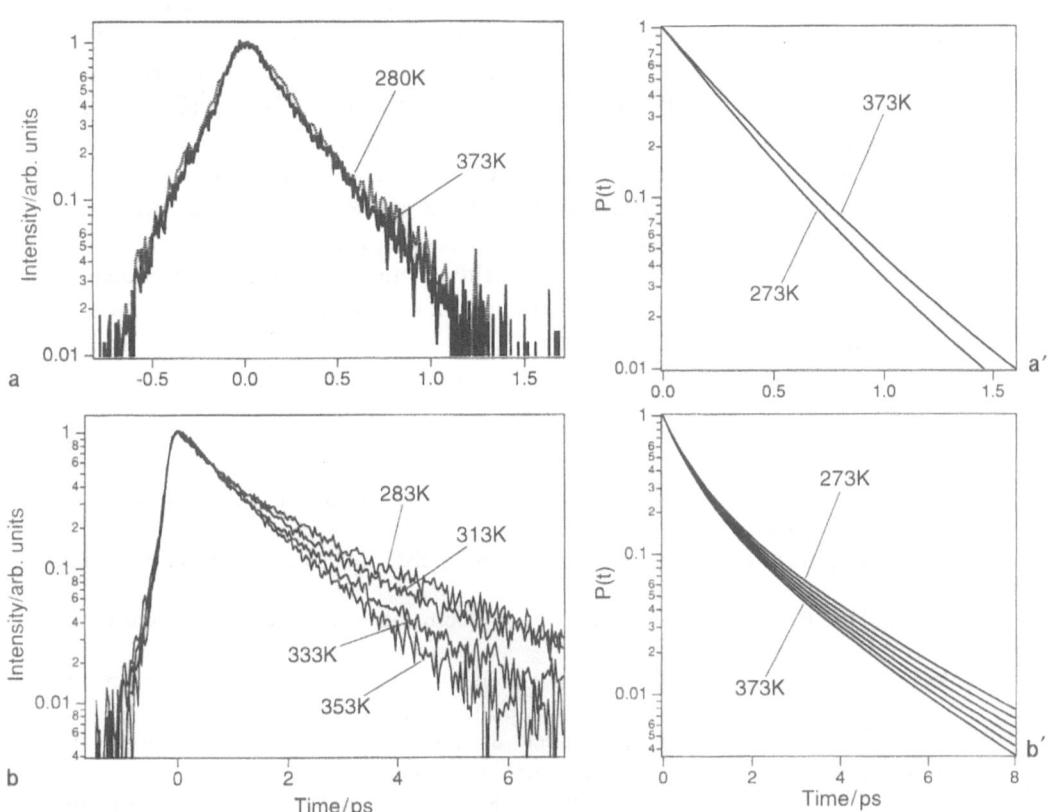

Fig. 3.3-5. Fluorescence decays of OX1 in *N,N*-dimethylamiline (*DMA*) (**a, a′**) and in aniline (*AN*) (**b, b′**). **a** Mono-exponential decay (280 fs) with no temperature dependence between 280 and 373 K. **b** Non-exponential decay with a clear temperature dependence at slower decay rates. **a′, b′** Simulation of the temperature dependence of fluorescence decays using a 2D model: **a′** $-\Delta G_0 = 0.248$ eV, $T = 273$ and 373 K (OX1 in *DMA*); **b′** $-\Delta G_0 = 0.0781$ eV, $T = 273, 293, 313, 353,$ and 373 K (OX1 in *AN*). (From [14], with permission)

Fig. 3.3-5(b′). The former is almost exponential, which coincides with the experimental result for OX1 in DMA. The latter corresponds to the ET of OX1 in AN, which produces nonexponential dynamics and the reaction becomes faster as the temperature increases. The major difference in the reaction dynamics seems to be the difference in $-\Delta G$ of 0.17 eV, which can also be calculated from the dielectric continuum model in solution [14]. In AN, the reaction has a potential energy barrier, and is affected by both slow motion on the X axis and fast motion on the q axis. In DMA, the reaction is very simple and can take place very rapidly along the q axis by exponential kinetics without traveling along the X axis. Thus, the difference in the energy gap may be the major reason for the different ET kinetics in AN and DMA.

Ultrafast *intramolecular* ET has also been observed in a few molecules. Metal–metal charge transfer that is faster than the solvation process has been found with the femtosecond pump–probe spectroscopic method in inorganic binuclear complexes such as $[(NH_3)_5Ru^{III}NCRu^{II}(CN)_5]^-$ and $[(NH_3)_5Ru^{III}NCFe^{II}(CN)_5]^-$ [19]. A similar ultrafast ET was observed with a polar organic intramolecular zwitterion [22]. The observed rate constants were analyzed in terms of the 2D ET model with a faster inertial-type solvent motion and a slower overdamped motion. Theoretical predictions are generally in good agreement with experimental data. These results also indicate that both the strong vibronic coupling between reactant and product and the inertial component of nuclear motion have a role in the ET dynamics of these compounds [23].

References

1. Marcus RA, Sutin N (1985) Electron transfer in chemistry and biology. Biochem Biophys Acta 811:265–322
2. Closs GL, Miller JR (1988) Intramolecular long-distance electron transfer in organic molecules. Science 240:440–447
3. Heitele H (1993) Dynamic solvent effect on electron-transfer reactions. Angew Chem Int Ed Engl 32:359–377
4. Yoshihara K, Tominaga K, Nagasawa Y (1995) Effects of solvent dynamics and vibrational motions in electron transfer. Bull Chem Soc Jpn 68:696–712
5. Jortner J (1976) Temperature dependent activation energy for electron transfer between biological molecules. J Chem Phys 49:4860–4867
6. Kemnitz K, Nakashima N, Yoshihara K (1988) Electron transfer by isolated rhodamine B molecules adsorbed on organic single crystals. A solvent-free model system. J Phys Chem 92:3915–3925
7. Miller JR, Calcaterra LT, Closs GL (1984) Intramolecular long-distance electron transfer in radical anions. The effects of free energy and solvent on the reaction rates. J Am Chem Soc 106:3047–3049
8. Miller JR, Beitz JV, Huddleston RK (1984) Effect of free energy on rates of electron transfer between molecules. J Am Chem Soc 106:5057–5068
9. Bixon M, Jortner J, Cortes J, Heitele H, Michel-Beyele ME (1994) Energy gap law for nonradiative and radiative charge transfer in isolated and in solvated supermolecules. J Phys Chem 98:7289–7299

10. Caspar JV, Kober EM, Sullivan BP, Meyer TJ (1982) Application of the energy gap law to the decay of charge-transfer excited states. J Am Chem Soc 104:630–632
11. Asahi T, Ohkohchi M, Mataga N (1993) Energy gap dependencies of the charge recombination process of ion pairs produced by excitation of charge-transfer complexes: Solvent polarity effects. J Phys Chem 97:13132–13137
12. Frohlich H (1949) Theory of dielectrics. Oxford University Press, London
13. Maroncelli M (1993) The dynamics of solvation in polar liquids. J Mol Liq 57:1–37
14. Nagasawa Y, Yartsev AP, Tominaga K, Johnson AE, Yoshihara K (1994) Temperature dependence of ultrafast intermolecular electron transfer faster than the solvation process. J Chem Phys 101:5717–5726
15. Jortner J, Bixon M (1987) Intramolecular vibrational excitations accompanying solvent-controlled electron transfer reactions. J Chem Phys 88:167–170
16. Kosower EM, Huppert D (1983) Solvent motion controls the rate of intramolecular electron transfer in solution. Chem Phys Lett 96:433–435
17. Kang TJ, Jarzeba W, Barbara PF, Fonseca T (1990) A photodynamical model for the excited state electron transfer of bianthryl and related molecules. Chem Phys 149: 81–95
18. Agmon N, Hopfield JJ (1983) Transient kinetics of chemical reactions with bounded diffusion perpendicular to the reaction coordinate: Intramolecular processes with slow conformational changes. J Chem Phys 78:6947–6959
19. Sumi H, Marcus RA (1986) Dynamical effects in electron transfer reactions. J Chem Phys 84:4894–4914
20. Walker GC, Akesson E, Johnson AE, Levinger NE, Barbara PF (1992) Interplay of solvent motion and vibrational excitation in electron-transfer kinetics: Experiment and theory. J Phys Chem 96:3728–3736
21. Kobayashi T, Takagi Y, Kandori H, Kemnitz K, Yoshihara K (1991) Femtosecond intermolecular electron transfer in diffusionless weakly polar systems: Nile blue in aniline and N,N-dimethylaniline. Chem Phys Lett 180:416–422
22. Nagasawa Y, Yartsev AP, Tominaga K, Bisht PB, Johnson AE, Yoshihara K (1995) Dynamical aspects of ultrafast intermolecular electron transfer faster than solvation dynamics. J Phys Chem 99:653–662
23. Tominaga K, Killer DAV, Johnson AE, Levinger NE, Barbara PF (1993) Femtosecond experiments and absolute rate calculations on intervalence electron transfer of mixed-valence compounds. J Chem Phys 98:1228–1243

3.4
Proton Transfer

Michiya Itoh

Excited-state proton transfer (ESPT) plays a vital role in many systems that are very important to any description of molecular interactions, not only in the field of photochemistry but also in photobiological phenomena. Inter- and intramolecular ESPT, as well as energy and electron transfer, also provide us with fundamental concepts in the design of the photochromic systems found in functional polymers and other molecular assemblies. In 1952, Förster [1] initiated research in the field of intermolecular ESPT with his study of 2-naphthol. He demonstrated that pK_a^* is the hydroxyl group of 2-naphthol decreased significantly in the singlet excited state compared with the ground state. Further, Weller [2] found that methyl salicylate (MS) presented an unusually large Stokes shifted fluorescence due to intramolecular ESPT. Since their pioneering work on ESPT, numerous investigations have been carried out in both the liquid and the solid state.

In addition to steady-state spectroscopy, nanosecond, picosecond, and even femtosecond time-resolved fluorescence and absorption spectroscopy has revealed the dynamic processes of ESPT. Recent supersonic jet spectroscopy has also demonstrated the dynamics of ESPT not only in intramolecular H-bonding systems, but also in jet-cooled intermolecular H-bonding in solvent clusters [3, 4]. This chapter is concerned only with ESPT from conventional spectroscopy in solutions, as described in the literature [5, 6]. Figure 3.4-1 shows diagrams of three typical types of potential energy curve for the ground and excited states as functions of the proton transfer coordinate (Q_H) in an asymmetric proton transfer system. Inter- and intramolecular ESPT is discussed in terms of these potential energy curves in Sects. 3.4.1 and 3.4.2.

The kinetics of excited-state proton transfer are described briefly below. The reaction scheme for ESPT and decay kinetics is

$$
\begin{array}{ccc}
S_1 & \underset{k_{-PT}}{\overset{k_{PT}}{\rightleftharpoons}} & S_1' \\[2mm]
h\nu \big\Updownarrow k_d & & h\nu' \big\Updownarrow k_d' \\[2mm]
S_0 & \rightleftharpoons & S_0'
\end{array}
$$

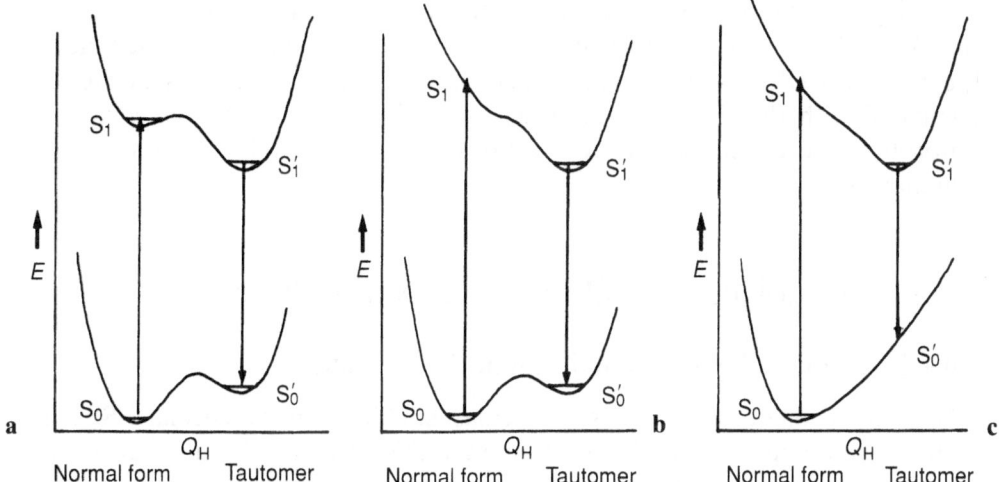

FIG. 3.4-1. Schematic potential energy diagrams showing the ground and excited states as functions of the proton transfer coordinates (Q_H). **a**, **b**, and **c** are three typical cases of asymmetric proton transfer systems. S_0 and S_1 are the ground and singlet excited states, respectively, in the normal form, and S_0' and S_1' are the corresponding states of the tautomer form

where S_1 and S_1' are the singlet electronic excited states of the normal and isomeric tautomer forms of ESPT, respectively, and k_{PT} and k_{-PT} are the ESPT and back proton transfer rate constants, respectively. The radiative rate constants of both states are expressed by k_r and k_r', and k_d and k_d' are their decay rate constants, respectively. The transient intensities of the normal (I) and tautomer (I') fluorescences are given by the following conventional equations:

$$I = k_r K \left\{ e^{-\lambda_1 t} + m e^{-\lambda_2 t} \right\} S_0 \tag{3.4-1}$$

$$I' = k_r' F \left\{ e^{-\lambda_1 t} - e^{-\lambda_2 t} \right\} S_0 \tag{3.4-2}$$

$$K = \left(\Delta_1 - \lambda_2 \right) / \left(\lambda_1 - \lambda_2 \right) \tag{3.4-3}$$

$$F = k_{PT} / \left(\lambda_1 - \lambda_2 \right) \tag{3.4-4}$$

$$m = \left(\lambda_1 - \Delta_2 \right) / \left(\Delta_1 - \lambda_2 \right) \tag{3.4-5}$$

$$\Delta_1 = k_d + k_{PT} \tag{3.4-6}$$

$$\Delta_2 = k_d' + k_{-PT} \tag{3.4-7}$$

$$\lambda_{1,2} = \left\{ \left(\Delta_1 + \Delta_2 \right) \pm \left[\left(\Delta_1 - \Delta_2 \right)^2 + 4 k_{PT} k_{-PT} \right]^{1/2} \right\} / 2 \tag{3.4-8}$$

Taking account of the single exponential decay of the normal fluorescence (I) in standard inter- and intramolecular ESPT, which corresponds to the rise time of the tautomer (I'), it was estimated that $m \ll 1$, and thus k_{-PT} is negligible compared with the ESPT rate constant k_{PT}. Time-resolved fluorescence and absorption spectra in subsequent sections are expressed in terms of these equations.

3.4.1 Intermolecular ESPT and Relaxation Processes

3.4.1.1 Aromatic Hydroxyl and Amino Compounds

2-Naphthol is one of the simplest molecules exhibiting intermolecular ESPT in aqueous solutions. Förster [1] showed that this molecule is more acidic in the singlet excited state than in the ground state. This effect is known as the Förster cycle. The different pK values in 2-naphthol were determined by Weller: $pK_a = 9.5$ and $pK_a^* = 3.09$ [2]. In 1968, Cohen and Marcus [7] adapted the theory he initially proposed for electron transfer reactions to the case of proton transfers. Since this pioneering work, ESPT has been studied in various H-bonding systems, e.g., phenanthrylamine and naphthylamine in aqueous solutions [6].

Robinson and co-workers [8] used picosecond spectroscopy to study the ESPT of 2-naphthol in water/methanol mixtures. They reported that the proton transfer rate increased as the temperature increased, and decreased as the methanol concentration increased. They reported that the activation energy of proton transfer is $3.45 \, \text{kcal mol}^{-1}$ in pure water. The Markov random walk theory indicates that a water cluster containing 4 ± 1 molecules is the proton acceptor.

The size of the molecules and the specific structure of the water clusters apparently determine the rate of ESPT in aqueous media. The effects of the size and structure of the solvent clusters on the role of the proton acceptor have recently been investigated with the aid of supersonic jet spectroscopy. This is discussed in detail in Chap. 5.2.

3.4.1.2 Aromatic Compounds with Bifunctional Groups and Their Dimers

Aromatic compounds with bifunctional groups, such as 7-hydroxyquinoline, 7-hydroxyflavone, and 3-hydroxyxanthone, show very different spectroscopic features from those of naphthol and aminonaphthalene. 7-hydroxyquinoline (7-HQ), a typical example of an aromatic compound with bifunctional groups, exhibits intermolecular ESPT in alcohol solutions [9]. A methanol solution of 7-HQ shows a UV ($\lambda_{max} = 378 \, \text{nm}$) and a large Stokes-shifted ($\lambda_{max} = 530 \, \text{nm}$) fluorescence, which can be attributed to the normal and tautomer fluorescence, respectively. The potential energy curve of 7-HQ in a methanol solution is shown in Fig. 3.4-1(a). The concentration dependence of the fluorescence and absorption spectra demonstrate that two H-bonded methanol molecules are required in the intermolecular ESPT of 7-HQ.

A transient absorption spectrum of a methanol solution of 7-HQ was observed at λ_{max} = 420–430 nm with a decay time of 3.5 μs (first laser excitation), and this was believed to be the absorption spectrum of the ground-state tautomer generated by intermolecular ESPT. A second laser excitation (440 nm) of the transient absorption band showed the tautomer fluorescence, which was two-step laser-induced fluorescence (TS-LIF). Plots of TS-LIF intensities versus delay times between the first and second laser pulses were consistent with the transient absorption decay time of the ground-state tautomer. The first observation of the ground state tautomer in the relaxation of ESPT in 7-HQ/methanol was further confirmed using the thermal lensing effect.

Other typical aromatic compounds with bifunctional groups, e.g., 7-hydroxyflavone (7-HF) and 3-hydroxyxanthone (3-HX) [10], showed ESPT leading to the formation of the excited-state tautomer and anion forms in methanol solutions. Picosecond fluorescence, transient absorption, and TS-LIF demonstrated that the tautomer and anion forms were generated from 1:2 and 1:1 methanol H-bonded complexes of these compounds, respectively, in the excited state (Fig. 3.4-2).

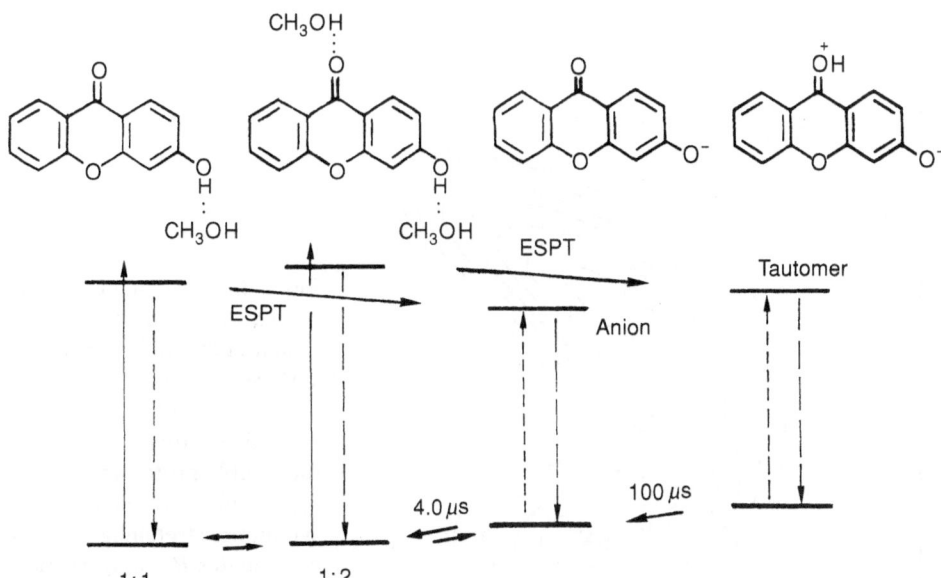

Fig. 3.4-2. Reaction scheme of excited-state proton transfer (*ESPT*) and relaxation in the 3-hydroxyxanthone/methanol system. (From [10], with permission)

3.4.1.3 Dimers of 7-Azaindole and 1-Azacarbazole as Aromatic Compounds with Bifunctional Groups

Kasha [11] investigated excited-state double proton transfer in the 7-azaindole (7-AI) dimer, which is a simple model of a hydrogen-bonding base pair in biological molecules. Numerous other investigations of double proton transfer in the 7-AI dimer using nano- and picosecond fluorescence in solutions [12] and in a supersonic free jet [13] have also been reported. When a 7-AI dimer formed in a nonpolar solution is excited to the S_1 state, the dimer exhibits a large Stokes-shifted visible fluorescence (450–500 nm) with a decay time of 3.0 ns in addition to the normal UV fluorescence (λ_{max} = 350 nm). The visible fluorescence was ascribed to a tautomer simultaneously generated by double proton transfer in the excited-state dimer of 7-AI. Very fast excited-state double proton transfer takes place in the almost coplanar structure of the dimer. The transient absorption

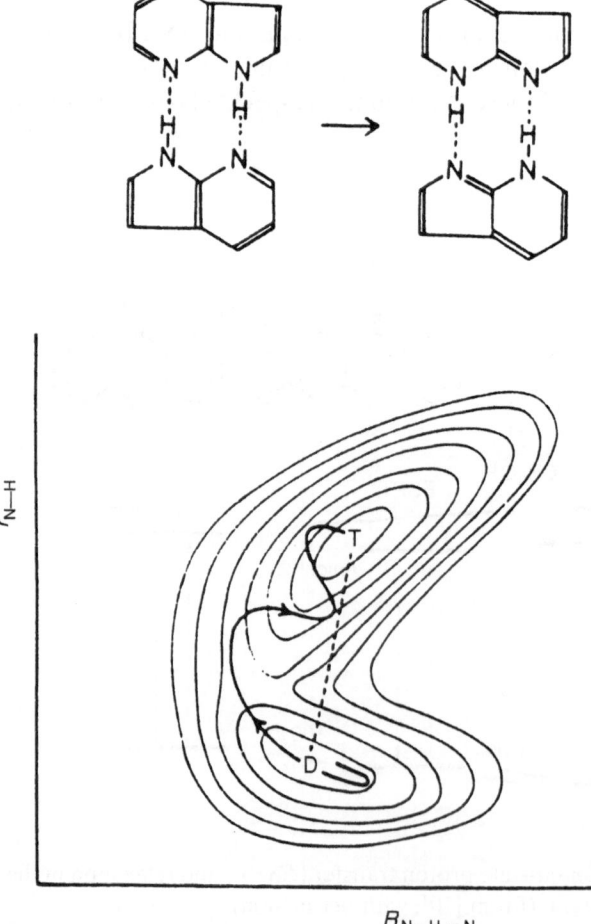

Fig. 3.4-3. Schematic representation of the S_1-state potential energy surface of a model hydrogen-bonded base pair plotted as a function of the double proton transfer, r_{N-H}, and intermolecular symmetric stretching coordinates R_{N-H-N}. D and T indicate the dimer and tautomer, respectively. (From [16], with permission)

spectrum of the dimer shows a band at $\lambda_{max} = 480$ nm with a decay time of $19\,\mu$s, which was ascribed to the transient absorption band of the ground-state dimer tautomer.

In contrast, the second laser excitation of the transient absorption band showed a red-shifted fluorescence (TS-LIF) spectrum of approximately 20–30 nm. The TS-LIF also showed rise and decay times of approximately 1.0 ns and 2.1 ns, respectively. After a series of experiments, including variable delay plots of TS-LIF, this TS-LIF was ascribed to the monomer tautomer, which was generated in a fast dissociation reaction by the second laser excitation of the ground-state dimer tautomer [14]. Further, supersonic jet spectroscopy, including picosecond time-resolved fluorescence of dimers of 7-AI and 1-azacarbazole, demonstrated that the symmetric stretching vibration in the hydrogen bond strongly encourages double proton transfer, which is ascribed to a dynamic coupling of proton motion in intermolelcular vibration [15] (Fig. 3.4-3).

3.4.2 Intramolecular ESPT and Relaxation Processes

3.4.2.1 Symmetric Intramolecular ESPT in a Matrix and in a Supersonic Jet

ESPT in typical symmetric proton transfer systems has been reported for 9-hydroxyphenalenone (9-HP) and tropolone (TRP) [17]. Two possible structures for these compounds arise from the transfer of a hydrogen atom between the two oxygen atoms. The physical properties of these structures are identical except for tunneling splitting in the molecular spectrum. The isomerization is represented by the double minimum potential energy curve shown in Fig. 3.4-4. Since the first observation of the S_1–S_0 absorption spectrum of TRP in the vapor phase, Aleves and Hollas [17] have reported that the electronic origin band exhibits a doublet structure as an isotopic effect of splitting. They conclude that the doublet structure arises from tunnelings at the zero-point levels of the ground and excited states.

Ito and co-workers [18] demonstrated vibrational bands of doublet structures arising from proton tunneling of intramolecular H-bonding in jet-cooled TRP. The difference in vibrational splitting between the excited and ground states was reported to be $19\,\text{cm}^{-1}$, which decrease to $2.3\,\text{cm}^{-1}$ on deuterium substitution. Comparable vibrational splittings, $21 \pm 1\,\text{cm}^{-1}$ and $7 \pm 1\,\text{cm}^{-1}$, were reported in an Ne matrix at 4 K of TRP and deuterium-substituted TRP.

3.4.2.2 Asymmetric Intramolecular ESPT in Nonhydrogen-bonding Solvents and in a Supersonic Jet

MS, 3-hydroxyflavone, 2-(2′-hydroxyphenyl)benzothiazole, and related compounds belong to the asymmetric intramolecular ESPT system [5] whose potential energy curves are shown in Fig. 3.4-1(b) and (c). All these compounds show

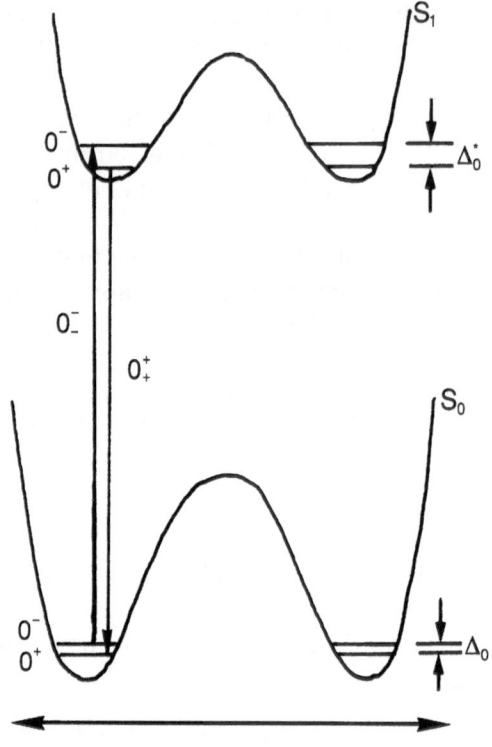

FIG. 3.4-4. Schematic diagrams of a symmetric double minimum potential in tropolone. (From [18], with permission)

a large Stokes-shifted fluorescence. In the case of MS, a dual fluorescence (λ_{max} = 340 and 450 nm) was observed [2]. The latter red-shifted fluorescence was assigned to an emission from the excited-state tautomer (S_1'). Numerous investigations into the ESPT and relaxation of MS and related compounds in solutions, in rare gas matrices and also in a supersonic jet [19], have been reported. Recent experiments with pico- and femtosecond spectroscopy have suggested that ESPT and ground-state reverse proton transfer (GSRPT) to the normal form occur within a few picoseconds. These features of ESPT and GSRPT in MS and related compounds seem to agree with the potential energy curve shown in Fig. 3.4-1(c).

Other typical asymmetric H-bonding systems exhibiting intramolecular ESPT are 2-(2′-hydroxyphenyl)benzothiazole (HBT) and -triazole, and their related compounds. Since some of these compounds, such as 2-(2′-hydroxy-5′-methylphenyl)benzotriazole (TIN, trade name Tinvin P), are used to diminish the photodegradation of polymers, considerable research effort has been put in to elucidating their photophysical deactivation mechanisms in solution, in polymers, and in crystals. These compounds usually show a large Stokes-shifted fluorescence in solution. The absorption spectra are due to the stable ground-state normal form (S_0), while the large Stokes-shifted fluorescence is associated with the keto tautomer (S_1') generated by ESPT. Very fast rates of ESPT are monitored by a rise in tautomer fluorescence. This was initially done by picosecond spectroscopy, but more recently the femtosecond pump or probe experiments have been used [20]. In these compounds, ESPT was believed to take place in a planar conformation of two aromatic systems with an intact intramolecular hydrogen bond. The keto tautomer is deactivated by internal conversion to its ground state (S_0') with $\tau = 150$fs, followed by the GSRPT with $\tau = 500$fs in tetrachloroethane.

In nonhydrogen-bonding solutions of HBT, however, Fujiwara and Itoh [21] suggested that *cis* and *twisted* ground-state conformers involving two aromatic rings are involved in the relaxation process of HBT. The decay times of these tautomers were 9.4 and 53 μs, respectively, at room temperature. Becker *et al.* [22] also suggested similar long-lived *cis* and *twisted* ground-state tautomers (zwitterions) of this compound in nonpolar solvents. The involvement of long-lived tautomers in GSRPT was suggested by time-resolved infrared spectroscopy [23]. On the other hand, triplet-state ESPT has been reported in some intramolecular H-bonding systems such as HBT and 7-hydroxy-1-indanone [24].

Other simple asymmetric H-bonding systems exhibiting intramolecular ESPT are 3-hydroxyflavone (3-HF) and related compounds. 3-HF exhibits a large Stokes-shifted fluorescence, which Kasha [11] ascribed to intramolecular ESPT. In the early stages of nano- and picosecond fluorescence studies of ESPT, a very slow rise-time for tautomer fluorescence was reported in nonpolar solutions of 3-HF, while a very fast rise-time (<10ps) was observed in 3-hydroxychromone (3-HC), which has no phenyl group at the 2-position of 3-HF [25]. These differences in ESPT behavior were ascribed to the torsional motion of the phenyl group in conjunction with the proton transfer. The remarkably different features of the ESPT in these compounds has been the subject of much controversy. Recent supersonic jet fluorescence studies of ESPT in 3-HC, 3-HF, and 2-(2′-naphthyl)-3-hydroxychromone (2-NHC) have demonstrated remarkably different upper limits of rate constants in these compounds (1.77×10^{12}s^{-1} for 3-HC, 6.5×10^{11}s^{-1} for 3-HF, and 1.54×10^{11}s^{-1} for 2-NHC) [26].

On the other hand, transient absorption and TS-LIF of 3-HF and related compounds has demonstrated that very long-lived ground-state tautomers are involved in the relaxation process. The involvement of long-lived ground-state tautomers in the relaxation of ESPT in 2-NHC has been demonstrated by transient absorption, TS-LIF [27], and time-resolved infrared spectroscopy [28].

References

1. Förster T (1951) Fluoreszenz organischer verbindungen. Vandenboeck & Ruprecht, Göttigen
2. Weller A (1955) Fluorescence of salicylic acid and related compounds. Naturwissenschaften 42:175–176
3. Kappens M, Leutwyler S (1988) Molecular beams of clusters. In: Scoles G (ed) Atomic and molecular beam method, vol 1. Oxford University Press, New York
4. Smalley RE, Wharton L, Levy DH (1977) Molecular optical spectroscopy with super- sonic beams and jets. Acc Chem Res 10:139–145
5. Caldin FF, Gold V (1975) Proton-transfer reactions. Chapman and Hall, London
6. Mataga N, Kubota T (1970) Hydrogen bonding complexes. In: Molecular interactions and electronic spectra. Dekker, New York, pp 293–369
7. Cohen AO, Marcus RA (1968) Slope of free energy plots in chemical kinetics. J Phys Chem 72:4249–4256
8. Lee J, Griffin RD, Robinson GW (1985) 2-Naphthol: A simple example of proton transfer effected by water structure. J Chem Phys 82:4920–4925
9. Itoh M, Adachi T, Tokumura K (1984) Time-resolved fluorescence and absorption spectra and two-step laser excitation fluorescence of excited-state proton transfer in a methanol solution of 7-hydroxyquinoline, J Am Chem Soc 106:850–855
10. Mukaihata H, Nakagawa T, Kohtani S, Itoh M (1994) Picosecond and two-step LIF studies of excite-state proton transfer in 3-hydroxyxanthone and 7-hydroxyflavone methanol solutions: Reinvestitation of tautomer and anion formations. J Am Chem Soc 116:10612–10618
11. Kasha M (1986) Proton-transfer spectroscopy. J Chem Soc, Faraday Trans 82:2378–2392
12. Barbara PF, Walsh PK, Brus LE (1989) Picosecond kinetics and vibrationally resolved spectroscopic studies of intramolecular excited-state hydrogen atom transfer. J Phys Chem 93:29–34
13. Fuke K, Yoshiuchi H, Kaya K (1984) Electronic spectra and tautomerism of hydrogen-bonded complexes of 7-azaindole in a supersonic jet. J Phys Chem 88:5840–5844
14. Tokumura K, Watanabe Y, Udagawa M, Itoh M (1987) Photochemistry of a transient tautomer of the 7-azaindole H-bonded dimer studied by two-step laser excitation fluorescence measurements. J Am Chem Soc 109:1346–1350
15. Fuke K, Tsukamoto K, Misaizu F, Kaya K (1991) Picosecond measurements of the vibrationally resolved proton transfer rate of the jet-cooled 1-azacarbazole dimer. J Chem Phys 95:4074–4080
16. Fuke K, Kaya K (1989) Dynamics of double-proton-transfer reaction in the excited-state model hydrogen-bonded base pairs. J Phys Chem 93:614–621
17. Alves ACP, Hollas JM (1973) Near-ultraviolet absortion spectrum of tropolone va- por: II Vibrational analysis. Mol Phys 25:1305–1314
18. Tomioka Y, Ito M, Mikami N (1983) Electronic spectra of tropolone in a supersonic free jet. Proton tunneling in the S_1 state. J Phys Chem 87:4401–4405

19. Felker PM, Lambert WR, Zewail AH (1982) Picosecond excitation of jet-cooled hydrogen-bonded systems: Dispersed fluorescence and time-resolved studies of methyl salicylate. J Chem Phys 77:1603–1605

20. Wiechmann M, Port H, Frey W, Larmer F, Elsasser T (1991) Time-resolved spectroscopy on ultrafast proton transfer in 2-(2′-hydroxy-5′-methylphenyl)benzotriazole in liquid and polymer environments. J Phys Chem 95:1918–1923

21. Fujiwara Y, Itoh M (1985) Transient absorption and two-step laser excitation fluorescence studies of photoisomerization in 2-(2-hydroxyphenyl)benzoxazole and 2-(2-hydroxyphenyl)benzothiazole. J Am Chem Soc 107:1561–1565

22. Becker RS, Lenoble C, Zein A (1988) A comprehensive investigation of the photophysics and photochemistry of salicylideneaniline and derivatives of phenylbenzothiazole, including solvent effects. J Phys Chem 91:3509–3517

23. Yuzawa T, Takahashi H, Hamaguchi H (1993) Submicrosecond time-resolved infrared study on the structure of photoinduced transient species of salicylideneaniline in acetonitrile. Chem Phys Lett 202:221–226

24. Chou PT, Martinez NL, Studer SI (1991) Role of triplet states in the reverse proton transfer mechanism of 7-hydroxy-1-indanone. J Phys Chem 95:10306–10310

25. Itoh M, Tokumura K, Tanimoto Y, Okada Y, Takeuchi H, Obi K, Tanaka I (1982) Time-resolved and steady-state fluorescence studies of excited-state proton transfer in 3-hydroxyflavone and 3-hydroxychromone. J Am Chem Soc 104:4146–4150

26. Ito A, Fujiwara Y, Itoh M (1992) Intramolecular excited-state proton transfer in jet-cooled 2-substituted 3-hydroxychromones and their water clusters. J Chem Phys 96:7474–7482

27. Itoh M, Fujiwara Y, Matsudo M, Higashikata A, Tokumura K (1990) Transient absorption and two-step laser-induced fluorescence studies of the intramolecular excited-state proton transfer and relaxation process in 2-naphthyl-3-hydroxychromones. J Phys Chem 94:8146–8152

28. Itoh M, Yuzawa T, Mukaihata H, Hamaguchi H (1995) Time-resolved infrared study of the relaxation process of intramolecular excited-state proton transfer in 2-naphthyl-3-hydroxychromone. Chem Phys Lett 233:550–554

3.5
Chemical Transformation

Tadashi Sugawara

Chemical transformation is the process of converting conveniently available substances into products of high value. It is therefore crucial to control the behavior of any reactive intermediates involved in chemical transformation in order to obtain the desired product in a highly efficienc manner.

Among reactive intermediates, one-centered diradicals such as carbenes [1] or nitrenes [2] play a significant role in chemical transformations. This is because these species exhibit characteristic electronic structures and reactivities that depend on their spin multiplicities. Chemical transformations, which take place through such reactive intermediates, can be utilized in many ways, as shown in Fig. 3.5-1. Carbene homologues are found in interstellar material [3], and carbene species make an important contribution to the formation of other unusual molecules [4]. The utility of carbenes and nitrenes in synthetic chemistry cannot be over stated. Highly strained or highly reactive compounds such as propellanes and cyclobutadienes, are synthesized through carbene intermediates. The addition reaction of carbenes is used as an efficient tool for modifying the C_{60} skeleton [5]. Transition metal carbene complexes are often used as a catalyst in olefin metathesis [6]. In the field of materials science, these highly reactive intermediates are also used as building blocks in high-spin organic molecules or organic magnetic material [7–9]. One application to functional materials is the use of the Wolff rearrangement of ketocarbene in photoresist processes [10]. An insertion reaction of aryl nitrenes is frequently used in a branching process in polymer chemistry [11], and as a photoaffinity label [12]. The anticancer activity of calicheamicin is derived from the highly reactive σ-1,4-diradical, which has a canonical structure of dicarbene. Interestingly, the diradical is formed by intramolecular cyclization of the enediyne moiety of calicheamicin [13].

It is important to understand the intrinsic nature of these species, not only to improve such functionalities, but also to explore novel functionality based on chemical transformations derived from carbenes, nitrenes, and related species. This chapter describes the electronic structures and reactivities of carbenes and nitrenes, including direct detection of highly reactive intermediates. The basic concepts of novel applications of such species at the molecular level are presented.

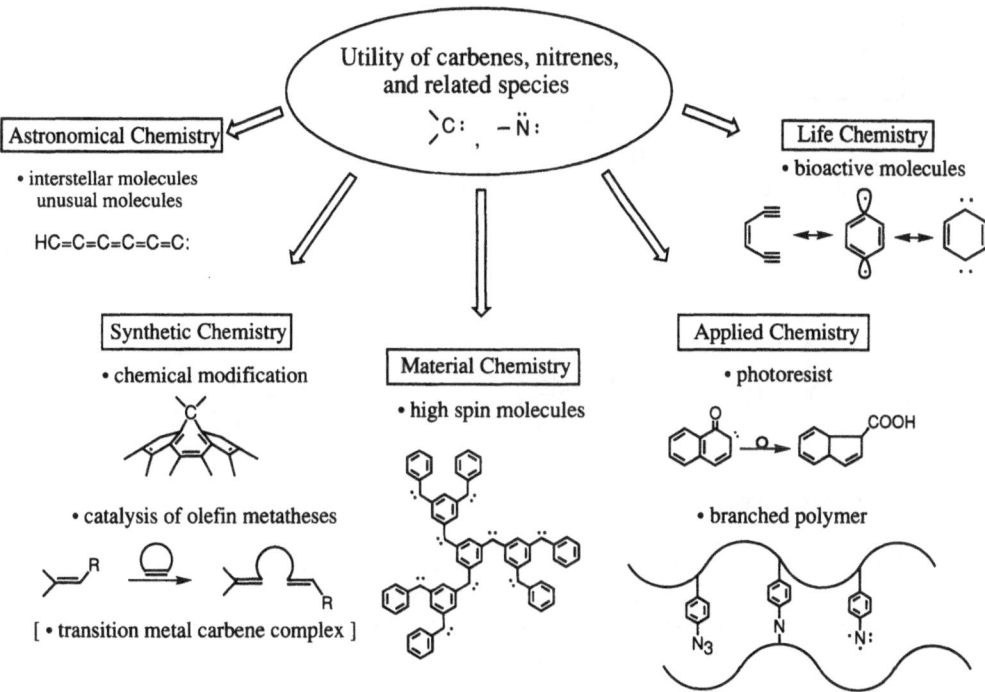

FIG. 3.5-1. The utility of carbenes, nitrenes, and related species

3.5.1 Molecular and Electronic Structure of Carbenes and Nitrenes

The electronic structure of methylene is characterized by the availability of nearly degenerated σ and π orbitals (Fig. 3.5-2). Two electrons occupying a low-lying σ-type orbital leads to singlet spin multiplicity. With two parallel spins occupying σ and π orbitals separately, the electronic structure leads to triplet spin multiplicity. Since the energy difference ($\Delta\varepsilon$) between these orbitals is usually smaller than the exchange interaction ($K_{\sigma\pi}$), a triplet becomes the ground state for carbenes. The energy difference (ΔE_{TS}) between the triplet and singlet methylene was found to be 38 kJ mol^{-1} by far infrared laser magnetic resonance spectroscopy, 37 kJ mol^{-1} by visible absorption and magnetic rotation spectroscopy, and 39 kJ mol^{-1} by photoelectron spectroscopy of CH_2. The energy difference was calculated theoretically as 39 kJ mol^{-1}. This agreement between experimental and theoretical values has only been established after many years of controversy [14]. When heteroatoms with lone pairs of electrons are directly attached to the divalent carbon atom, the ground state spin multiplicity of the carbenes becomes a singlet. The ground spin multiplicity and the energy difference (ΔE_{TS}) between the triplet and singlet states of some carbenes have been calculated theoretically

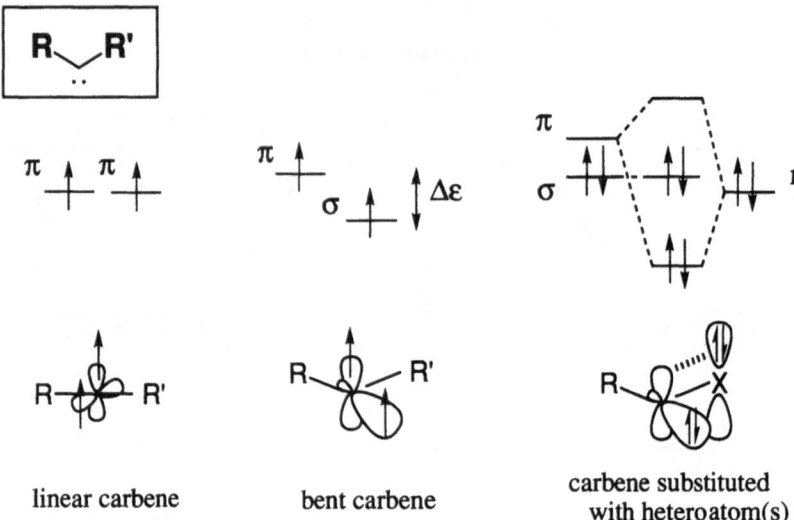

linear carbene   bent carbene   carbene substituted
                                               with heteroatom(s)

FIG. 3.5-2. Electronic structure of linear, bent, and heteroatom-substituted carbenes. A linear carbene has singly-occupied degenerated π-orbitals (*left*). The degeneracy of π-orbitals is removed in a bent carbene (*middle*). Orbital interaction between the empty π-orbital of carbene and the filled n-orbital of a heteroatom stabilizes the singlet configuration of a heteroatom-substituted carbene (*right*)

and some are listed in Table 3.5-1 [15]. The electronic structure of carbenes is similar to that of transition metal complexes. Transition metal complexes with weak ligands have a high-spin ground state, whereas metal complexes with strong ligands have a low-spin ground state. Recently a singlet carbene stabilized by heteroatoms was isolated and its molecular structure was determined by X-ray crystallographic analysis [16].

The molecular structure of methylene also depends on its spin multiplicity ($^3B_1$, $^1A_1$). Although the bond distance of C—H is 1.1 Å for both $^3B_1$ and $^1A_1$, the bond angle of HCH is 134° for $^3B_1$ and 120° for $^1A_1$ [1]. Theoretical values are consistent with experimental data. The molecular structure and spin distribution of diphenylmethylene (DPM) have been determined based on the ENDOR experiments by Hutchison and Kohler [17]. DPM has a bent structure and the two phenyl rings have a twist angle of 54°. The sign of the electron spin density of carbon atoms alternates, i.e., a positive value shows an up-spin density and a negative value is a down-spin density. Sixty per cent of the π-spin is localized at the divalent carbon atom and the rest is distributed over two benzene rings, whereas more than 70% of the σ-spin is localized on the divalent carbon. The bent structure of the geminate bianthracenylidene is known to relax to a linear structure with *sp* hybridization of the divalent carbon [18]. This structural relax-

TABLE 3.5-1. Energy difference between triplet and sin-
glet states of carbenes $(\Delta E_{TS})^a$

Carbene	ΔE_{TS} (kJ mol^{-1})
HCH	77.0
HCF	-16
FCF	-144
HCCH$_3$	69.5
HCCN	80.4
NCCCN	75 ± 8

[a]The energy difference is calculated in terms of STO-3G restricted Hartree-Fock method
(RHF) for singlet carbenes and
STO-3G for triplets. A negative
value indicates that the singlet
state is more stable than the triplet.

ation is presumably caused by the steric effect of hydrogen atoms at the *peri* positions.

Nitrene has a linear molecular structure [N—H bond distance: 1.034 Å (experimental), 1.040 Å (calculated) for $^1\Delta$; 1.0362 Å (experimental), 1.045 Å (calculated) for $^3\Sigma$], and has triplet ground spin multiplicity [ΔE_{TS} = 150 kJ mol^{-1} (experimental), 163 kJ mol^{-1} (calculated)]. While the molecular symmetry causes a degeneracy of the singlet state ($^1\Delta$), this degeneracy is removed by alkyl- or arylnitrenes. The energy difference between the triplet and singlet phenylnitrene has recently been reinvestigated theoretically by Borden and co-workers [19], and their calculated value ($\Delta E_{TS} = 76$ kJ mol^{-1}) agrees with experimental data [20].

3.5.2 Generation of Carbenes and Nitrenes from Diazo and Azide Compounds

The generation of methylene from diazomethane has been theoretically explained by Yamabe et al. based on the state correlation diagram [21]. For the elimination of nitrogen, the least-motion path conserving C_{2v} symmetry throughout the reaction is energetically unfavorable, and thermal decomposition is initiated through a bent out-of-plane motion, whereas elimination at the lowest excited state is initiated through a bent in-plane path (Cs).

The energy profile of the decomposition of diphenyldiazomethane is not fully understood, mainly because of the non-fluorescent character of diazo compounds. Based on the absorption spectrum of diphenyldiazomethane, its $n\pi^*$ state is located at 210 kJ mol^{-1}, and the $\pi\pi^*$ state is 320 kJ mol^{-1} above S_0. The quantum yield of decomposition is 0.12 when UV light is used for irradiation, and is 3.2×10^{-4} for visible light. The singlet carbene generated from the singlet excited state of the diazo precursor undergoes an internal conversion to the lowest singlet state of diphenylmethylene (S_1 of DPM), and it interconverts to the ground triplet state (T_0 of DPM). A photoacoustic measurement of diphenyldiazomethane shows that the decomposition of the diazo precursor to the triplet DPM and a nitrogen molecule is an almost isoenergetic process, $\Delta H = 0 \pm 7.1\,\mathrm{kJ\,mol^{-1}}$ [22].

The electronic structure of azides and the thermal generation of nitrenes from azides have been reported by Bock and Dammel [23]. In the cases of methyl azide and vinyl azide, synchronous rearrangement occurs to give methanimine or 2H-azirine, respectively.

3.5.3 Reaction Patterns of Carbenes and Nitrenes in Solution

The reaction scheme of diphenylmethylene (DPM) was reported by Bethell *et al.* [24] and is shown in Fig. 3.5-3. The singlet DPM (S) reacts with *iso*-propanol to give diphenylmethyl *iso*-propylether (E) or with the remaining diazo compound to give azin (A), whereas the triplet (T), which is in equilibrium with the singlet, gives diphenylmethane (M) by removing the tertiary hydrogen of *iso*-propanol followed by a hydrogen atom transfer. The triplet DPM also reacts with dissolved oxygen at a diffusion-controlled rate to give benzophenone through an intermediate carbonyl oxide [25, 26].

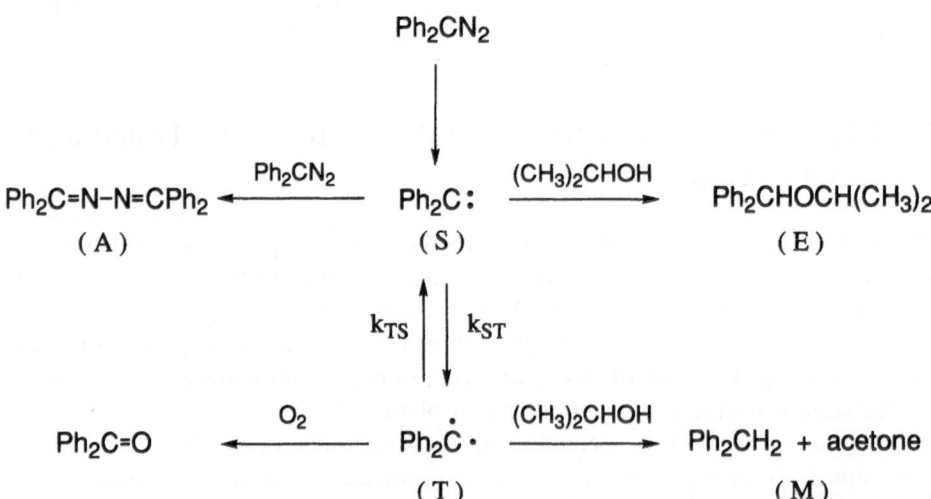

FIG. 3.5-3. Reaction scheme of singlet and triplet diphenylmethylene (*DPM*). *S*, singlet carbene; *T*, triplet carbene; *A*, azine derivative; *E*, ether derivative; *M*, diphenylmethylene

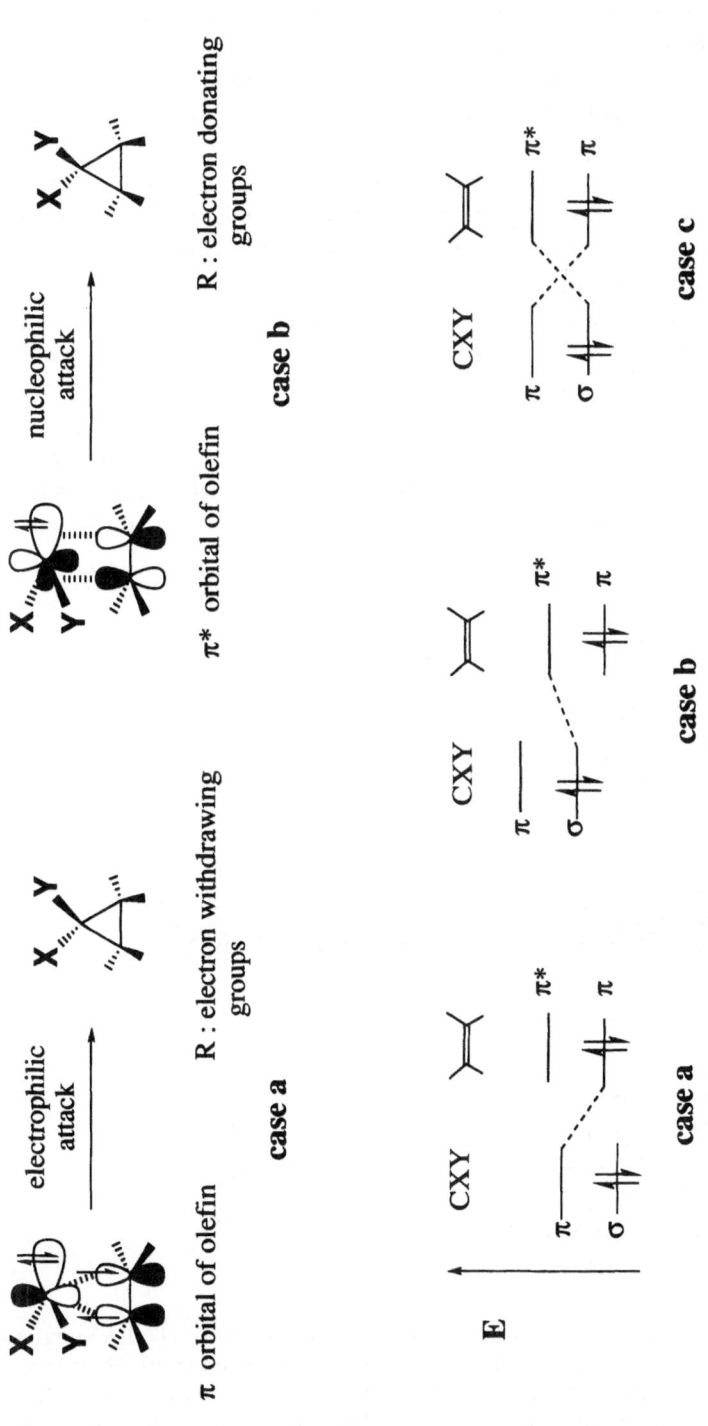

FIG. 3.5-4. Orbital interaction in a cyclopropane formation by a singlet carbene and an olefin. *CXY*, carbene substituted by *X*, *Y* groups. **Case a** Electrophilic attack. The empty π-orbital of a carbene interacts with the filled π-orbital of an olefin. **Case b** Nucleophilic attack. The filled σ-orbital of a carbene interacts with the unoccupied π^*-orbital of an olefin. **Case c** Ambiphilic attack. The empty π- and the filled σ-orbital of a carbene interact with the π- and the π^*-orbitals of an olefin, respectively

Since singlet DPM has both a filled σ orbital and an empty π orbital, the reactivity of DPM is governed by frontier orbitals which play a major role during reaction with a substrate. In the reaction between DPM and alcohol, the empty π orbital of DPM interacts with a lone pair of electrons on the oxygen in alcohol giving a zwitter-ionic intermediate, which in turn rearranges to give the ether via proton transfer.

The formation of cyclopropanes is one of the most useful carbene reactions. The reactivity of substituted methylenes with olefins is explained in terms of the frontier orbital interaction, which is crucial in the addition reaction path [27]. Orbital interaction in electrophilic carbenes takes place between the empty π orbital of carbenes and the filled π orbital of olefins, while the interaction between the filled σ orbital of carbenes and the empty π^* orbital of olefins is important for nucleophilic carbenes. Both types of orbital interactions may be involved in ambiphilic carbenes (Fig. 3.5-4).

Skeletal rearrangement is another feature of the reactivity of carbenes (Fig. 3.5-5). Generally, the activated σ bond attaching the alpha carbon to the carbenic carbon migrates to the carbenic center [28]. Back donations of the σ electrons of

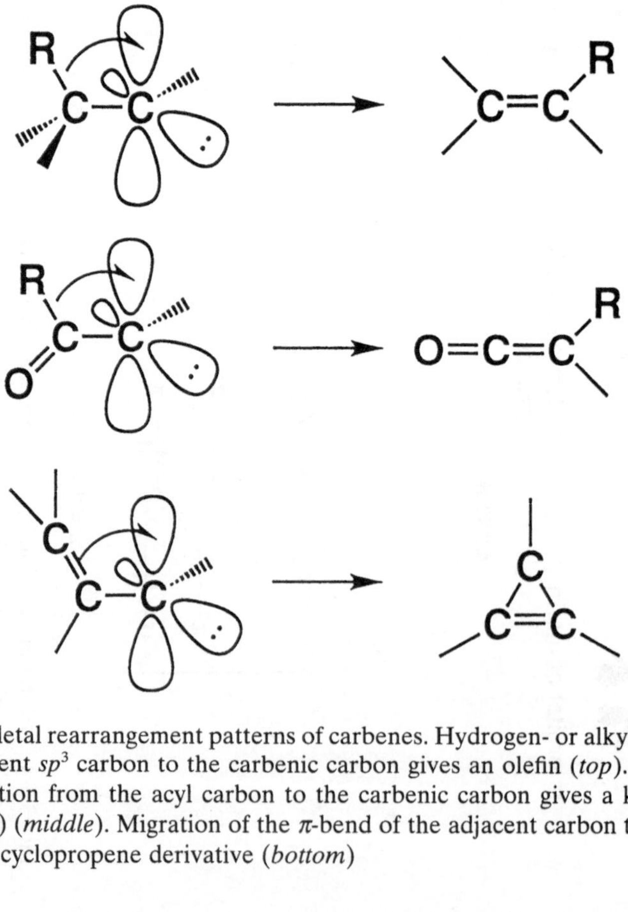

FIG. 3.5-5. Skeletal rearrangement patterns of carbenes. Hydrogen- or alkyl(R)-migration from the adjacent sp^3 carbon to the carbenic carbon gives an olefin (*top*). Hydrogen- or alkyl(R)-migration from the acyl carbon to the carbenic carbon gives a ketene (Wolff-rearrangement) (*middle*). Migration of the π-bend of the adjacent carbon to the carbenic carbon gives a cyclopropene derivative (*bottom*)

the carbenic carbon to the alpha carbon form a double bond leading to ethylene and ketene derivatives (Wolff-rearrangement), while π-bond migration to the carbenic carbon gives cyclopropene derivatives.

The absolute rate constants of these individual processes can be determined by time-resolved laser spectroscopy. Assignment of the absorption peaks of carbenes is based on the spectra of carbenes obtained in glassy matrices. The reactivity of DPM in solution was extensively studied by Closs and Rabinow [29] and more recently by Turro, Eisenthal, Sciano, and others [30, 31]. They analyzed the kinetic data, assuming equilibrium between the singlet and triplet DPM, and determined absolute rate constants for the kinetic processes in which singlet and triplet carbenes are involved. Using these kinetic data, the energy difference between the singlet and triplet states is estimated to be $21\,\text{kJ}\,\text{mol}^{-1}$. Recently, stable triplet carbenes which have a long half-life in solution have been reported. For example, the half-life of the triplet didurylcarbene was measured as $210\,\text{ms}$ in a degassed benzene solution with an intial concentration of diazo precursor of $1.0 \times 10^{-4}\,\text{M}$ [32]. The half-life increased to $2\,\text{s}$ after deuteration of the methyl groups of the duryl moieties.

The reactivity of phenylnitrene is shown in Fig. 3.5-6. The singlet nitrene is in equilibrium with its valance isomer azacycloheptatetraene, and the latter reacts with diethylamine to give a good yield, of the azepine derivative. The singlet nitrene inserts into the C—H bond of the substrate to give the aniline derivative in the case where the electrophilicity is activated by the electron-withdrawing substituent (e.g. p-CN). The triplet nitrene resulting from this intersystem cross-

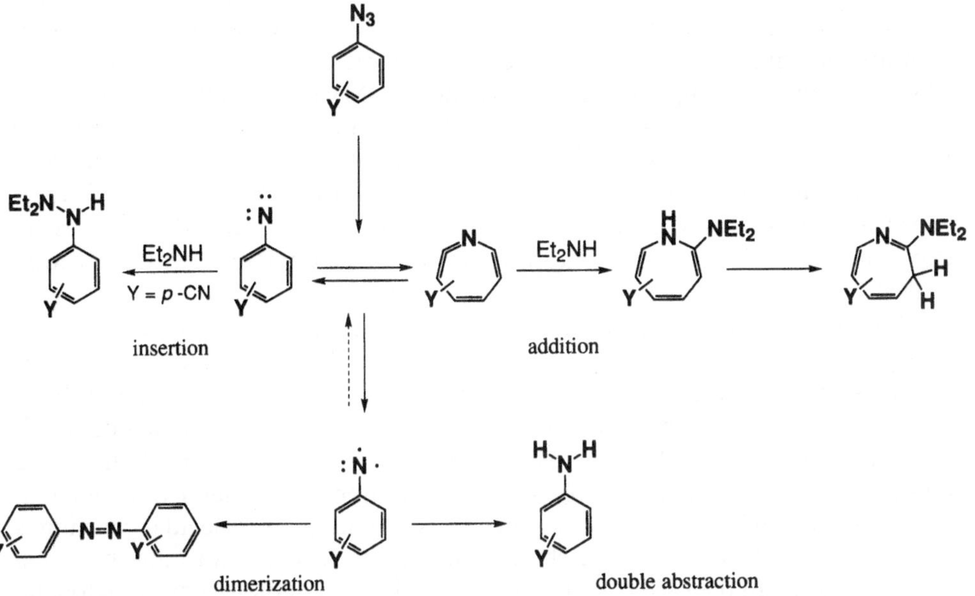

FIG. 3.5-6. Reaction scheme of singlet and triplet phenylnitrene with a substituent (Y)

ing gives aniline through double abstraction of hydrogen. The triplet nitrene is dimerized to give the azo compound.

Rate constants for phenylnitrene have also been determined by laser photolysis. The rate constant of formation of 1-H azepine was found to be dependent on the concentration of amine. The rate constant was $1.2 \times 10^6 \mathrm{s}^{-1}$ in a hexane solution of phenylazide (3.6×10^{-3} M) and diethylamine (0.19 M), and increased to $9.1 \times 10^6 \mathrm{s}^{-1}$ at an amine concentration of 9.29 M [33].

3.5.4 Decomposition of Diazo and Related Compounds Coupled with Electron Transfer

If spontaneous elimination of nitrogen occurs from the cation or anion radical of a diazo compound, these species should give the carbene cation radical or the carbene anion radical, respectively. These exotic species may be of use in chemical transformations which lead to more new applications. Radical cations and anions of diazo compounds are known to be generated by one-electron oxidation or reduction, both chemically and electrochemically. For example, when a degassed solution of diphenyldiazomethane in CH_2Cl_2 containing 0.1 M n-Bu$_4$NBF$_4$ was electrolyzed at +1.3 V (*vs.* Ag/AgCl) in an electron spin resonance (ESR) cavity at −30°C, the ESR spectrum of the σ-type cation radical (a_N = 1.72 and 1.01 mT) is obtained. Introduction of a t-butyl group, however, leads to a reversal of its electronic structure, and the ESR spectrum of the π-radical species (a_N = 0.39 and 0.34 mT, a_H(*ortho*) = 0.35, a_H(*meta*) = 0.08, a_{CH_3} = 0.67 mT) is detected [34].

We must now consider whether unimolecular loss of nitrogen from such species can give rise to carbene radical cations or anions. The parent (CH$_2$:) has been identified among the products of electric discharges through gaseous CH_4 and CH_2N_2. Oxidation of diphenyl diazomethane (DDM) is performed by copper(II) perchlorate to initiate a chain reaction giving tetraphenylethylene as the major product along with small amounts of azine, pH$_2$C=N—N=CPh$_2$, and other products. The chain carriers are cation radicals of both tetraphenylethylene and azine, which are formed by the ambient attack on the neutral DDM. There is no experimental evidence for the intermediacy of a free diphenylmethylene cation radical.

Although the electrochemical one-electron reduction of diphenyldiazomethane in polar aprotic solvents gives diphenylmethane and azine as the major products, the unimolecular loss of nitrogen from Ph$_2$CN$_2$ to give the carbene anion radical has not been confirmed [35]. Electrochemical reduction of 9-diazofluorene has also been examined. It was first reported that the fluorenylidene anion radical is formed by elimination of nitrogen from the anion radical of 9-diazofluorene. This was later revised, and the bimolecular reaction is now considered to occur before the elimination of nitrogen, forming the anion radical of azine. In the case of reductive decomposition of diethyl diazomalonate, however, the kinetic data support the intermediacy of the carbene anion species [36].

Oxidative or reductive decomposition of aryl azide has not yet been satisfactorily explained. The electron affinity (EA) of phenylnitrene (C_6H_5N) was found to be 1.45 ± 0.02 eV by photoelectron spectroscopy [37], and these authors also investigated the electronic structure of the phenylnitrene anion radical. The phenylnitrene anion radical has been reported to react with phenyl azide to give 1.4-diphenyltetrazine, and the latter decomposes to give an anion radical of the azo compound.

Electron-transfer-induced decomposition of diazo or azide compounds in an organized molecular assembly by an electron-relay mechanism may prove to be extremely interesting from the viewpoint of obtaining high concentrations of carbenes or nitrenes.

3.5.5 Spectroscopic Study of Carbenes and Nitrenes in Matrices

A matrix isolation technique is often used in spectroscopic studies of the molecular structure and valence isomerization of carbenes or nitrenes [38] because the lifetime of such highly reactive intermediates is increased significantly when they are generated in matrices at cryogenic temperatures.

When phenyldiazomethane is photolyzed in rigid glass at cryogenic temperatures, the intense ESR signal of triplet phenylmethylene is detected. The ESR spectrum of triplet carbenes is interpreted in terms of the zero-field parameters D (0.405 cm^{-1}) and E (0.0194 cm^{-1}). The D parameter provides a measure of the interaction of two magnetic dipoles. The value is reversibly proportional to r^3, where r is the average separation of two parallel spins. The E parameter is a measure of the deviation from the cylindrical distribution of two parallel spins.

IR spectra of the photolyzed sample showed a characteristic absorption band at 1825 cm^{-1}. Since this band is assignable to a cumulative double bond, the species is ascribed to cycloheptatetraene (Fig. 3.5-7) [39]. The spectroscopic data suggest that the singlet phenylmethylene efficiently isomerizes to its valence isomer, cycloheptatetraene, on the singlet energy surface, competing with the intersystem crossing to the triplet carbene. Photolysis of naphthyldiazomethane in an argon matrix gives a cyclopropene-type isomer which has characteristic absorption bands at $v = 1765$ and 1755 cm^{-1} prior to the formation of the cycloheptatetraene-type isomer. When diazoketone (2085 and 1702 cm^{-1}) is photolyzed at 12 K in an argon matrix, IR absorption bands of ketocarbene (1665 cm^{-1}) and ketene (2127 cm^{-1}) are detected, suggesting that an efficient rearrangement of ketocarbene to ketene occurs.

The zero-field parameters of phenylnitrene have been found to be $D \cong 1$ cm^{-1} and $E \cong 0$ cm^{-1}. This value suggests that the average distance between two parallel spins is much shorter than that of phenylmethylene, and that the distribution of unpaired electrons is not significantly distorted from cylindrical symmetry. Although the behavior of phenylnitrene is basically the same as that of phenylmethylene, an interesting rearrangement from phenylnitrene to pyridylcarbene is also reported.

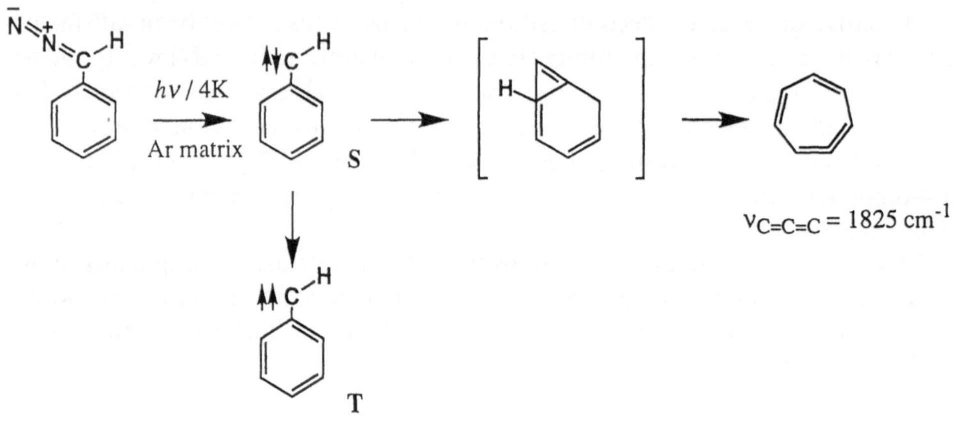

FIG. 3.5-7. Valence isomerization and intersystem crossing of phenylmethylene. $\nu_{C=C=C}$ denotes the stretching vibrational frequency of the C=C=C group

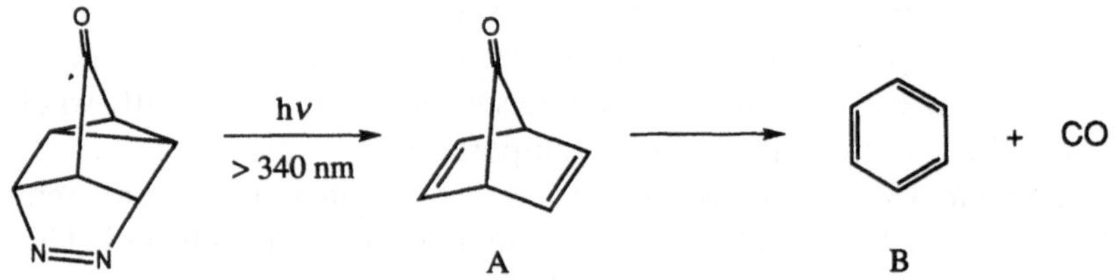

Matrix	Temperature	Product A	Product B	
Toluene-d_8	77 K	Major	Minor	
3-Methylpentane	10 K	Minor	1	
N_2	10 K	Minor	2.9	
Ar	10 K	Trace	6.5	

FIG. 3.5-8. Effect of a matrix on the product distribution obtained from the cryogenic photolysis of tetracyclic azo compound. The relative yield of benzene is shown under three conditions

   The effect of matrices on scavenging the excess energy of photochemically generated reactive intermediates is clearly demonstrated by the product distribution obtained by the photolysis of the tetracyclic azo compound [40], as shown in Fig. 3.5-8. When the azo compound is photolyzed in an *iso*-pentane glass, 7-oxonorbornadiene is isolated, while photolysis of the precursor in an argon matrix gives benzene and carbon monoxide (CO). This contrasting result indicates that a matrix of a complex molecular shape is a better energy scavenger to remove excess energy from the photochemically generated bicyclic compound which, in turn, is a precursor for generating benzene and CO.

3.5.6 Chemical Reaction of Carbenes and Nitrenes in Condensed Phases

Although the lifetimes of matrix-isolated carbenes and nitrenes are extremely long at cryogenic temparatures, the matrix-isolated species reacts at elevated temperatures. However, the reactivity of carbenes and nitrenes in glassy matrices is very different from that in fluid media. This is because of the extremely high viscosity of the reaction environment. Most of the photogenerated carbenes undergo intersystem crossing prior to reacting as a singlet species, and the resultant triplet carbene abstracts hydrogen from the matrix molecule to give an intermolecular radical pair, which in turn recombines to give the final product. The abstraction of hydrogen is believed to occur through the tunneling mechanism. When the carbene is surrounded by carbenes or by the remaining diazo compounds, they react with each other to give ethylene derivatives or azine. The reactivity of phenylnitrene in matrices is similar. The temperature-dependence of the product distribution of phenylnitrene in solution has been examined over a wide temperature range. The distribution of the products is similar to that in the glassy matrix with decreasing temperature.

The reactivity of arylnitrene in crystals of the corresponding azides exhibits topochemical reactivity and extremely high kinetic stability. For example, the half-life of p-carboxyphenylnitrene is 2 weeks at room temperature [41]. This extraordinarily high stability is explained by tight packing and by the inert environment around the univalent nitrogen (Fig. 3.5-9). In general, the crystal structure of aryl azide is characterized by the face-to-face arrangement of azido groups based on the dipolar interaction. In the case of the p-carboxyl derivative, the azide array is sandwiched by aromatic rings above and below. This formation of the azo compound is confirmed by the crystal structure described above.

The decay rates of arylnitrenes at the *meta-* or *para-*position with various substituents were measured in a crystalline environment. Although unique azo compounds were obtained in all cases examined, their thermodynamic parameters varied over a wide range, as shown in Table 3.5-2. This tendency cannot be explained by assuming that the rate-determining step is in the recombination process of two nitrenes, because the collapse of two nitrenes should proceed with almost no barriers. Thus the rate-determining step is a diffusional process in the crystals. The difference in thermodynamic parameters is ascribed to the different molecular packing of these azides. Although the general feature of the face-to-face arrangement of azide groups is maintained, the one-dimensional array of azide groups stacks perpendicular to the molecular plane, forming a two-dimensional sheet of azide groups. The small enthalpy of activation and large negative entropy of activation observed in cases other than p-carboxy derivatives is interpreted in terms of a cooperative motion in the formation of the azo

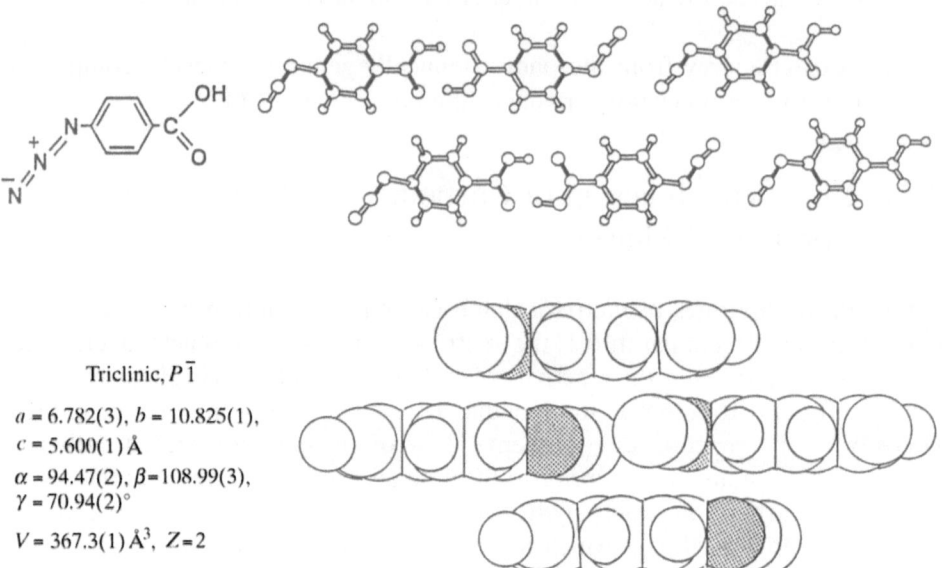

Triclinic, $P\bar{1}$

$a = 6.782(3)$, $b = 10.825(1)$,
$c = 5.600(1)$ Å
$\alpha = 94.47(2)$, $\beta = 108.99(3)$,
$\gamma = 70.94(2)°$

$V = 367.3(1)$ Å^3, $Z = 2$

FIG. 3.5-9. *p*-Carboxyphenylnitrene generated in the host crystal of *p*-carboxyphenyl azide (the *black ball* represents the univalent nitrogen). $a,b,c,\alpha,\beta,\gamma$, cell parameters of a single crystal of *p*-carboxyphenyl azide. V, cell volume; Z, number of molecules in the unit cell

TABLE 3.5-2. Thermodynamic parameters of the decay of arylnitrenes in crystals[a]

	$t_{1/2}$ (at 20°C)	$\Delta H^{\neq}$ (kJ mol^{-1})	$\Delta S^{\neq}$ (J mol^{-1} K^{-1})
$\cdot\ddot{N}$—⟨⟩—COOH	10 days	95	−44
$\cdot\ddot{N}$—⟨⟩ COOH	<<1 min	29	−203
$\cdot\ddot{N}$—⟨⟩—NHCOCH$_3$	100 min	33	−205
$\cdot\ddot{N}$—⟨⟩—N−COCH$_3$ CH$_3$	<<1 min	18	−234
$\cdot\ddot{N}$—⟨⟩—NO$_2$	10 hours	29	−197
$\cdot\ddot{N}$—⟨⟩—CN	~5 min	37	−177

[a] $t_{1/2}$, half-life; $\Delta H^{\neq}$, activation enthalpy; $\Delta S^{\neq}$, activation entropy.

compound within the two-dimensional sheet. The fact that the reaction is controlled by diffusional processes in crystals will be useful in the design of solid-state reactions in general [42].

3.5.7 Decomposition of Diazo Compounds Coupled with Energy Transfer

If a molecule has multi-chromophores which are weakly coupled electronically, the electronic energy in the excited state migrates as an exciton within the chromophores. The triplet energy transfer in triptycene is one famous example. When a diazobenzyl chromophore is connected at the *meta*-position, the electronic features may correspond to the case described above. When 1,3-bis(α-diazobenzyl)benzene is photolyzed in host crystals of benzophenone or in a rigid glass at cryogenic temperatures, the quintet signals from the *meta*-substituted dicarbene are observed immediately after the UV irradiation has started. The rate of formation of the quintet signals shows a linear dependence on the light intensity. It has been concluded from these findings that the quintet species is formed by a one-photon process [43]. The one-photon mechanism is explained by efficient intramolecular energy migration between diazo chromophores, coupled with the elimination of nitrogen. However, photolysis of 1,3-bis(α-diazobenzyl)benzene in solution proceeds in a stepwise manner through a monodiazo–monocarbene species that is formed by elimination of one nitrogen molecule from the bisdiazo compound by one photon. In a fluid solution, the excess energy of the photoexcited monodiazo–monocarbene is rapidly dispersed by translational or overall rotational modes. Under such circumstances, the biscarbene species cannot be formed by one photon.

The photodecomposition of diazo compounds in crystals provides an intermolecular counterpart of a similar phenomenon. The elimination of nitrogen from the diazo compound may be coupled with an intermolecular energy transfer, because the photochemical decomposition of diphenyldiazomethane is an isoenergetic process [22]. ESR spectroscopy can selectively detect clusters of carbenes which are magnetically coupled in different sizes. Therefore close observation of ESR spectra of carbene clusters should provide an insight into the mechanism of the photochemical generation of carbenes in the host crystals of the diazo compound. The photolytic behavior of bis(4-methoxyphenyl)-diazomethane has been extensively studied [44] because it was found that photolysis of the diazo crystals exhibits ESR signals of the ground-state quintet species which is derived from the dimeric carbene cluster coupled ferromagnetically. The initial rate constant of formation of the quintet signals was monitored as a function of the light intensity at 20 K. The number of photons was found to be 1.14 for light with a wavelength of 254 nm, and 1.65 with a wavelength of 340 nm. This result suggests that the one-photon process contributes significantly to the formation of the quintet species in irradiation at 254 nm, and that the two-photon process predominates in irradiation at 350 nm. This phenomenon may be interpreted by assuming that after removal of one nitrogen molecule from a diazo molecule, the excess energy can be efficiently transferred to the adjacent diazo molecule, resulting in the formation of the second carbene. The dependence of the ratio Q/T on the wavelength may open up the possibility of developing interesting applications such as a photoimage with higher dimensionality.

3.5.8 Photochemical Generation of High-Spin Polycarbenes and Polynitrenes

It is more than 20 years ago that the first ground-state quintet dicarbene was reported by Itoh [7] and Wasserman *et al.* [45], independently but almost at the same time. An ESR spectrum of dicarbene generated from the bisdiazo compound, which was doped in a single benzophenone crystal, provided the quintet spin multiplicity in the ground state. The mechanism of the spin alignment of π-spins can be explained by the topochemical nature of the π-system, in which a

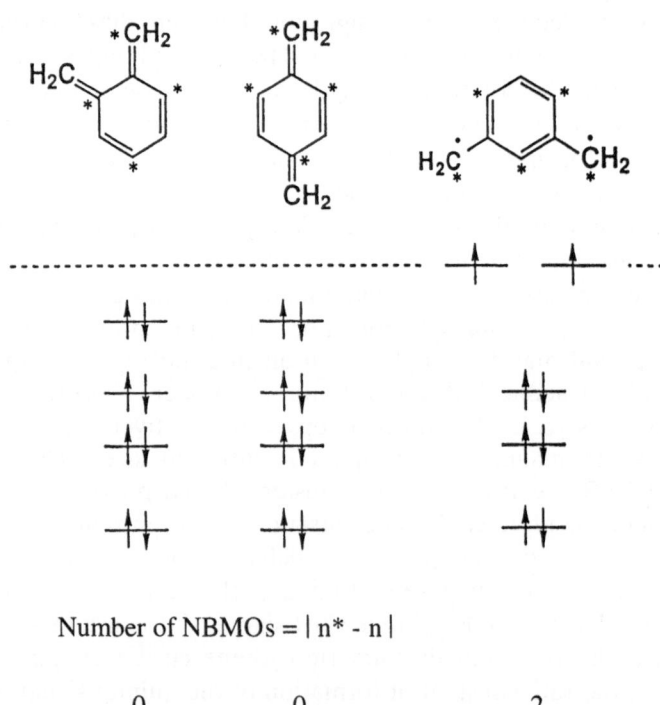

Number of NBMOs = | n* - n |

0   0   2

FIG. 3.5-10. Molecular orbital diagram of *ortho-*, *meta-*, and *para-* xylylenes. *Asterisks* are marked on every other carbon atom. *NBMO*, non-bonding molecular orbital. The number of NBMOs can be calculated by subtracting n (number of unasterisked carbon atoms) from $n*$ (number of asterisked carbon atoms)

benzene ring is connected to two methylene radicals at the *meta*-position. When the methylene radicals are introduced at the *ortho* or *para* positions, the species can be expressed as an *ortho*- or *para*-quinoid structure according to the valence bond picture, but in the *meta* case no quinoid structure can be drawn. When the electronic structure of these species is shown by a molecular orbital (MO) picture, the *meta* isomer is found to have two degenerated MOs (Fig. 3.5-10). Therefore two electron spins occupy each orbital separately, resulting in the triplet ground state. In the case of *meta*-dicarbene, the two additional σ-spins at the carbenic carbons also become parallel to the π-spins based on Hund's rule. Thus the quintet ground state spin multiplicity of *meta*-dicarbene is explained by this characteristic electronic structure.

Linear polycarbenes and star-burst-type polycarbenes [46] are also generated from the corresponding polydiazo compounds, and the high-spin multiplicity of these species has been clarified by magnetic susceptibility measurements and ESR spectroscopic studies. The quintet 1,3-dinitrenobenzene and the septet 1,3,5-trinitrenobenzene have also been reported. For example, tetracarbene was generated from the tetradiazo compound doped in a benzophenone crystal by photolysis in the cryostat of a magnetic balance. The spin multiplicity of tetracarbene has been shown by Sugawara *et al.* [47] to be a nonet in terms of the

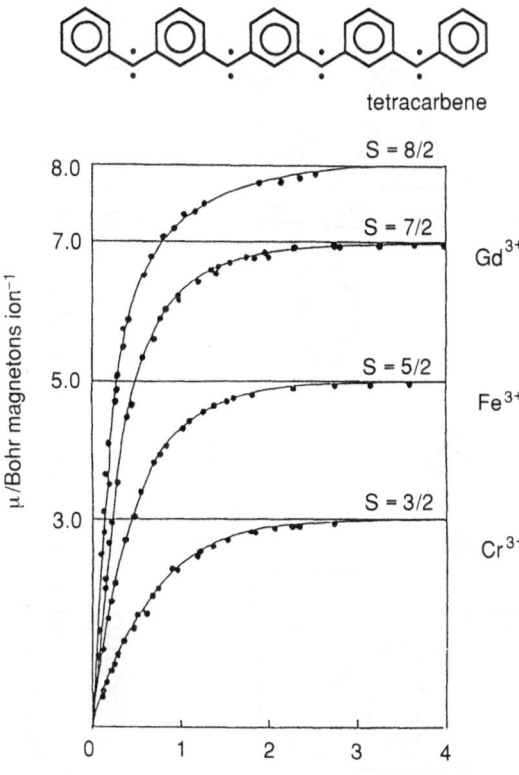

tetracarbene

FIG. 3.5-11. Plot of the magnetic moment (μ_B per ion) of transition metal ions together with tetracarbene of various spin quantum numbers (S) versus temperature-normalized external magnetic field B/T (T/K)

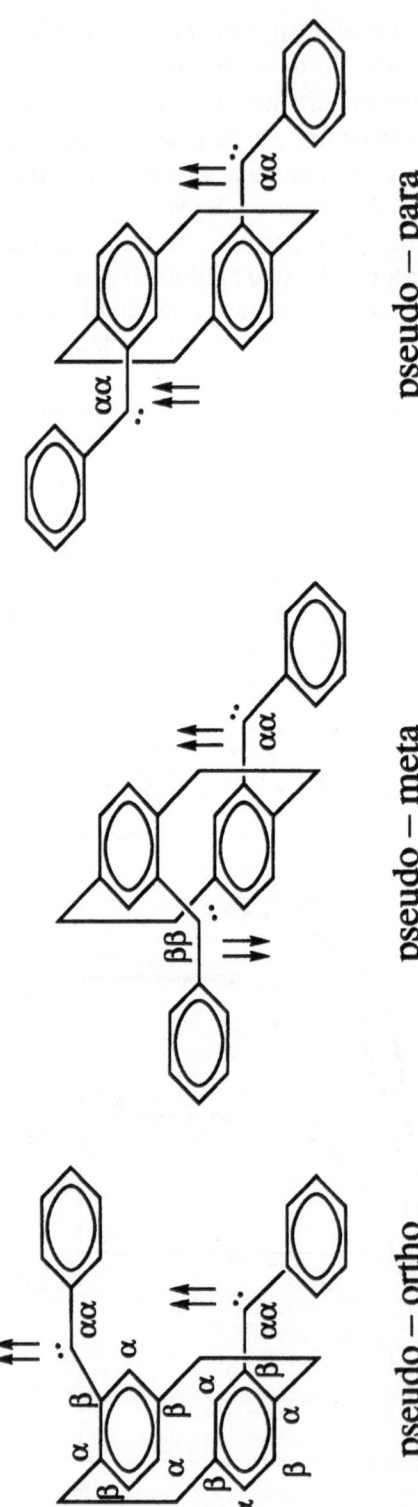

pseudo – ortho   **pseudo – meta**   **pseudo – para**

FIG. 3.5-12. Spin alignment in [2.2]paracyclophane-dicarbene. α represents a carbon atom with an up-spin density, and β a carbon atom with a down-spin density

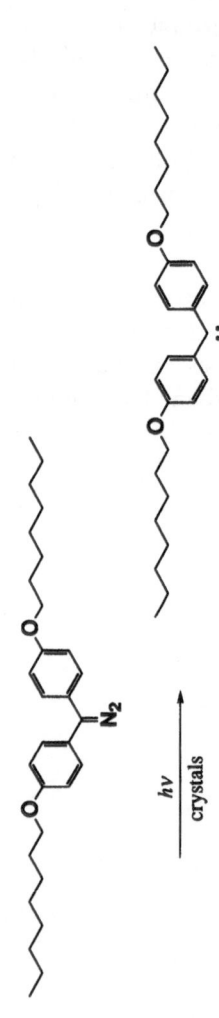

FIG. 3.5-13. Ferromagnetically coupled carbene cluster having a Weiss temperature of +2K, generated in crystals of bis[*p*-(octyloxy)phenyl]diazomethane

temperature-dependence of its magnetic susceptibility and the field-dependence of its magnetization. This result is consistent with an interpretation based on ESR data [48]. Figure 3.5-11 compares the magnetization of tetracarbene with those of transition metal ions. Fe(III) has five parallel spins in d orbitals, giving rise to the S = 5/2 state, and Gd(III) with seven parallel spins has the S = 7/2 state. However, tetracarbene shows an even higher spin state. This means that the spin multiplicity of organic molecules, when they are appropriately designed, can exceed even that of transition metals ions. The spin alignment in *meta*-polyphenylmethylene (MPPM) has been described as follows. The σ-spin at the divalent carbon has a ferromagnetic coupling with a π-spin on the same carbon. Since the up π-spin is found again on the next divalent carbon atom through the strong spin correlation in the conjugated π-system, all the σ-spins on the divalent carbons align in the same direction through their coupling with the parallel π-spins. Thus the mechanism of high-spin ordering in poly-carbene is very unusual when compared with that in pre-existing ferromagnetic materials.

When bulk magnetism is considered, carbene species also provide an important spin system for the study of intermolecular ferromagnetic coupling. The isomers of [2.2]paracyclophane-dicarbene have been designed and generated as model compounds in which to examine ferromagnetic intermolecular magnetic interactions (Fig. 3.5-12) [49]. For example, the ESR spectrum of the *pseudo-ortho* isomer shows intense ground-state quintet signals, while the *meta* isomer gives thermally populated triplet signals, indicating that this isomer has the ground singlet state. This is experimental evidence for McConnell's theory of intermolecular spin alignment.

When photolysis of polycrystals of bis(*p*-methoxyphenyl)diazomethane is carried out at cryogenic temperatures, the ground-state quintet signal is observed through ferromagnetic intermolecular interaction [44] as described in Sect. 3.5.8. In the case of bis[*p*-(octyloxy)phenyl]diazomethane in particular, the photolyzed sample shows a positive Weiss temperature of 2 K. This result indicates that the introduction of long alkyl chains at the *para*-positions is a favorable arrangement which causes two-dimensional ferromagnetic coupling among apin-distributed benzene rings (Fig. 3.5-13) [50]. Thus the spin system is interpreted as two-dimensional metamagnetic system. The active development of organic ferromagnets will be described in Sect. 4.1.2.

References

1. Jones M Jr, Moss RA (eds) (1973–1975) Carbenes. Wiley, New York, 2 vol
2. Lwowski W (ed.) (1970) Nitrenes. Wiley, New York
3. Brown RD, Pullin DE, Rice EHN, Rodler M (1985) The infrared spectrum and force field of C_3O. J Am Chem Soc 107:7877–7880
4. Wentrup C (1984) Synthesis of unusual molecules by flash vacuum pyrolysis of hetero-cyclic compounds. Lect Hetro Chem 7:91–94

5. Suzuki T, Li Q, Khemani KC, Wudl F, Almarson Ö (1991) Systematic inflation of Buckminsterfullerene C_{60}: Synthesis of diphenyl fulleroids C_{61} to C_{66}. Science 254:1186–1188

6. Fu GC, Nguyen ST, Grubbs RH (1993) Catalytic ring-closing metathesis of functionalized dienes by a ruthenium carbene complex. J Am Chem Soc 115:9856–9857

7. Itoh K (1963) Electron spin resonance of an aromatic hydrocarbon in its quintet ground state. Chem Phys Lett 1:235–238

8. Mataga N (1968) Possible "ferromagnetic states" of some hypothetical hydrocarbons. Theor Chim Acta 10:372–376

9. Iwamura H (1990) High-spin organic molecules and spin alignment in organic molecular assemblies. Adv Phys Org Chem 26:179–253

10. Pacansky J (1980) Recent advances in the photodecomposition mechanisms of diazooxides. Polym Eng Sci 20:1049–1053

11. Reiser A, Leyshon LJ, Johnson L (1971) Effect of matrix rigidity on the reactions of aromatic nitrenes in polymers. Trans Faraday Soc 67:2389–2396

12. Knowles JR (1972) Photogenerated reagents for biological receptor-site labeling. Acc Chem Res 5:155–160

13. Walker S, Murnick J, Kahne D (1993) Structural characterization of a calicheamicin-DNA complex by NMR. J Am Chem Soc 115:7954–7961

14. Schaefer HF III (1986) Methylene: A paradigm for computational quantum chemistry. Science 231:1100–1107

15. Baird NC, Taylor KF (1978) Multiplicity of the ground state and magnitude of the T_1–S_0 gap in substituted carbenes. J Am Chem Soc 100:1333–1338

16. Arduengo AJ III, Harlow RL, Kline M (1991) A stable crystalline carbene. J Am Chem Soc 113:361–363

17. Hutchison CA Jr, Kohler BE (1969) Electron nuclear double resonance in an organic molecule in a triplet ground state. Spin densities and shape of diphenylmethylene molecules in diphenylethylene single crystals. J Chem Phys 51:3327–3335

18. Wasserman E, Kuck VJ, Yager WA, Hutton RS, Greene FD, Abegg VP, Weinshenker NW (1971) Electron paramagnetic resonance of 9,9′-Dianthrylmethylene. A linear aromatic ground-state triplet methylene. J Am Chem Soc 93:6335–6337

19. Hrovat DA, Waali EE, Borden WT (1992) Ab initio calculations of the singlet–triplet energy difference in phenylnitrene. J Am Chem Soc 114:8698–8699

20. McDonald RN, Davidson SJ (1993) Electron photodetachment of the phenylnitrene anion radical: EA, ΔH_f°, and the singlet–triplet splitting for phenylnitrene. J Am Chem Soc 115:10857–10862

21. Yamabe S, Minato T, Osamura Y (1980) Theoretical study of photochemical reactions: Electron assignment and the state correlation diagram. Int J Quantum Chem 18:243–250

22. Peters KS (1994) Time-resolved photoacoustic calorimetry: From carbenes to proteins. Angew Chem Int Ed Engl 33:294–302

23. Bock H, Dammel R (1987) The pyrolysis of azides in the gas phase. Angew Chem Int Ed Engl 26:504–526

24. Bethell D, Stevens G, Tickle P (1970) The reaction of diphenylmethylene with isopropyl alcohol and oxygen: The question of reversibility of singlet–triplet interconversion of carbenes. J Chem Soc Chem Commun 792–794

25. Sugawara T, Iwamura H, Hayashi H, Sekiguchi A, Ando W, Liu MTH (1983) Time-resolved absorption spectroscopic detection of 10,10′-dimethyl-10-silaanthracen-9(10H)-one oxide. Chem Lett 1261–1262

26. Casal HL, Sugamori SE, Sciano JC (1984) Study of carbonyl oxide formation in the reaction of singlet oxygen with diphenyldiazomethane. J Am Chem Soc 106:7623–7624

27. Moss RA (1980) Carbenic selectivity in cyclopropanation reaction. Acc Chem Res 13:58–64

28. Nickon A (1993) New perspectives on carbene rearrangements: Migratory aptitudes, bystander assistance, and geminal efficiency. Acc Chem Res 26:84–89

29. Closs GL, Rabinow BE (1976) Kinetic studies on diarylcarbenes. J Am Chem Soc 98:8190–8198

30. Eisenthal KB, Turro NJ, Aikawa M, Butcher JA Jr, Du Puy C, Hefferson G, Hetherington W, Korenowski GM, McAuliffe MJ (1980) Dynamics and energetics of the singlet-triplet interconversion of diphenylcarbene. J Am Chem Soc 102:6563–6565

31. Griller D, Nazran AS, Sciano JC (1984) Reaction of diphenylcarbene with methanol. J Am Chem Soc 106:198–202

32. Tomioka H, Okada H, Watanabe T, Hirai K (1994) An extremely long-lived triplet carbene: Reactivity, optical absorption spectrum, and kinetics of highly congested diarylcarbenes. Angew Chem Int Ed Engl 33:873–875

33. Schrock AK, Shuster GB (1984) Photochemistry of phenyl azide: chemical properties of the transient intermediates. J Am Chem Soc 106:5228–5234

34. Ishiguro K, Sawaki Y, Izuoka A, Sugawara T, Iwamura H (1987) ESR study on the σ- and π-radical cations formed by one-electron oxidation of phenyldiazomethanes. J Am Chem Soc 109:2530–2531

35. Bethell D, Parker VD (1982) Intermediates in the decomposition of aliphatic diazo-compounds. Part 17. Formation and reaction of diazodiphenylmethane anion radical in solution. J Chem Soc Perkin Trans II 841–849

36. Galen DAV, Young MP, Hawley MD, McDonald RN (1985) Kinetic evidence for the formation of the carbene anion radical $(EtO_2C)_2C^{-}$. J Am Chem Soc 107:1465–1470

37. Travers MJ, Cowles DC, Clifford EP, Ellison GB (1992) Photoelectron spectroscopy of the phenylnitrene anion. J Am Chem Soc 114:8699–8701

38. Chapman OL (1979) Photochemistry of diazocompounds and azides in argon. Pure Appl Chem 51:331–339

39. McMahon R, Abelt CJ, Chapman OL, Johnson JW, Kreil CL, LeRoux JP, Mooring AM, West PR (1987) 1,2,4,6-cycloheptatetraene: The key intermediate in arylcarbene interconversions and related C_7H_6 rearrangements. J Am Chem Soc 109:2456–2469

40. LeBlanc BF, Sheridan RS (1988) Observation and substituent control of medium-dependent hot-molecule reactions in low-temperature matrices. J Am Chem Soc 110:7250–7252

41. Mahé L, Izuoka A, Sugawara T (1992) How a crystalline environment can provide outstanding stability and chemistry for arylnitrenes. J Am Chem Soc 114:7904–7906

42. McBride JM (1983) The role of local stress in solid-state radical reactions. Acc Chem Res 16:304–312

43. Itoh K, Konishi H, Mataga N (1968) Optical absorption and luminescence spectra of a ground-state quintet hydrocarbon molecule. J Chem Phys 48:4789–4790

44. Sugawara T, Tukada H, Izuoka A, Murata S, Iwamura H (1986) Magnetic interaction among diphenylmethylene molecules generated in crystals of some diazodiphenyl-methanes. J Am Chem Soc 108:4272–4278

45. Wasserman E, Murray RW, Yager WA, Trozzolo AM, Smolinsky G (1967) Quintet ground state of m-dicarbene and m-dinitrene compounds. J Am Chem Soc 89:5076–5078
46. Nakamura N, Inoue K, Iwamura H (1992) Synthesis and characterization of a branched-chain hexacarbene in a tridecet ground state. An approach to superparamagnetic polycarbenes. J Am Chem Soc 114:1484–1485
47. Sugawara T, Bandow S, Kimura K, Iwamura H, Itoh K (1986) Magnetic behavior of nonet tetracarbene as a model for one-dimensional organic ferromagnets. J Am Chem Soc 108:368–371
48. Teki Y, Takui T, Itoh K, Iwamura H, Kobayashi K (1986) Preparation and ESR detection of a ground-state nonet hydrocarbon as a model for one-dimensional organic ferromagnets. J Am Chem Soc 108:2147–2156
49. Izuoka A, Murata S, Sugawara T, Iwamura H (1987) Molecular design and model experiments of ferromagnetic intermolecular interaction in the assembly of high-spin organic molecules. J Am Chem Soc 109:2631–2639
50. Sugawara T, Murata S, Kimura K, Iwamura H, Sugawara Y, Iwasaki H (1985) Design of molecular assembly of diphenylcarbenes having ferromagnetic intermolecular interaction. J Am Chem Soc 107:5293–5294

Part IV
Structure and Properties of Molecules and Molecular Systems

4.1
Electric and Magnetic Properties

YUSEI MARUYAMA

A molecule is the most elementary unit of a substance which still retains the intrinsic nature or function of that substance. Molecules are composed of atoms, but the nature of a molecule is usually very different from that of the component atoms. On the other hand, the nature or function of a molecular solid is primarily determined by its constituent molecules in the zero-order approximation. The structure and properties of molecules in a condensed state are usually very similar to those in the isolated state. The differences between the two states are sometimes very small but are sometimes significantly large. The degree of these differences depends on the interactions between the molecules in the condensed phase. The main intermolecular interaction is usually of the van der Waals' type, which is much weaker than valence bond or ionic interactions. A weak interaction between molecules leads to a large separation between molecules, and as a consequence electrons are usually localized inside a molecule even in the condensed phase.

Conversely, electronic states in conventional metallic or semiconducting substances are characterized by delocalized electronic energy bands throughout the bulk phase. The bands can be fairly wide, ~10 eV or more, due to strong interatomic interactions. In principle, these energy bands dominate the properties of such solid materials. However, in the case of molecular materials, the width of such electronic energy bands is usually much narrower, less than 1 eV and sometimes even less than the thermal energy. A one-electron energy band picture, therefore, is hardly applicable to the analysis of the electronic state of molecular materials. Other electronic effects are also crucially important, such as electron–electron repulsion (electron correlation), electron–phonon coupling (vibration or rotation), electronic disorder which leads to the localization of electrons, and electronic polarization of the lattices in molecular materials. Moreover, the magnitude of these energies is usually within a similar range, and this makes the problem very difficult to solve. Almost all these effects are temperature- or pressure-dependent and many are also time-dependent. The dominant factor can easily be changed by quite small perturbations. Thus electronic phase-transitions may frequently take place in molecular materials following a change in temperature or pressure.

The shapes of conventional molecules are usually linear or planar; 3-dimensional shapes are rare. Any anisotropy in molecular shape may lead to anisotropy in intermolecular interactions, and the low-dimensional nature of the molecular solid may be apparent. A spherical molecule such as C_{60} can form a cubic lattice crystal, which may facilitate high temperature superconducting transitions in fullerides. However, even in C_{60} molecules, the electron distribution on the surface of the molecule is not homogeneous, i.e., not spherically symmetric. Therefore the mutual orientation of molecules in the crystal lattice is not always identical, which causes inherent disorders in intermolecular interactions in the low-temperature crystal phase. The detailed orientation of molecules or very weak libration is sometimes crucial in determining the nature of crystals, and this type of feature must be considered when studying condensed molecular materials.

In this chapter electric, magnetic, optical, and thermal properties of molecular solids are investigated.

4.1.1 Electric Properties of Organic Molecular Solids: Organic Semiconductors, Conductors, and Superconductors

This section focuses on the electrical conductivity or resistivity of organic molecular solids. The original observation of electrical conduction in organic solids was presented as the concept of "organic semiconductors" by Akamatsu and Inokuchi in 1950 [1]. Typical organic semiconductors known at that time were such polycyclic aromatic hydrocarbons as anthracene, perylene, and violanthrone, and phthalocyanine was included by Eley in 1948 [2]. The essential feature in these substances is that they consist of π-electron systems, and it is known that a variety of functionalities of organic materials are caused by such systems.

A breakthrough with the discovery of organic metallic conductors was reported by Akamatu et al. in 1954 [3] when this property was found in perylene–bromine complexes. Charge transfer interactions between the two components are essential to the extraordinarily high conductivity of these binary compounds. The first observation of metallic conductivity was reported in 1973 by Heeger and co-workers [4] for tetrathiafulvalene–tetracyano-quinodimethane (TTF–TCNQ). Charge-carrier conduction bands are formed by the overlap of electron wavefunctions between molecules which are aligned in one dimension with molecular plane-to-plane stacking. The one-dimensional structure causes an inherent instability in the electronic band structure which results in an energy gap at the Fermi-energy level. This electronic phase transition is accompanied by lattice distortion through an electron–phonon coupling which stabilizes a charge density wave (CDW) insulating state. This type of metal–insulator transition usually occurs at a finite temperature, e.g., 53 K for TTF–TCNQ.

To stabilize metallic phases at low temperatures, there have been attempts to increase the dimensionality by introducting intercolumnar interactions or heteroatom–heteroatom interactions in directions parallel to each plane between TTF-type molecules. The replacement of sulfur atoms in the TTF moiety by selenium atoms, which have larger interaction radii and a smaller ionization potential, is effective in increasing dimensionality. High-pressure application is a physically effective way to increase dimensionality and to suppress the metal–insulator transition.

Electron–electron repulsive correlation is another crucial factor in realizing a metallic state in narrow electron energy band systems. The relationship between the on site (on one molecule) Coulomb repulsive energy (U) and the band-width (W) is especially important in influencing whether the material will be metallic or insulating. If the ratio U/W is much greater than unity, the material is a Mott-type insulator. Various strategies have been tried to reduce U, such as using larger sized molecules, or twin-type or multivalency molecules. The former approach was very successful using (bis)ethylenedithio–tetrathiafulvarene (BEDT–TTF) as an electron donor [5].

The other factor which makes a material nonmetallic is the localization of charge carriers, electrons or holes. There are many possible causes of localization, such as disorders in structure or energy, impurities, or electron–phonon interactions (polaron or self-trapped state). If these effects are strong enough, nonmetallic conduction may appear even if there is no distinct energy gap in the electronic band.

A necessary condition for a superconductor is that the material is metallic at some temperature that is low enough for superconductivity. Therefore it has been necessary to stabilize the metallic state or to suppress the metal–insulator transition in organic molecular solids to produce organic superconductors. Although there have been many attempts to clarify the origins of superconductivity in organic molecular solids, our understanding is still far from complete. Thus, the strategy for designing organic superconductors is similar to that used to achieve satisfactory organic metals. In this case, however, the first organic superconductor, tetramethyltetraselenofulvalene (($TMTSF)_2PF_6$), was found under high pressure, thus suppressing the metal–insulator transition that occurs at ambient pressure. Futher developments are discussed in Sect. 4.1.1.2.

4.1.1.1 Organic Semiconductors and Conductors

Over the years there have been many studies on organic semiconductors and conductors [6]. Typical single-component organic semiconductors consist of π-electron conjugated systems such as polycyclic aromatic compounds or phthalocyanines. The introduction of sulfur atoms into carbon atom skeletons was an important step in expanding knowledge of organic semiconductors. A new intermolecular interaction in heterocyclic molecular solids is due to sulfur–

sulfur σ-type contacts rather than π-electron overlaps between carbon atoms. A variety of new organic semiconductors has been synthesyzed based on TTF skeletons.

Bis[1,2,5] thiadiazolo-p-quinobis (1,3-dithiole) (BTQBT). An interesting study of single-component organic semiconductors was carried out by Yamashita *et al.* [7]. A typical example is bis [1,2,5] thiadiazolo-*p*-quinobis(1,3-dithiole) (BTQBT), which was synthesized with very strong intermolecular interactions in single-component molecular crystals. This new molecule has electron-withdrawing heterocycles which are fused to the skeleton of a very strong electron donor. 2,2′-*p*-quinobis(1,3-dithiole) (QBT).

(a) TTF   (b) BTQBT   (c) QBT

In addition to the advantage of strong intermolecular interactions (heteroatom–heteroatom contacts), extending the π-conjugation on the molecular plane of BTQBT may reduce the on-site Coulomb repulsion. Futhermore, this molecule is highly polarized because of intramolecular charge transfer, and accordingly it is expected to have a relatively small highest occupied molecular orbital–lowest unoccupied molecular orbital (HOMO–LUMO) gap. All these features may contribute to the high electrical conductivity of this material.

Single BTQBT crystals actually exhibited an unusually high conductivity of $1 \times 10^{-5}\,\mathrm{S\,cm^{-1}}$ at room temperature, with a thermal activation energy of $0.21\,\mathrm{eV}$. The Hall effect and its temperature-dependence were observed by Imaeda *et al.* [8]. The Hall mobility with a positive sign is $2.4\,\mathrm{cm^2\,V^{-1}\,s^{-1}}$ at $330\,\mathrm{K}$ and $6.3\,\mathrm{cm^2\,V^{-1}\,s^{-1}}$ at $175\,\mathrm{K}$, and this changes with temperature as $T^{-1.6}$. As far as I know, this is the first successful observation of the Hall effect for the dark current of single-component organic semiconductors. Further research along these lines may help to clarify the electrical conduction mechanisms in organic molecular solids, which are still in the "dark state".

(*DMe–DCNQI*)$_2$*Cu Complex*. The molecular structure of the bis (2,5-dimethyl-dicyanoquinonediimine) copper [(DMe–DCNQI)$_2$Cu] complex is shown below.

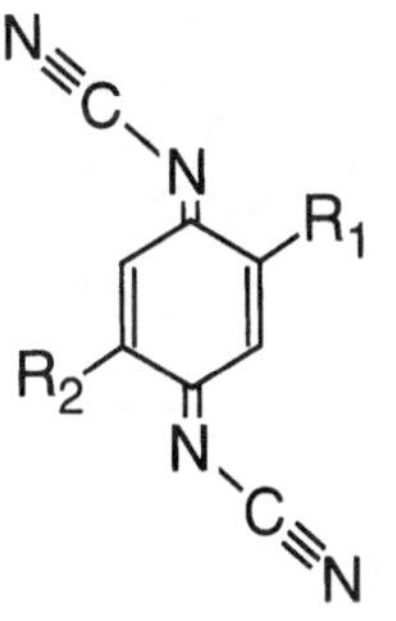

R$_1$,R$_2$-DCNQI

DMe–DCNQI is an electron-accepting molecule coordinating to Cu ions at the nitrogen atoms of cyanide groups. The DMe–DCNQI molecular planes overlap each other in the crystal to form a one-dimensional column structure, and copper ions connect these one-dimensional columns to form three-dimensional networks, as shown in Fig. 4.1-1. The copper ions are in a mixed valence state of 4/3+ average valence charge, which indicates that the copper d-orbitals are interacting with the π-orbitals (LUMO) of DMe–DCNQI molecules.

The three-dimensional electronic structure stabilizes the metallic state at low temperatures without metal–insulator transition. However, the metallic state is transformed to the insulating state under relatively low pressure (>100 bar), and around the critical pressure region "metal-insulator-metal" reentrant behavior was observed with decreasing temperature [9] (Fig. 4.1-2). Similar metal–insulator transition and reentrant behavior was discovered with the substitution of the hydrogen atoms in DMe–DCNQI by deuterium, or the substitution of ^{12}C in cyanide groups by ^{13}C (Fig. 4.1-2.) These isotope exchanges may cause a qualitatively similar effect to the pressure-effect, which is called a chemical pressure effect.

The theoretical details of the strongly correlated electron systems of (DMe–DCNQI)$_2$Cu crystals have been studied by Fukuyama [10].

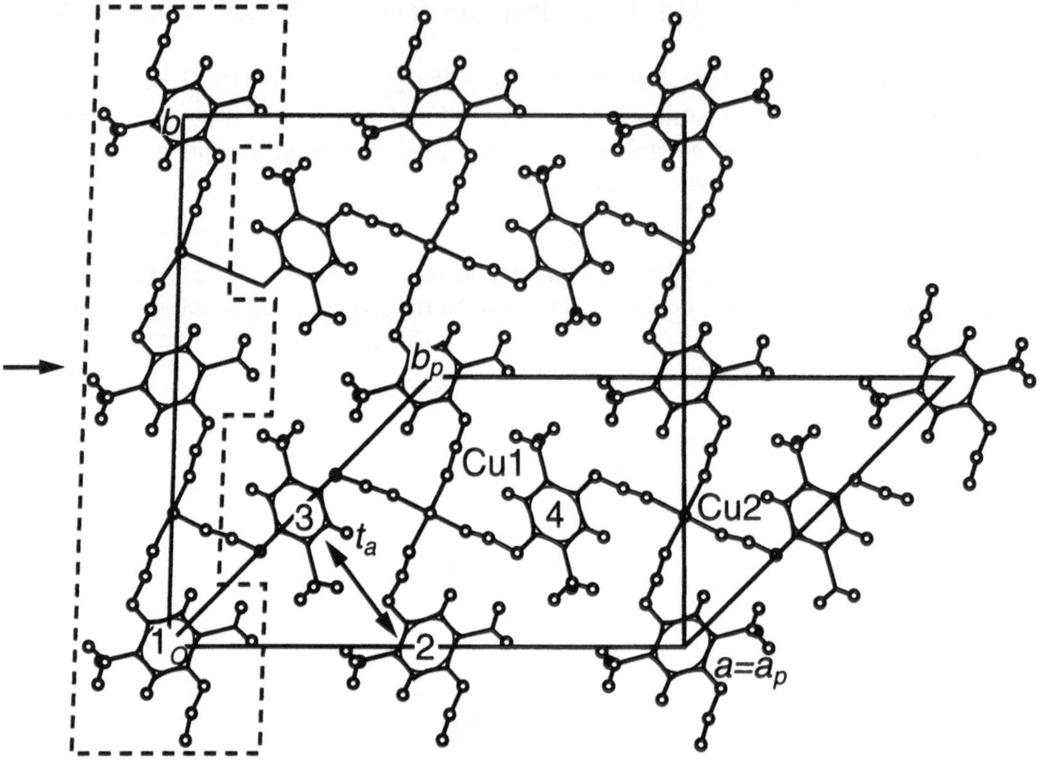

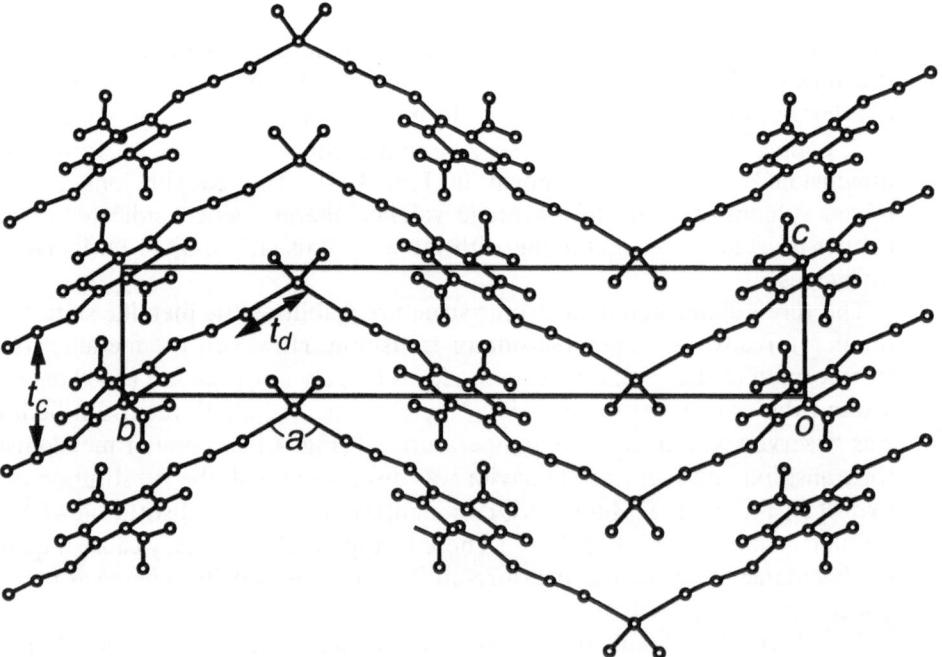

$a_p = a, b_p = (a + b + c)/2, c_p = c$

Fig. 4.1-1. Crystal structure of (DMe–DCNQI)$_2$Cu

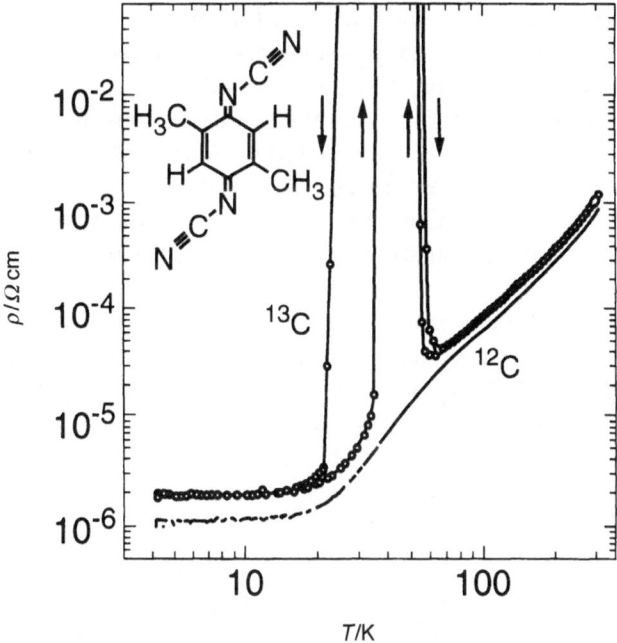

FIG. 4.1-2. Reentrant metal–insulator transition by isotope substitution in (DMe–DCNQI)$_2$Cu

4.1.1.2 Organic Superconductors

Altempts have been made to stabilize metallic states at low temperatures by enhancement of dimensionality and reduction of on-site Coulomb repulsion energies. The first organic superconductor, discovered by Jerome et al. [11], was (TMTSF)$_2$X (X = PF$_6$, $T_c = 0.9$ K at 12 kbar). The TMTSF complex forms a quasi-one-dimensional columnar structure, and the intercolumnar Se–Se interaction is essential to stabilize the metallic state. Although in the case of the $X = PF_6$ compound, some effort is necessary to supress the metal–insulator transition, the $X = ClO_4$ compound was found to be superconducting at 1.4 K under ambient pressure owing to the stronger intercolumnar interactions in this case.

(BEDT–TTF)$_2$X compounds, investigated by Saito *et al.* [5], have shown much more successful results. The larger size of the π-electron conjugation of BEDT–TTF may reduce the on-site Coulomb repulsion, and the two-dimensional (not quasi-one-dimensional) crystal structure (Fig. 4.1-3) is very advantageous in the metallic phase.

The structure and function of the insulating layer X in (BEDT–TTF)$_2$X are also important for the two-dimensional metallic character of these compounds. In the cases of $X = Cu(NCS)_2$, $X = Cu(CN)[N(CN)_2]$, $X = Cu[N(CN)_2]Br$, and $X = Cu[N(CN)_2]C$, the superconducting critical temperatures T_c are relatively high (10.4, 12.3, 11.2, and 12.5 K, respectively). The highest T_c is 13.1 K under 0.3 kbar for $X = Cu[N(CN)_2]Br$. The pressure-dependence of T_c for $X = Cu(NCS)_2$,

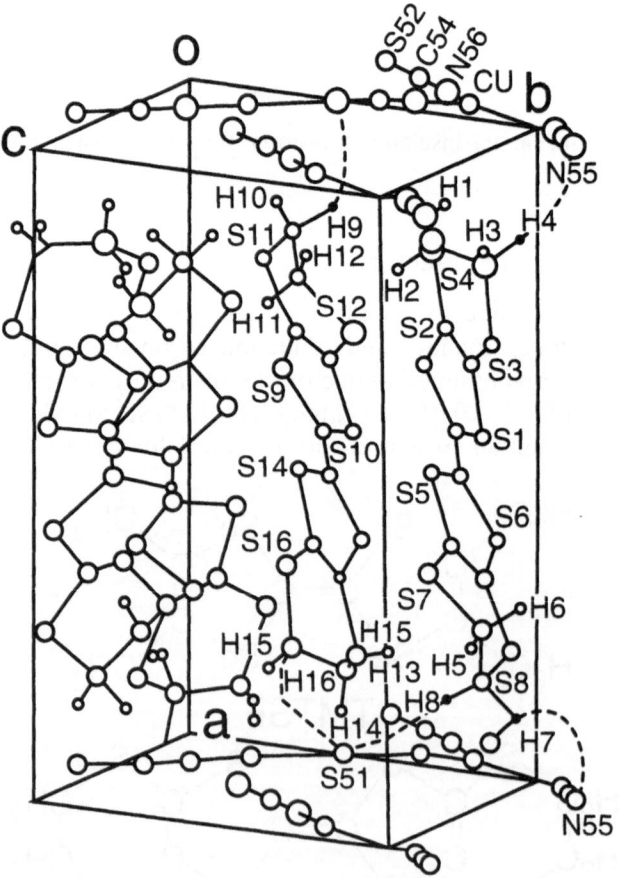

FIG. 4.1-3. Crystal structure of ((bis)ethylenedithio–tetrathiafulvarene)$_2$Cu(NCS)$_2$ ((BEDT–TTF)$_2$Cu(NCS)$_2$)

however, is slightly negative. The isotope effect for T_c is also negative for $(BEDT–TTF)_2Cu(NCS)_2$ when the peripheral hydrogen atoms are replaced by deuterium atoms. Conversely, substitution of ^{12}C atoms in the inner TTF core by ^{13}C atoms showed almost no isotope shift in T_c. The magnetic field effects for metallic states have been very extensively studied to reveal the 2-dimensional nature and Fermi-surface structures of these materials. $(BEDT–TTF)_2X$ compounds may be classified as type-II superconductors, but their magnetic behavior below T_c is not yet clarified.

4.1.1.3 Fullerenes

The discovery of new cage-type carbon atom clusters (C_{60}, or fullerenes) opened a new horizon in the field of molecular materials [12]. After the invention of a mass-production method for fullerenes by carbon contact-arc discharge [13], the discovery of superconductivity at 18 K in K_3C_{60} [14] strongly accelerated the study of fullerenes.

The electrical conductivity of C_{60} single crystals is strongly and negatively dependent on the ambient oxygen pressure, and is believed to be around $10^{-5}\,S\,cm^{-1}$ at 295 K in oxygen-free pure C_{60} single crystals. The thermal activation energy of conductivity is 0.26 eV, which indicates that a pure C_{60} crystal is a molecular semiconductor. The charge-carrier mobility in pure C_{60} single crystals was measured by two methods: time-of-flight [15] and the Hall effect [16]. The drift mobilities obtained by the former method were $1.7\,cm^2\,V^{-1}\,s^{-1}$ for holes and $0.5\,cm^2\,V^{-1}\,s^{-1}$ for electrons and both are temperature-independent over a range of temperatures below (50–250 K) and above (260–300 K) the phase transition point at around 260 K. At around 260 K the mobility increased step-wise by 70% with decreasing temperature. It is known that changes in molecular rotational freedom, and accordingly in the crystal structure, take place at this temperature, which suggests that the transport of electrons or holes may couple with molecular rotations. The Hall mobility with a positive sign is reported to be $0.15\,cm^2\,V^{-1}\,s^{-1}$, and again it is almost independent of temperature between 290 K and 570 K.

C_{60} crystals doped with alkali metals or alkaline-earth metals, M_xC_{60}, have resulted in a new family of molecular metals and superconductors which have a much higher T_c than any organic superconductor known so far. The dominant stoichiometry that yields the metallic state is $X = 3$, and the corresponding alkali metal-doped complexes, make a systematic family of molecular superconductors, the fullerides (Fig. 4.1-4). The highest T_c in this family is 33 K for Cs_2RbC_{60}, as shown in Table 4.1-1. However, it was recently reported that Cs_3C_{60} stabilized under pressure may have a T_c of approximately 40 K. The intrinsic nature of the metallic state and the mechanisms of superconductivity in fullerides are not yet clear. In the Bardeen–Cooper–Schrieffer (BCS) regime a possible theoretical approach based on intramolecular vibration-mediated Cooper pairing was proposed, taking into account the strongly correlated electron system and relevant phonons and Coulomb energy windows [17]. However, other approaches such

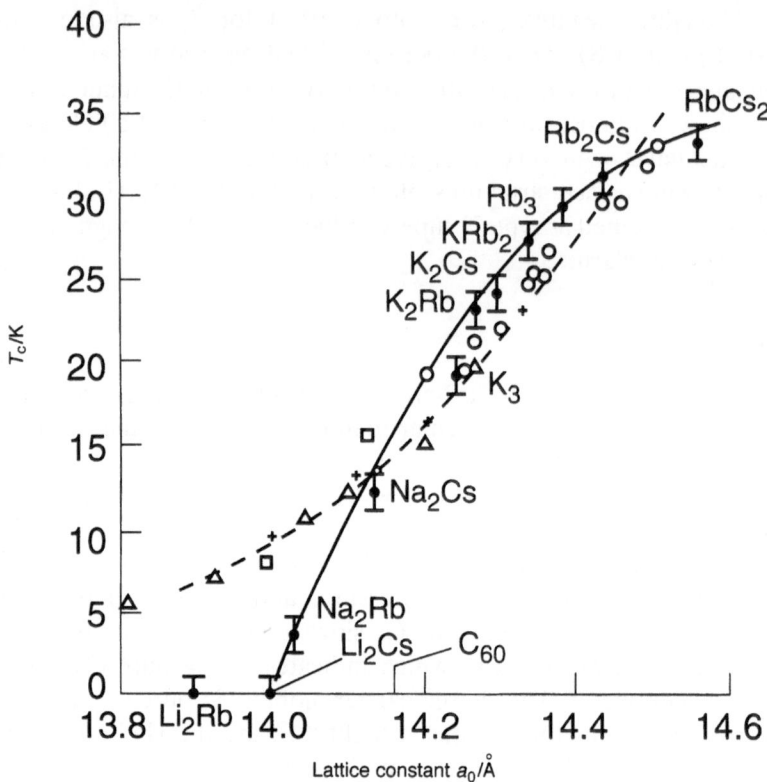

FIG. 4.1-4. Correlation between T_c and a_0 in alkali metal–C_{60} superconductors

TABLE 4.1-1. Superconducting T_c values for exohedrally doped C_{60}

M_xC_{60}	$T_c(K)$	M_xC_{60}	$T_c(K)$
Li_2RbC_{60}	<2	K_2RbC_{60}	23
Li_2CsC_{60}	<2	K_2CsC_{60}	24
Na_2RbC_{60}	3.5	K_3C_{60}	19.3
Na_2CsC_{60}	12	$RbCs_2C_{60}$	33
$Na_2Cs(NH_3)_4C_{60}$	29.6	Rb_2KC_{60}	27
$NaK_2(NH_3)_xC_{60}$	11	Rb_2CsC_{60}	31.3
$NaRb_2(NH_3)_xC_{60}$	13	Rb_3C_{60}	29
$Na_xN_yC_{60}$	13	Cs_3C_{60}	40
$(NaH)_xC_{60}$	15	Ca_5C_{60}	8.4
KRb_2C_{60}	23	Ba_6C_{60}	7
$K_{1.5}Rb_{1.5}C_{60}$	22.2	Sr_6C_{60}	4

as charge-transfer pairing could be more general and relevant to molecular superconductors.

4.1.2 Magnetic Properties of Organic Molecular Solids

In the study of the electronic structures of molecular solids, observation of their magnetic properties is indispensable for a full understanding of their electronic states, including electron-spin states. The magnetic properties of organic semiconductors or conductors have been investigated to elucidate their electronic structure in parallel with their electrical properties. Normal organic semiconductors show diamagnetism or thermally activated paramagnetism because of their closed-shell electronic structure. However, charge-transfer complexes or some neutral radicals may have unpaired or nonzero spin moment. Thus, organic conductors based on charge-transfer complexes have shown a variety of magnetic properties, such as Pauli paramagnetism, antiferromagnetism, or ferromagnetism.

Metal–insulator transitions associated with CDW or SDW (spin density wave) states are always accompanied by characteristic changes in magnetism, and it is very important to study the temperature-dependence of magnetism at the same time as that of electrical conductivity.

Another very important subject in this field is the creation or synthesis of molecular ferromagnets, which is the subject of this section.

A molecule is, in general, a diamagnetic, closed-shell species, but there are two unusual families of organic molecules whose members are stable, open-shell entities. One consists of molecules containing neutral, localized unpaired electron-containing functional groups (free radicals), and the other consists of charged species (radical ions). In the former category are the so-called neutral radicals, the nitroxyls (e.g. nitronyl nitroxide), triphenyl verdazyls, sterically hindered hydrazyls [e.g., diphenylpicrylhydrazyl (DPPH)], and sterically hindered phenoxyls (galvinoxyl). In the last group are the radical ions which result from single-electron transfer from the π-molecular orbitals of an electron donor molecule which has a relatively low ionization potential, or from transfer of an electron to the π-molecular orbitals of an electron acceptor molecule which has a relatively high electron affinity.

As radical molecules, neutral or charged, have unpaired electrons and show paramagnetism, they might be expected to be ferromagnetic materials. However, once these paramagnetic species are in the solid state, they are usually Curie–Weiss paramagnets. If the interactions between spins in different molecules are strong enough to become ordered at some temperatures, the antiferromagnetically ordered states usually dominate the ferromagnetic states, since the bonding states between molecules are usually more stable than the antibonding states. Thus, the next problem was how to make the bonding states unstable, or make two unpaired spin states orthogonal. Success was achieved with nitronyl nitroxide radicals, as shown in Table 4.1-2 [16]. In the crystal,

TABLE 4.1-2. Molecular structure of typical ferro-magnetic nitroxide radicals

	p-NPNN, $T_c = 0.65\,\text{K}$[a]
	TMAO, $T_c = 1.48\,\text{K}$
	BTEMPO, $T_c = 0.3\,\text{K}$

[a] From [16].

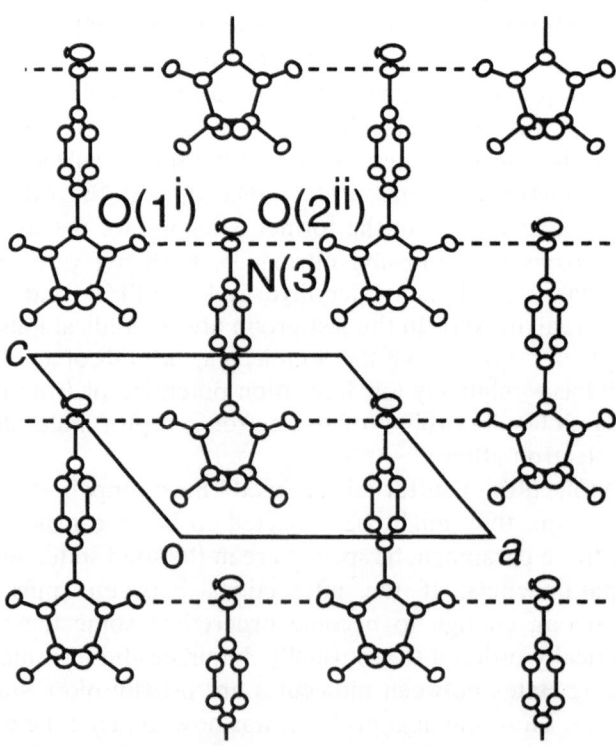

FIG. 4.1-5. Molecular arrangement in a p-nitronyl nitroxide crystal

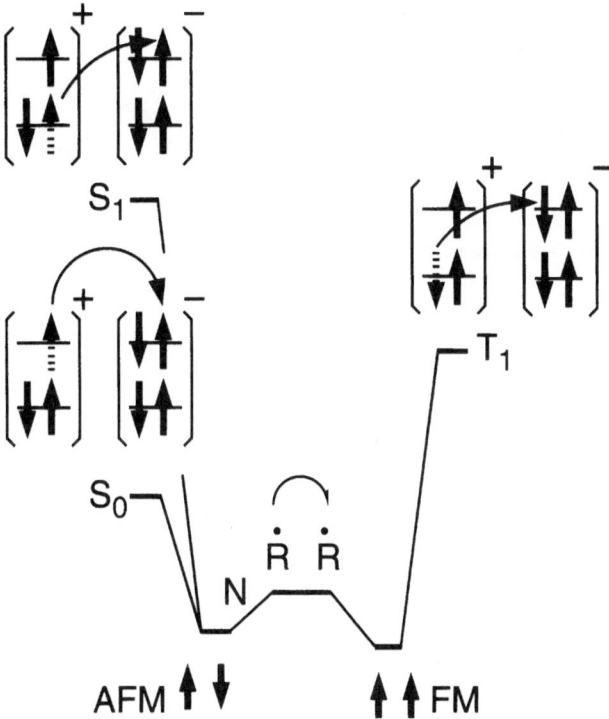

Fig. 4.1-6. Three charge transfer configurations in the radical dimer which lead to a ferromagnetic (*FM*) state or an antiferromagnetic (*AFM*) state

unpaired spin states on –NO groups tend to be localized and widely separated owing to crystal stability, as shown in Fig. 4.1-5. Thus, an almost orthogonal relation between two unpaired spin states in neighboring molecules does occur in this case. Further, large on-site Coulomb repulsion may contribute to the stabilization of the triplet or ferromagnetic states in this system. Some schematic spin configurations which stabilize a ferromagnetic state relative to an antiferromagnetic state are shown in Fig. 4.1-6. The charge transfer process between radical molecules (dimers) is essential in this mechanism. The crystal phase of p-nitronyl nitroxide (α-phase) shows ferromagnetic ordering below 0.65 K [18].

Other systematic approaches have also been proposed based on: (a) charge-transfer complexes, (b) very high spin polycarbene and polyradical solids, and (c) co-crystallization of high- and low-spin molecules in a ferrimagnetic array, as described in the review article by Wudl and Thompson [19]. In case (a), the essential point is that the triplet states which appear in the charge transfer scheme may stabilize the ferromagnetic ordering in the solid state, as shown below.

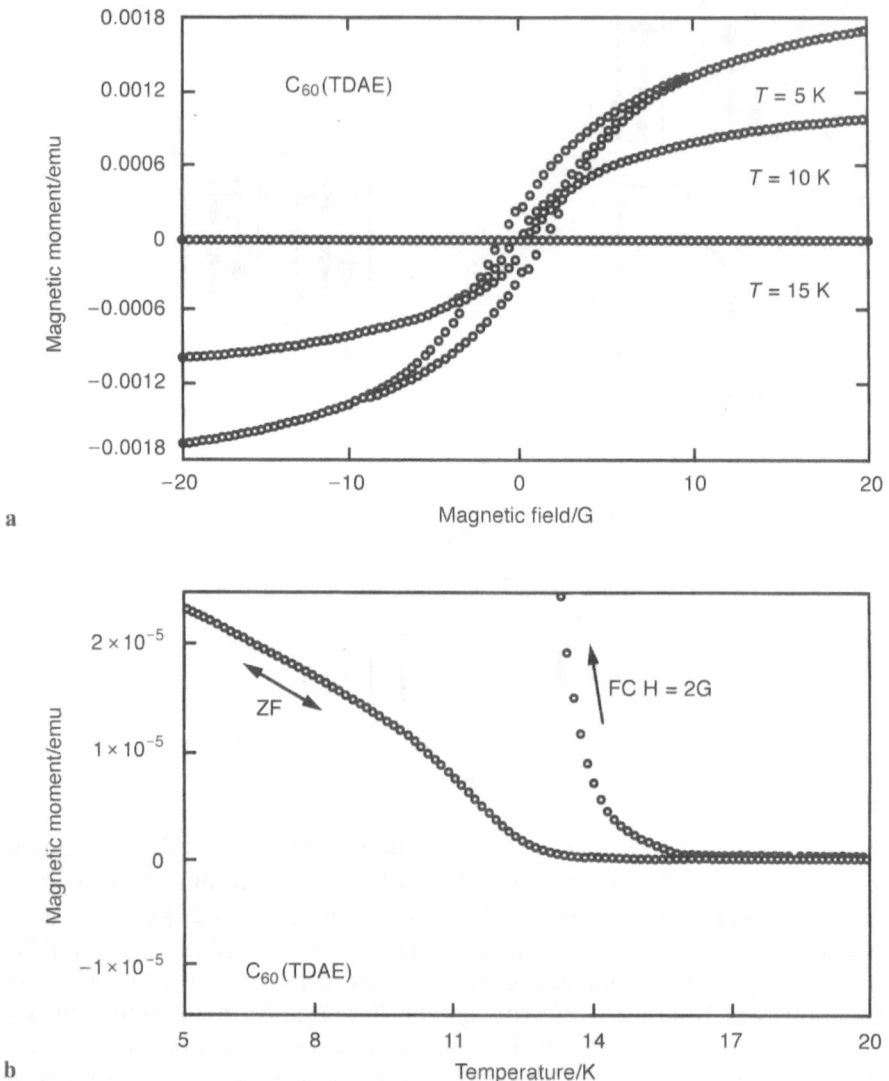

FIG. 4.1-7. Spontaneous magnetization in C_{60}–tetrakis(dimethylamino)-ethylene (*TDAE*)

The C_{60} tetrakis(dimethylamino)-ethylene (TDAE) complex could be a molecular ferromagnet in this scheme [20]. Although extensive studies of this system have revealed behavior such as spontaneous magnetization below 13 K and the hysteresis curve in isothermal magnetization (Fig. 4.1-7) [21], whether the nature of the magnetism is really ferromagnetic or spin-glass-like is still unclear. Recent experiments by two research groups [22, 23] are expected to show that the quest for confirmation of ferromagnetic spins in this material will not be successful.

Recently, more promising high-T_c systems have been reported. One is a transition metal ion complex, $V[Cr(CN)_6]_{0.86} \cdot 2.8H_2O$, which showed a T_c of 315 K [24]. The other is a charge-transfer complex containing tetracyanoquinodimethane (TCNQ), such as ($Me_4N^+F_4TCNQ^- \cdot \frac{1}{2}F_4TCNQ$), which showed an S-shaped magnetization curve at room temperature [25]. As well as these two systems, many other attempts will be made to find a high-T_c molecular ferromagnet. This subject will be very important in the field of molecular solid-state science in the next few years.

References

1. Akamatu H, Inokuchi H (1950) On the electrical conductivity of violanthrone, isoviolanthrone, and pyranthrone. J Chem Phys 18:810–811
2. Eley DD (1948) Phthalocyanines as semiconductors. Nature 162:819
3. Akamatu H, Inokuchi H, Matsunaga Y (1954) Electrical conductivity of the perlene–bromine complex. Nature 173:168–169
4. Coleman LB, Cohen MJ, Sandman DJ, Yamagishi FF, Garito AF, Heeger AJ (1973) Superconducting fluctuations and the Peierls instability in an organic solid. Solid State Commun 12:1125–1132
5. Saito G, Enoki T, Inokuchi H (1982) A novel behavior of electrical resistivity in a new two-dimensional organic metal, (BEDT–TTF)$_2$ClO$_4$ (1,1,2-trichloroethane)$_{0.5}$. Chem Lett 1345–1348
6. Inokuchi H, Sano M, Maruyama Y, Sato N (1989) Proceedings of the OJI international seminar on organic semiconductors—40 years. Mol Cryst Liq Cryst 171:1–356
7. Yamashita Y, Tanaka S, Imaeda K, Inokuchi H (1991) Tetrathio-derivatives of p-quinodimethanes fused with 1,2,5-thiadiazoles. A novel type of organic semiconductor. Chem Lett 1213–1216
8. Imaeda K, Yamashita Y, Li Y, Mori T, Inokuchi H, Sano M (1992) Hall-effect observation in the new organic semiconductor bis(1,2,5-thiadiazolo)-p-quinobis(1,3-dithiole) (BTQBT). J Mater Chem 2:115–118
9. Tomic S, Jerome D, Aumuller A, Erk P, Hunig S, von Schultz JU (1988) Pressure–temperature phase diagram of the organic conductor (DM–DCNQI)$_2$Cu. Synth Met 27:B281–B288
10. Fukuyama H (1993) The origin of reentrant metallic state in (DCNQI)$_2$Cu. J Phys Soc Jpn 62:1436–1438
11. Jerome D, Mazaud A, Ribault M, Bechgaard K (1980) Superconductivity in a synthetic organic conductor (TMTSF)$_2$PF$_6$. J Phys Lett 41:L95–L98

12. Kroto HW, Heath JR, O'Brien SC, Curl RF, Smally RE (1985) C_{60}: Buckminsterfullerene. Nature 318:162–163
13. Kratschmer W, Lamb LD, Foctiropoulos K, Huffman DR (1990) Solid C_{60}: a new form of carbon. Nature 347–358
14. Hebard AF, Rosseinsky MJ, Haddon RC, Murphy DW, Glarum SH, Palstra TTM, Ramirez AP, Kortan AR (1991) Superconductivity at 18 K in potassium-doped $C6_0$. Nature 350:600–601
15. Frankevich E, Maruyama Y, Ogata H (1993) Mobility of charge carriers in vapor-phase grown C_{60} single crystal. Chem Phys Lett 214:39–44
16. Arai T (1994) Electrical transport properties of single crystal C_{60}. Doctoral thesis, Tokyo University
17. Ivanov V, Maruyama Y (1995) Disorder and phonon windows for superconductivity in doped fullerenes. Physica C 247:147–155
18. Kinoshita M, Turek P, Tamura M, Nozawa K, Shiomi K, Nakazawa Y, Ishikawa M, Takahashi M, Awaga K, Inabe T, Maruyama Y (1991) An organic radical ferromagnet. Chem Lett 1225–1228
19. Wudl F, Thompson JD (1992) Buckminsterfullerene C_{60} and organic ferromagnetism. J Phys Chem Solids 53:1449–1455
20. Allemand PM, Khemani KC, Koch A, Wudl F, Holczer K, Donovan S, Gruner G, Thompson JD (1991) Organic molecular soft ferromagnetism in a fullerene C_{60}. Science 253:301–303
21. Suzuki A, Suzuki T, Whitehead RJ, Maruyama Y (1994) Evidence of spontaneous magnetic order in the C_{60} complex with tetrakis(dimethylamino)ethylene. Chem Phys Lett 223:517–520
22. Suzuki A, Suzuki T, Maruyama Y (1995) Magnetic and electrical behaviors of C_{60} (TDAE) single crystal. Solid State Commun 96:253–257
23. Blinc R, Mihailovic PCD, Venturini P, Omerzu A, Arcon D, Pokhodnia K (1995) Single crystal ESR of the organic ferromagnet TDAE-C_{60}. In: Kuzmany H, Fink J, Mehring M, Roth S (eds) Physics and chemistry of fullerenes and derivatives. World Scientific, pp 485–488
24. Ferlay S, Mallah T, Ouahes R, Veillet P, Verdaguer M (1995) A room-temperature organometallic magnet based on Prussian blue. Nature 378:701–703
25. Sugimoto T, Tsujii M, Matsuura H, Hosoito N (1995) Weak ferromagnetism below 12 K in a lithium tetrafluorotetracyanoquinodimethanide salt. Chem Phys Lett 235:183–186

4.2
Optical Properties

YOSHINORI TOKURA (4.2.1, 2)
KAZUHIKO SEKI (4.2.3, 4)

The optical and photoelectronic properties of molecular systems essentially reflect the individual electronic states of constituent molecules. Nevertheless, some important modifications show up occasionally in optical and photoelectron spectra of molecular solids due to intermolecular interaction or the delocalization effect of electrons extended over molecular units. In the ordinary optical spectra, the Coulombic effect, or the final-state interaction, is important in the optical excitation process because of the localized or low-dimensional nature of electronic structures. This Coulombic effect results in the dominant role of the excitons, as explained in the following sections. Photoelectron spectroscopy, however, gives one-particle spectra which contain a lot of information about the π- or σ-valence electrons extending over the molecular units. In some cases, spectral variation reveals important information about the local configuration of molecular units which would otherwise be difficult to establish.

After a brief review of the basic features of optical spectra, this section describes the application of optical, photoelectron, and X-ray absorption spectroscopy to some functional molecular systems.

4.2.1 Basic Optical Properties of Molecular Solids

The optical spectrum near the absorption edge in a solid is produced mainly by exciton absorption. The exciton, the concept of which is described in detail in Sect. 2.3.1, is a composite particle composed of an electron and a hole which are bound together by Coulomb interaction. In the case of inorganic semiconductors and ionic crystals, the one-electron band picture generally holds good, and their optical properties can basically be understood in terms of interband transitions. The exciton spectra show up as modifications of the interband transition spectra near the absorption edge. In particular, the exciton in inorganic semiconductors, which is called the Wannier–Mott exciton, forms a hydrogen-like series of eigenstates with a large exciton Bohr radius. In contrast, the optical properties of a molecular solid are almost always dominated by the transitions of the excitons, i.e., the so-called Frenkel excitons, in which the electron and hole are tightly bound on a single molecular unit, or at most within adjacent molecules.

The variations of exciton spectra in prototypical molecular solids are now considered, along with the strength and anisotropy (dimensionality) of π-electron interactions between molecular units. Optical absorption spectra in most of the molecular crystals which are composed of identical π-molecules show similar features to those of isolated molecules. With an increase in intermolecular inter-action, however, considerable modifications become apparent as generic features of molecular excitons. As an example, the optical reflectance spectra of an anthracene crystal (monoclinic) is shown in Fig. 4.2-1. The anthracene crystal forms a monoclinic lattice with two molecules in the unit cell. Thus, the polarized reflectance (or absorption) spectra of the (001) face show the differently polarized components, as indicated by $0-0_b$ and $0-0_a$ in Fig. 4.2-1. The observed

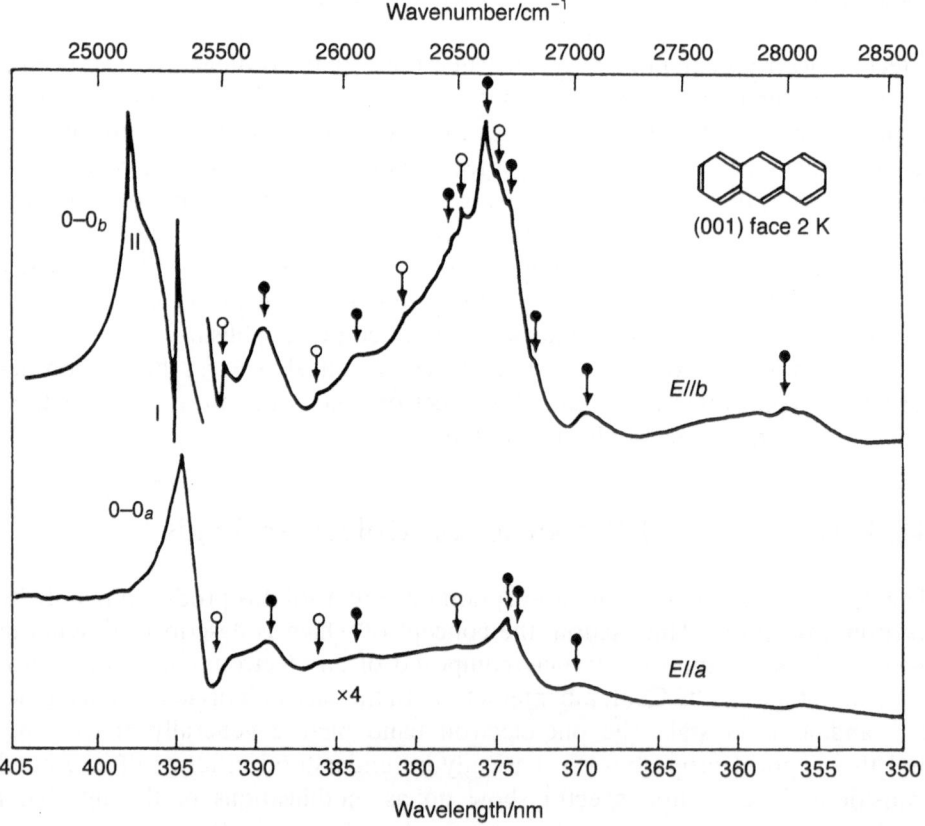

FIG. 4.2-1. Polarized reflectance spectra of excitons in an anthracene crystal at 2 K. *Open* and *solid circles* indicate two-particle edges and destabilized vibrons, respectively. *0–0$_a$* and *0–0$_b$* indicate the respective Davydov components of the 0–0 exciton. *I* and *II* represent sharp reflection minima corresponding to the surface exciton states

polarization-dependent splitting of the exciton state is called Davydov splitting, and arises from the interaction between molecular excitations on the adjacent inequivalent-site molecules. Each transition dipole can be represented by a ± combination of the transition dipoles of the inequivalent-site molecules (two in the case of anthracene crystals) in the unit cell. A Davydov splitting of around $220\,cm^{-1}$ represents an exciton transfer interaction in the [110] direction. As for the $0-0_b$ component, sharp reflectance minima, I_b and II_b, are discernible in the reflectance band of the $0-0_b$ exciton. Philpott and Turlet [1] have assigned these structures to the surface exciton states on the (001) surface.

The molecular excitation embedded in the crystal lattice shows two types of modifications compared with the case of an isolated molecule. One is the excitation transfer, or excitonic effect, which results in Davydov splitting as observed in anthracene spectra, for example. Another modification is the so-called site-shift effect, which originates from the transition dipole–dipole interaction with surrounding molecules and makes the molecular excitation energy lower than that for the isolated molecule. The origin of surface excitons I and II observed in anthracene crystals is the lack of site-shift energies in the anthracene molecules in the first and second surface layer, respectively.

The site-shift energy ($D < 0$) plane-wise [1] is

$$D = D_0 + 2\sum_{n=1}^{\infty} D_n \qquad (4.2\text{-}1)$$

where D_0 (<0) is the contribution from the same (001) plane as the molecule, and D_n (<0) is the contribution from the nth nearest (001) plane. The magnitude of D_n rapidly decreases with n. Surface excitons I and II have energies higher than the bulk exciton by $-(D_1 + D_2 + D_3 + \ldots)$ and $-(D_2 + D_3 \ldots)$, respectively. The latter is much smaller than the former, which is why surface exciton II is observed close to the energy position of the bulk exciton. The $(001)_a$ counterpart of surface exciton I, (I_a), is barely discernible as a small peak in the a-polarized reflectance spectra. The poor exciton dispersion ($<5\,cm^{-1}$) along the [001] direction also favors the existence of the surface exciton series. In fact, the Davydov splitting of surface exciton I almost coincides with that of the bulk exciton, reflecting the two-dimensional exciton band within the (001) plane.

On the high-energy side of the 0–0 excitons, a number of fine structures show up in the reflection spectra due to the coupling of the exciton with molecular vibrations. Vibronic exciton states are categorized into two types of excitations: the one-particle state, or so-called vibron, which is composed of a molecular exciton and vibrational quantum on the same molecular site, and the two-particle (or many-particle) state, which involves a bare exciton and molecular vibrational quantum (quanta) on different molecular sites. Such a molecular picture is, of course, not accurate when the exciton transfer interaction is strong enough to decompose the vibron into the two-particle state. The sharp features denoted by open circles in Fig. 4.2-1 correspond to the lower band edges of the two-particle (or three-particle) states associated with various molecular vibrational quanta [1]. These energy positions are the same for b- and a-polarization, and precisely coincide with the sum of the $0-0_b$ exciton energy (exciton band edge) and the vibrational energy (energies). On the other hand, the rather broad peaks marked with solid circles are remnants of the vibrons which were destabilized as a result of the decomposition process of exciton transfer interaction.

The vibron state, i.e., the two-particle complex, can be a well-defined crystal excitation in cases where such a state is split off from the corresponding two-particle band. In naphthalene, for example, the B_{1g} molecular vibration $(509\,cm^{-1})$ shows a frequency reduction $(86\,cm^{-1})$ on the electronically excited molecule due to the quadratic coupling effect [2, 3]. Thus, an attractive interaction works between the molecular exciton and the vibrational quantum, and results in the formation of a stable complex, i.e., the localized vibron, well below the corresponding two-particle edge. Such an isolated vibron absorption with minimal Davydov splitting is observed prototypically for naphthalene and benzene crystals.

In addition to molecular excitons that govern the absorption spectra of molecular crystals, intermolecular charge-transfer (CT) excitations, or CT excitons, exist at the higher energy side of the molecular exciton but well below the interband transitions. In general, these CT excitons (as well as the interband transitions) have little (if any) oscillator strength. However, in cases where the CT interaction is large or the energy separation between the molecular and CT excitons is small, the mixing between the molecular and CT excitons becomes significant. In these cases, the original molecular exciton bears the character of the CT excitation, whereas the CT exciton can be seen in the ordinary absorption spectrum because it shares the character of the molecular excitation. (More precisely, mixing occurs only between the odd-parity states of the molecular and CT excitons and the even-parity CT excitons remain one-photon forbidden.) These CT excitations are believed to play an important role in the primary process of photocarrier generation.

As a typical example of CT excitons in molecular crystals, the spectra of absorption (the imaginary part of the dielectric constant, ε_2) and electroabsorption (the electromodulated part of ε_2, $\Delta\varepsilon_2$) for the β-crystal of Cu–phthalocyanine (Cu–Pc) [4] are shown in Fig. 4.2-2. The doubly degenerate lowest $\pi-\pi^*$ excitation (the so-called Q-band) of the Pc molecule splits into two bands in the crystal, denoted by A and B in Fig. 4.2-2, both of which are also subject to Davydov (i.e., polarization-dependent) splitting. On the higher energy side of the B-band, progressive vibronic absorption bands, C_n ($n = 0, 1, 2 \ldots$), occur. However, the vibrational spacing $(650\,cm^{-1})$ does not coincide with that of the neutral Pc molecule.

The electroabsorption signals shown in Fig. 4.2-2 can be obtained by Kramers–Kronig analysis of the electroreflectance spectra, which are recorded as the relative change $(\Delta R/R)$ in the reflectivity caused by the alternating external electric field. The Pc crystal possesses inversion symmetry, and hence the electromodulation signal is proportional to the square of the field amplitude. The electroabsorption signals are found at the energy positions corresponding to molecular excitons A and B. However, the Stark signal of the isolated Pc mol-

FIG. 4.2-2. Spectra of the imaginary part of the dielectric constant (ε_2) and electro-modulated components $\Delta\varepsilon_2$ of a β-Cu-phthalocyanine crystal at 77 K. *Shaded areas* represent charge-transfer excitons and their vibronic sidebands

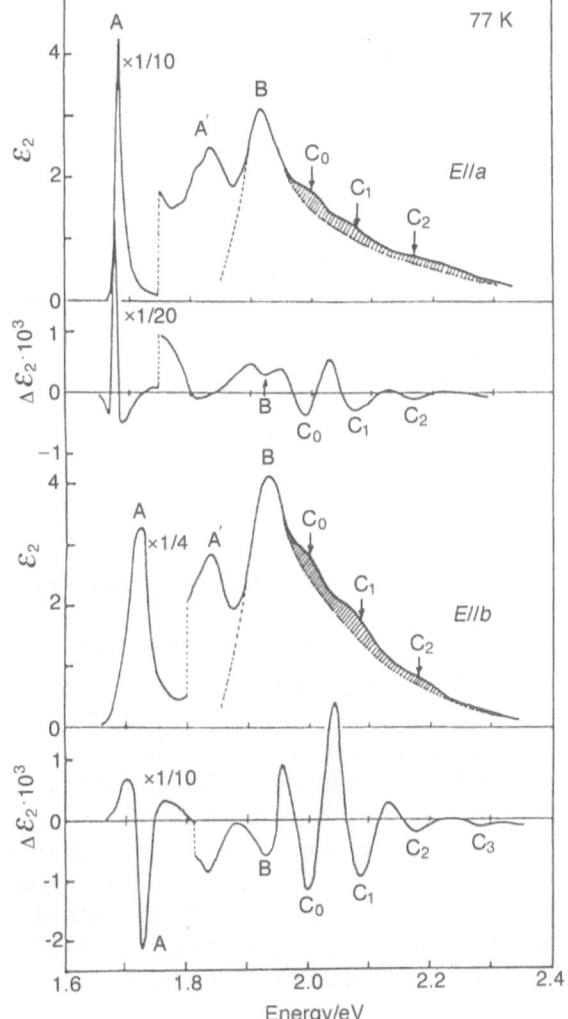

ecule is much smaller than that observed in the crystal, and the observed field effect is probably due to the intermolecular CT component of the A and B excitons. Strong electromodulation signals are observed corresponding to the C_n transitions, indicating that the C_n transitions are the s-CT (odd-parity) exciton and its vibronic sidebands, which are hybridized with the molecular excitons F_1 and F_2, as shown schematically in Fig. 4.2-3. The vibrational spacing (650 cm^{-1}) and the coupling strength observed for the CT exciton series are characteristic of the charge-transferred (ionic) Pc molecule. Thus, electromodulation spectroscopy applied to molecular or polymeric systems can be used to detect CT states or symmetry-forbidden states.

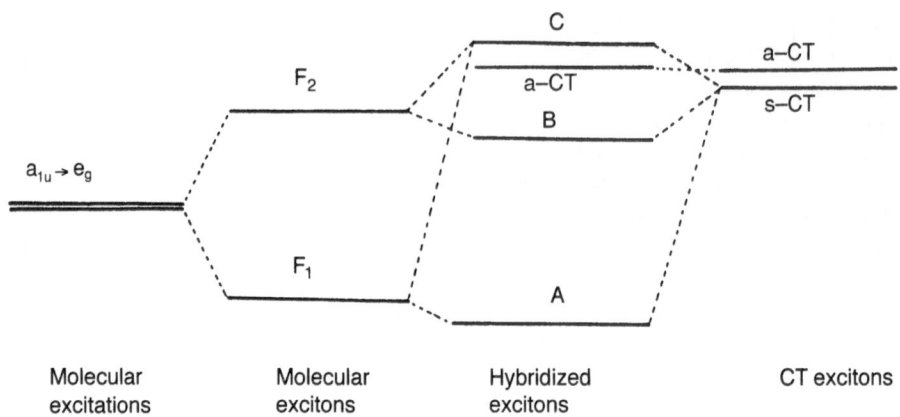

FIG. 4.2-3. The hybridization effect between the charge-transfer (*CT*) and molecular excitons in a β-Cu-phthalocyanine crystal. (The vibronic effect is not taken into account)

In the case of molecular compounds composed of electron donor (D) and acceptor (A) molecules, the CT exciton state is located at the lowest position and hence governs the low-lying optical response. Figure 4.2-4 shows the absorption and luminescence spectra of anthracene–PMDA (pyrometric dianhydride) crystals, a typical weak CT molecular compound. Anthracene–PMDA crystals show an almost neutral ground state, while the first excited (optically allowed) state is the singlet CT exciton in which an electron is transferred from the D molecule (anthracene) to the A molecule (PMDA). The zero-phonon line of the CT exciton is clearly observed for both the absorption and the luminescence spectra [5]. A rather broad absorption (or luminescence) band which follows the zero-phonon lines is ascribed to the progression of the phonon sideband (intermolecular vibration of several tens of cm^{-1}). The CT exciton is known to couple strongly with the lattice mode which modulates the CT interaction between the D and A molecules.

The even-parity CT exciton, which is the in-phase combination of the $A^0D^+A^-$ and $A^-D^+A^0$-type CT excitations, is optically forbidden, but is located at an energy $10\,cm^{-1}$ higher than the odd-parity CT exciton [6]. Such a forbidden exciton state is observed under an external electric field along the DA stacking direction [6] or in isotopically mixed crystals [7] in which the centrosymmetry of the crystal is perturbed. The energy separation between the odd- and even-parity CT excitons corresponds to the $k = 0$ splitting of the CT exciton band, and reflects the CT exciton dispersion width [6]. The CT exciton band width in anthracene–PMDA was experimentally estimated to be about $50\,cm^{-1}$ by coherent potential approximation (CPA) analysis of the CT exciton spectra of the isotopically mixed crystal [7].

One of the interesting CT exciton states observed in DA-type CT compounds is the condensation of CT excitons, or the exciton-string states. The permanent dipole of CT excitation produces an attractive interaction between adjacent CT

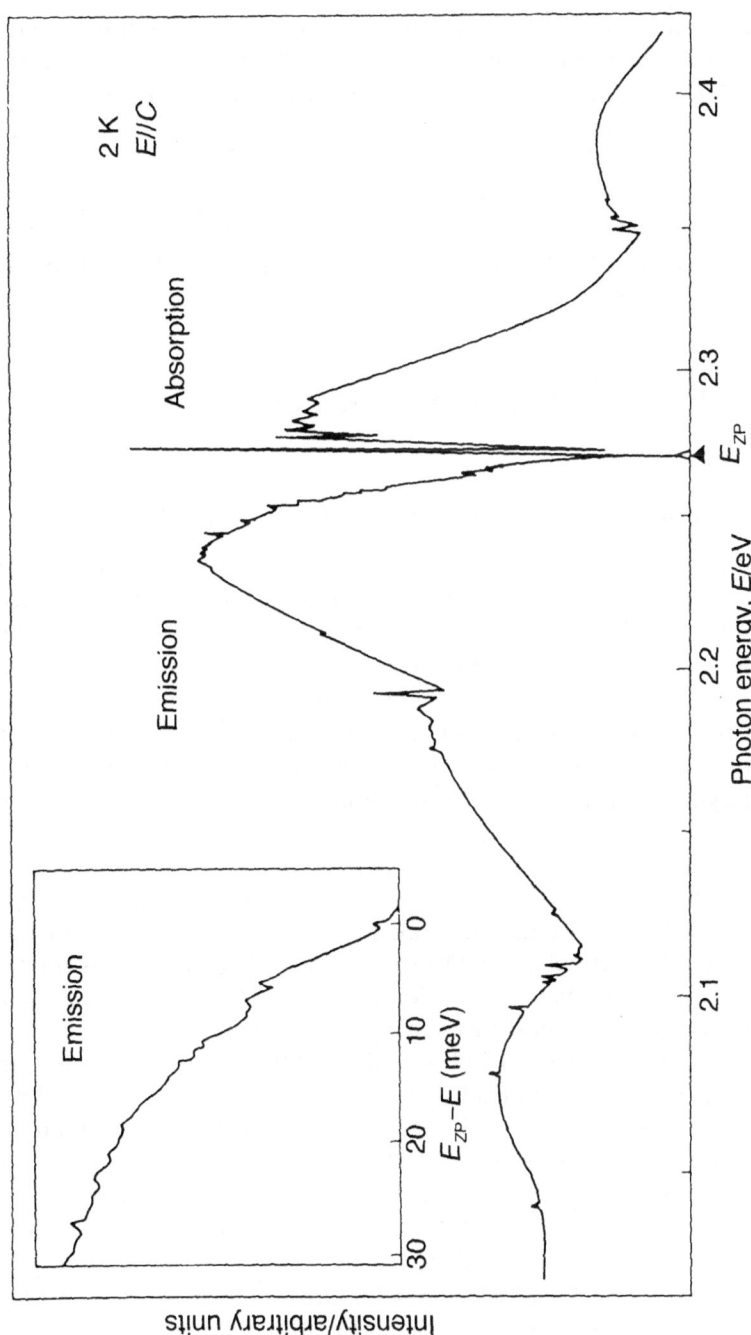

FIG. 4.2-4. Spectra of ε_2 (absorption) and emission of an anthracene–PMDA crystal at 2 K with light polarization along the stack axis. E_{zp} is the energy of a zero-phonon absorption. *Inset* is the enlarged spectrum of the emission near E_{zp} in which the zero-phono emission line is clearly shown

excitons [8]. Therefore, the CT excitons (number n) strung along the DA stack should have a lower energy and hence be more stable than the n independent CT excitons. Such a CT exciton string can be observed in anthracene–PMDA crystals by means of the picosecond pump–probe method, up to the $n = 3$ state [8]. The exciton string is analogous to the biexciton, or excitonic molecule [9], which is a two-exciton complex analogous to the hydrogen molecule and observed in inorganic semiconductors and insulators. However, the exciton string tends to form multiexciton states beyond the two-exciton complex because of its one-dimensional nature.

4.2.2 Optical Properties of Some Functional Molecular Systems

4.2.2.1 TTF-p-Chloranil with Collective Charge-Transfer Instability

The aforementioned attractive Coulombic interaction between CT excitations can lead to a sort of exciton-condensed state when the DA compound locates on the verge of the neutral–ionic (N–I) phase boundary. The ground state of a DA compound is categorized as either the nearly neutral (D^0A^0) state or the nearly ionic (D^+A^-) state. This phase boundary is governed by the competition between two quantities: the ionization energy for the DA pair ($D^0A^0 \rightarrow D^+A^-$) and the Madelung energy per DA pair for the D^+A^- ionic lattice. When the Madelung energy gain is larger (smaller) than the ionization energy loss, the I (N) ground state is realized. In a real system, the CT interaction between the DA pair hybridizes the genuine I and N states and hence the degree of CT or molecular ionicity ρ ($D^{+\rho}A^{-\rho}$) becomes fractional. Nevertheless, most actual compounds possess a fairly well-defined ground state, either the N (mostly $\rho < 0.3$) or the I (mostly $\rho > 0.6$) state. In particular, the DA compounds adjacent to the N–I phase boundary undergo phase transition between the N and I states when external pressure is applied or the temperature is changed [10, 11].

An example is tetrathiafulvalene (TTF)-p-chloranil (CA) [10–13], in which the N–I transition is caused by the application of hydrostatic pressure of 1.1 GPa or by decreasing the temperature below $T_{NI} = 82$ K. Figure 4.2-5 shows the temperature-dependence of reflectance spectra of TTF–CA crystals [12]. The reflectance peak around 0.8 eV polarized parallel to the stack axis (Fig. 4.2-5a) represents the CT excitons. The spectral shape of the CT exciton shows an abrupt change at T_{NI} due to spin-Peierls-like dimerization in the I-phase. Nevertheless, the energy of the CT exciton (0.8 eV) changes little upon NI transition, which is exactly as expected [10]. On the other hand, the reflectance structures polarized perpendicular to the stack axis (Fig. 4.2-5b) correspond to the intramolecular excitations (mainly on TTF molecules) and hence critically reflect the molecular valence (ionicity ρ). The molecular excitation (exciton) spectra undergo a marked change on NI transition, and analysis of the change in spectral position can be utilized to estimate the ground-state ionicity [12].

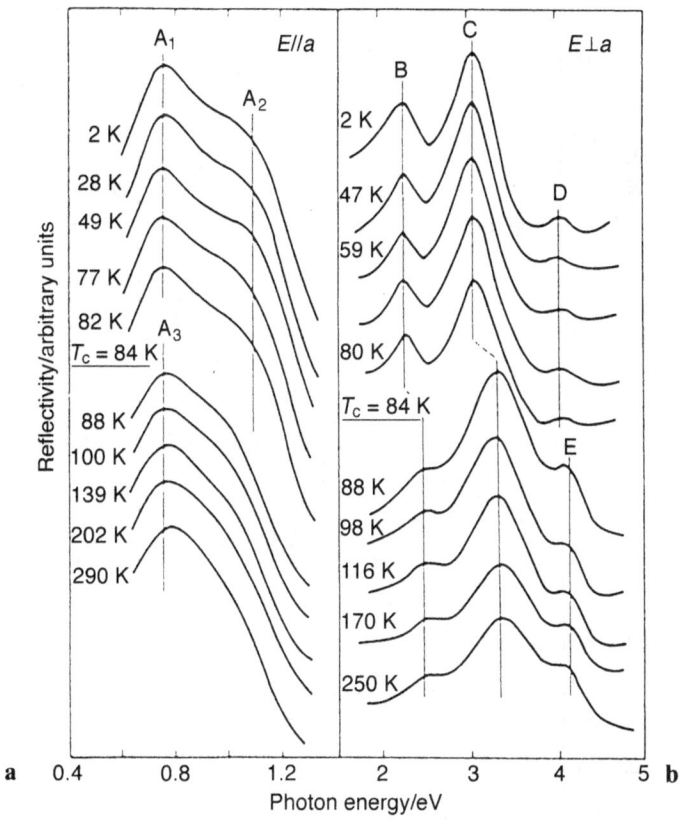

FIG. 4.2-5. Temperature-dependence of reflectance spectra for **a** the charge-transfer excitation bands (A_1, A_2, and A_3) and **b** the intramolecular excitation bands (B, C, D, and E) in TTF–chloranil. T_c is the neutral–ionic phase transition temperature: $T_c = 84$ K (82 K) in the warming (cooling) run

The instability of molecular valence in DA compounds near the NI phase boundary is also triggered by photoexcitation. For example, the CT exciton, which is photogenerated in the ionic DA stack, corresponds to the D^0A^0 pair. If the compound is on the verge of the NI phase boundary, such a local neutral state can be a nucleus around which the local I–N conversion evolves. In other words, a CT exciton with the characteristics of a neutral DA pair may undergo self-multiplication via string-type condensation of CT excitons. In this case, the excess energy needed to create metastable N-states (or a condensed state in CT excitons) may be supplied from the lattice relaxation energy of the initially photoexcited CT exciton, as theoretically argued by Toyozawa [14].

Figure 4.2-6 shows the photoreflectance spectra (i.e., photoinduced change in reflectance spectra) of TTF–CA at 77 K (I-phase) and 90 K (N-phase) [15]. When the crystal is in the I-phase, e.g., at 70 K, photoexcitation causes a large change in the molecular excitation spectra (polarized perpendicular to the stack axis), as

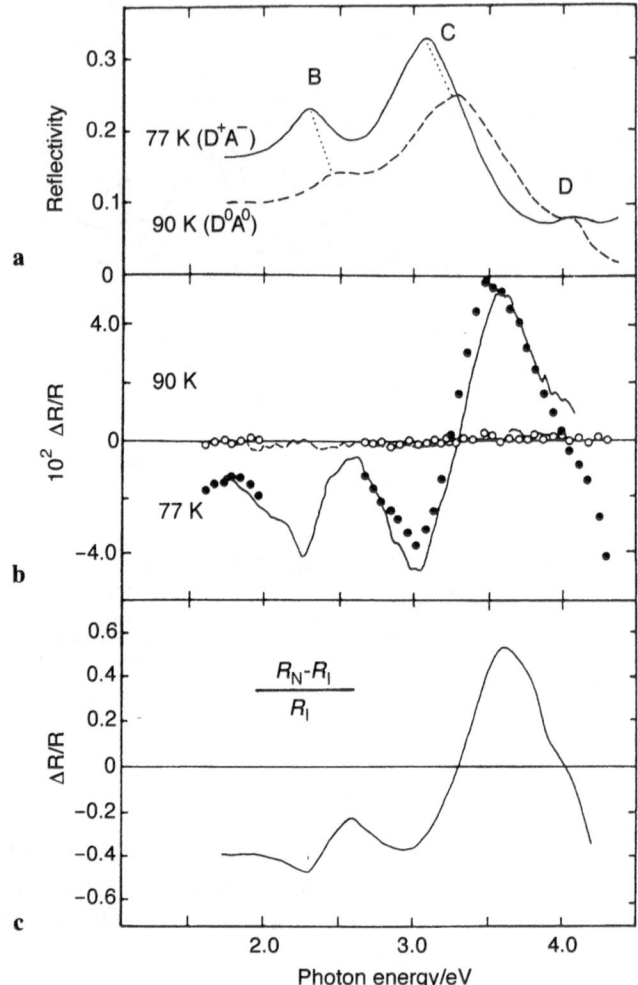

FIG. 4.2-6. **a** Reflectance spectra and **b** photoreflectance spectra for molecular excitation bands at 77 K (ionic phase) and 90 K (neutral phase) in a TTF–chloranil crystal. In **a**, *B*, *C*, and *D* represent the intramolecular excitation bands of TTF. **b** shows the results for pulse excitation shown by *open* (90 K) and *solid* (77) *circles*. *Solid* and *broken lines* represent the result by cw laser excitation. **c** is the calculated differential spectrum using the ionic phase (R_I) and neutral phase (R_N) spectra shown in **a**

shown for the cases of 10 ns pulse excitation (dots) and stationary (cw) excitation (solid line) in Fig. 4.2-6b. The photoreflectance spectra almost coincide with the calculated differential spectra (Fig. 4.2-6c) between the I-phase and the N-phase, indicating that photoexcitation or the resultant generation of the CT exciton in the I-phase induces the local N-phase region. From a comparison of the photoreflectance spectra (pulse excitation) with the calculated differential spectra, it is concluded [14] that a photogenerated CT exciton produces more than 80 pairs of neutral D^0A^0. Furthermore, the efficiency is only weakly dependent on temperature below T_c. This may be viewed as a kind of photoinduced phase transition via the self-multiplying, string-type condensation of CT excitons, as described above.

4.2.2.2 Photoinduced Phase Transition in Polydiacetylenes

The phenomena of photoinduced phase transition, a type of photochromism with extremely high efficiency and threshold (switching) behavior, may be observed in a wide class of organic systems. The necessary condition for the occurrence of the photoinduced phase transition is that free-energy minima corresponding to two electronic phases are nearly identical but separated by some potential barrier. In this case, photogenerated species such as polarons, self-trapped excitons, and solitons can trigger the first-order-like phase transition.

An organic system that shows well-defined photoinduced phase transition is a single crystal of polydiacetylene, PDA–4Un, $+RC-C{\equiv}C-CR{+}_x$, substituted with alkyl urethanes $(R=(CH_2)_4OCONH(CH_2)_{n-1}CH_3)$ [16]. In PDA–4Un, the urethane groups are connected to each other by hydrogen bonds which run parallel to the polymer backbone. In common with other PDAs, there are two spectroscopically distinct phases in the series of PDA–4Un: the so-called blue phase (A-phase) and the red phase (B-phase). Figure 4.2-7 shows the temperature-dependence of the electronic absorption spectra as well as Raman spectra for the backbone C=C stretching modes of the PDA–4U3 ($n = 3$) crystal [16]. The number attached to each spectrum in Fig. 4.2-7 indicates the temperature point at which the spectrum was measured in the course of the heating and cooling runs, as shown in Fig. 4.2-8. Each phase shows a distinct absorption peak due to the 0–0 exciton transition around 1.9 eV for the A-phase and around 2.3 eV for the B-phase, as typically seen in spectra 1 and 3, respectively, in Fig. 4.2-7. Each peak is associated with vibronic sidebands, denoted A′ or B′ (mainly in the C=C and C≡C stretching modes or their combination). In accordance with the change in the electronic spectra with the temperature-induced A–B transition, the Raman mode frequency of the C=C stretching mode shifts from 1450 cm^{-1} to 1500 cm^{-1}, as shown on the right-hand side of Fig. 4.2-7, indicating the change in the polymer backbone structure.

Figure 4.2-8 shows a trace of the reflectivity of a single crystal of PDA–4U3 with varying temperature (in the heating and cooling runs) monitored at the peak position (1.95 eV) for the A-phase. The trace shows a hysteresis loop with a width of around 60 K. The local free-energy minima corresponding to the two (A and B) phases are separated by a potential barrier, and the relative stability of each phase is exchanged at some temperature within the hysteresis loop, as shown schematically in the inset.

In the thermal cycle shown in Fig. 4.2-8, the sample was photoexcited at point 2 in the heating run and at point 4 in the cooling run with a single shot of pulsed dye-laser (10 ns width). The exciting photon energies from the pulse laser were 2.81 eV and 3.18 eV for points 2 and 4, respectively, and the excitation photon

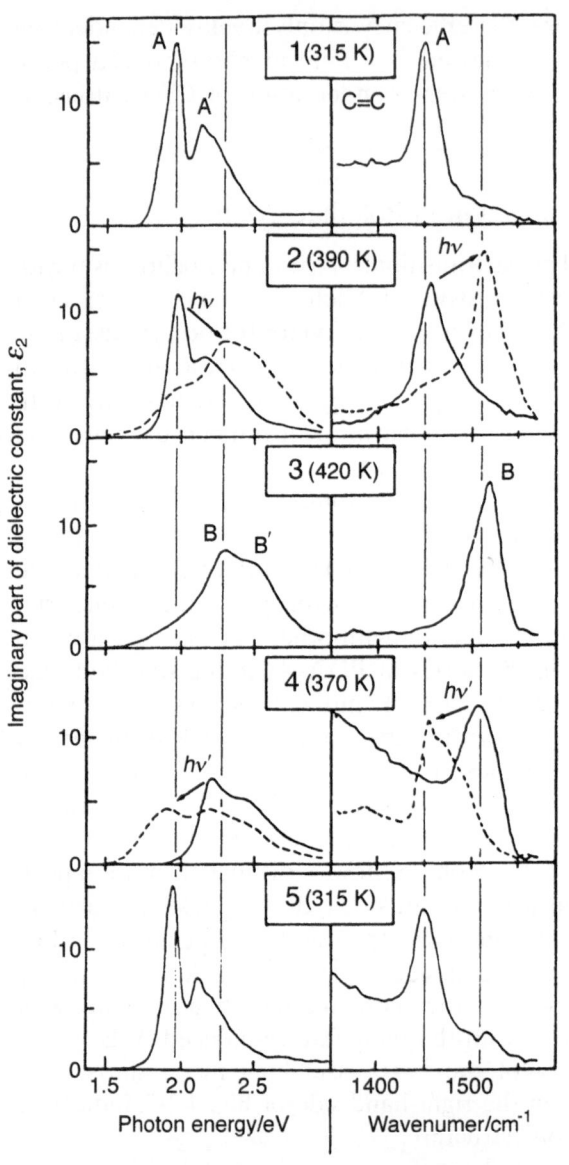

FIG. 4.2-7. Spectra of the imaginary part of the dielectric constant (ε_2) for the exciton absorption (*left*) and Raman spectra of the C=C stretching mode (*right*) at various temperatures (labeled 1–5) which are indicated on the hysteresis loop shown in Fig. 4.2-8. The *broken lines* in the spectra at temperature points 2 and 4 were obtained after photoexcitation by a single shot of a pulsed dye-laser. A (A′) and B (B′) represent the exciton absorption (vibronic sideband) and Raman band characteristic of the A- and B-phase, respectively

density was 5×10^{18} cm^{-3}. The exciton absorption (ε_2) and Raman spectra after photoexcitation are shown by the broken lines in Fig. 4.2-7. Changes in both absorption and Raman spectra show that nearly 100% of the A-phase is converted into the B-phase at point 2, and that about 50% of the B-phase is converted into the A-phase at point 4. The excitation intensity corresponds to one absorbed photon for every 140 repeated [$\xleftarrow{} $RC–C≡C–CR$\xrightarrow{}$] units. Such an extremely high photoconversion efficiency is ascribed to the collective nature of the phase transition between the two phases (A and B) whose free-energy levels are

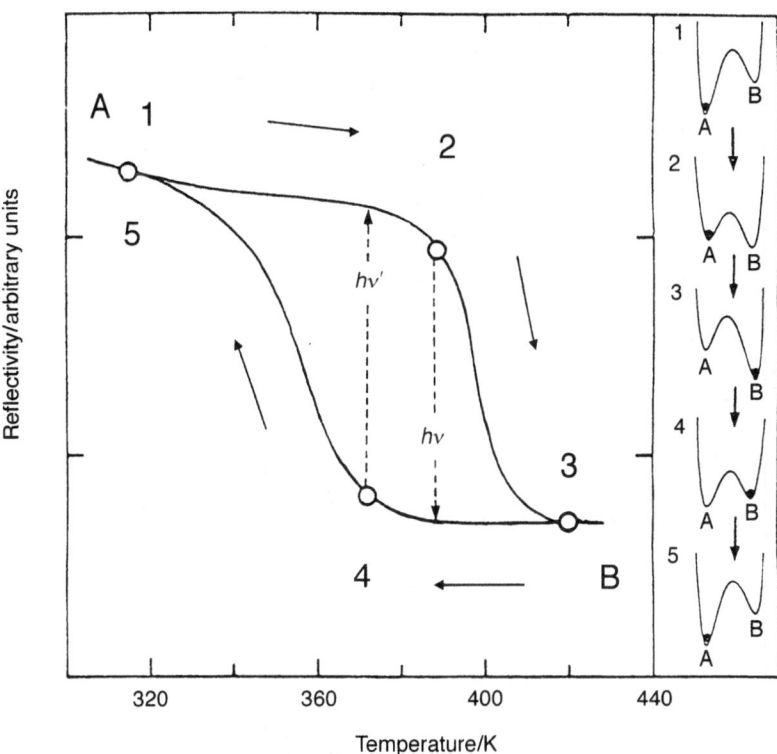

FIG. 4.2-8. Temperature-dependence of reflectivity in a poly-4U3 crystal at 1.95 eV and schematic diagrams of the free energy, with the minima corresponding to the A and B phases. The temperature points at which the reflectance and Raman spectra were measured are denoted by *open circles* and numbered in order of the heating and cooling processes

nearly degenerate in the hysteresis temperature region. The photoconverted fractions for the A → B and B → A phases are observed to depend nonlinearly on the excitation intensity, showing threshold behavior at around $3 \times 10^{18}\,\text{cm}^{-3}$ (corresponding to one absorbed photon per 230 repeated units). This implies that there is some collective interaction among photo-excited species, and a resultant critical size for the nucleus where local distortion evolves into macroscopic phase conversion.

Figure 4.2-9 shows the dependence of the phase-conversion fraction on the exciting photon energy both for the A → B and B → A phase changes. In the experiment, a single-shot laser pulse with a duration of about 10 ns was used, and the laser power was adjusted by varying the photon energy so that the absorbed photon density within the crystal was kept constant (around $6.5 \times 10^{18}\,\text{cm}^{-3}$). The excitation spectra for the converted fraction show threshold behavior at energies 0.4–0.5 eV higher than the absorption peak energies for both A → B and B → A

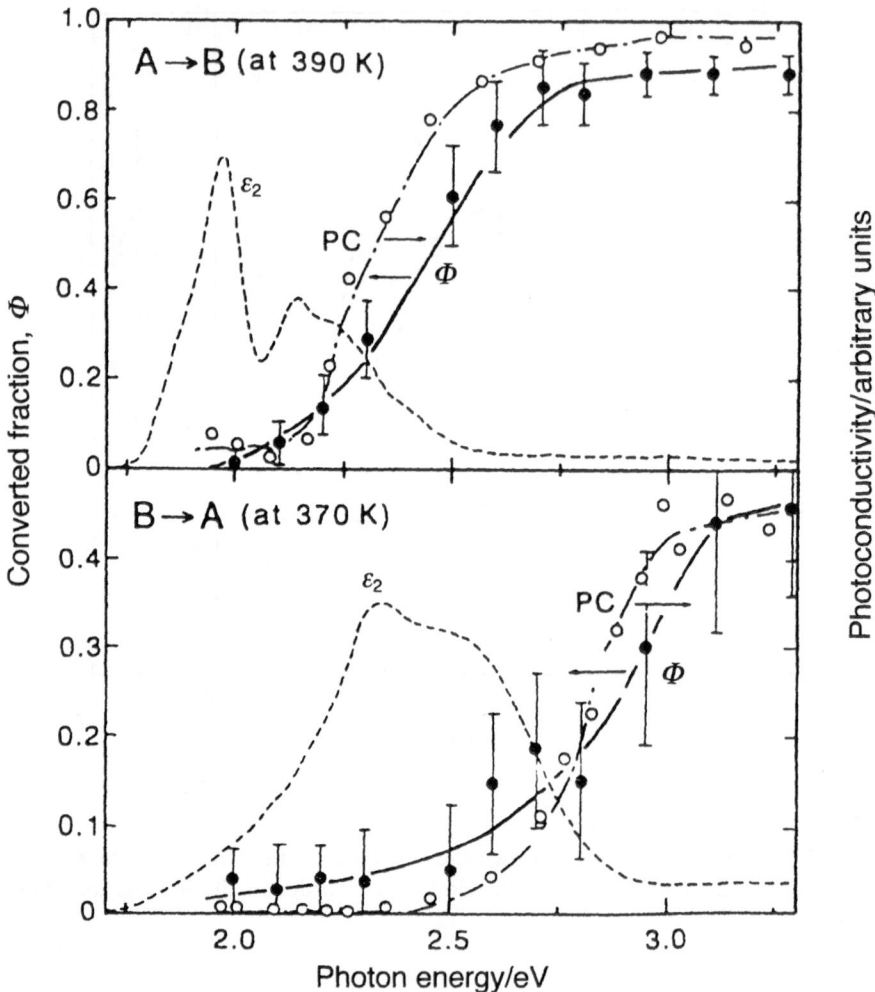

FɪɢG. 4.2-9. Dependence of the converted fraction of poly-4U3 crystal on the exciting photon energy (*solid circles*) for the photo-induced A → B (*upper*) and B → A (*lower*) phase transitions. Absorbed excitation photon density was kept constant, ~6.5 × 10^{18} cm^{-3}, with varying photon energy. *Open circles* and *broken lines* show the action spectra of photoconductivity (*PC*) and the exciton absorption spectra (ε_2 spectra) in the respective phases

phase changes. In particular, the efficiencies are very low in the energy region where the exciton transition strongly absorbs light. This is very different from the situation for laser-heating-induced phase changes. Figure 4.2-9 also shows action spectra of photocurrents measured on single crystals of PDA–4U3 at the same temperatures as the measurements of the photoinduced phase changes. The action spectra also show the threshold behaviors, including band-to-band transitions located at energies 0.4–0.5 eV higher than exciton transitions in both A and B phases of PDA–4U3. The observed parallel behaviors between the excitation spectra for the conversion fraction and the photoconductivity indicate that the photogenerated carriers are the microscopic source for photoinduced phase transitions. It is likely that the photocarriers in PDAs are subject to a strong and rapid lattice relaxation process as in other conjugated polymers, forming polaron or bi-polaron states. Such polaronic species have a role as domain walls transiently separating the A and B phases on the polymer backbones.

Making use of the characteristic that the A- and B-phase PDAs have different values of band gaps (E_g^A and E_g^B), we can bi-directionally control the phase transition by dichromatic excitation [16]. Figure 4.2-10 shows the cyclic change in the reflectance spectra of PDA–4U3 crystals at 375 K, near the middle of the hysteresis loop shown in Fig. 4.2-8. First, a single-shot excitation pulse at 2.58 eV (above E_g^A) will convert almost 100% of the A-phase into the B-phase, as is

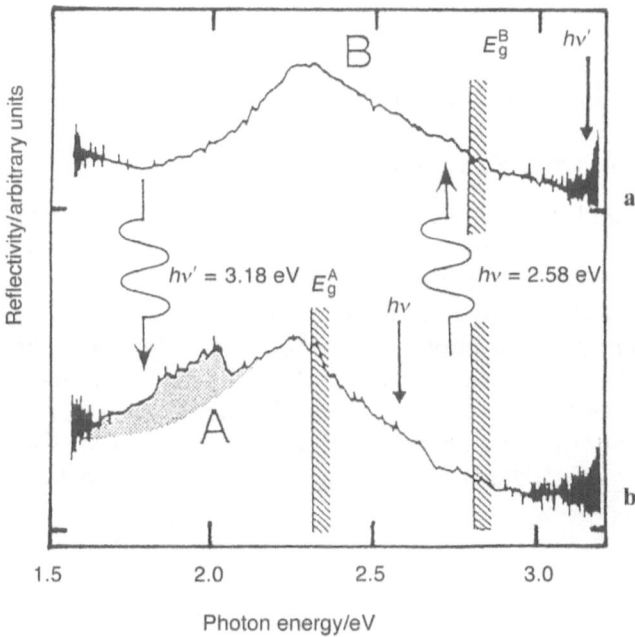

Photon energy/eV

FIG. 4.2-10. Repeated an bi-directional switching between the A- and B-phase of poly-4U3 crystal by dichromatic excitation: $hv' = 3.18$ eV and $hv = 2.58$ eV. **a** Reflectance spectrum characteristic of the B-phase converted by hv' excitation, and **b** the spectrum coverted from the B-phase by hv excitation. E_g^A (ca. 2.3 eV) and E_g^B (ca. 2.8 eV) represent the band-gap energies for the A- and B-phases, respectively. The *shaded area* under the curve indicates the partly restored A-phase band

shown in Fig. 4.2-10a. Next, a single-shot excitation pulse at 3.18 eV (above E_g^B) modifies the B-phase spectrum into the spectrum shown in Fig. 4.2-10b, in which an additional reflectance peak (hatched area) can be seen due to the partly (about 50%) converted A-phase region. Subsequently, the (A + B) phase crystal is irradiated again by a single pulse with a photon energy of 2.58 eV, i.e., higher than E_g^A but lower than E_g^B. The crystal completely reverts to the B-phase (Fig. 4.2-10a), with a spectrum which is identical to the one that was first converted from the A-phase virgin crystal. Thus, dichromatic excitation at 2.58 and 3.18 eV can bidirectionally and repeatedly switch the crystal phase between the A (50%) + B (50%) state and the genuine B (100%) state.

According to a time-resolved study [16], most of the photoinduced phase transition in polydiacetylene is completed within the duration (10 ns) of an exciting laser pulse. This initial process in which collective interactions between polanonic species seem to change a local distortion into a macroscopic phase change will be useful in research in to the possible photodetection of the dynamics of first-order phase transitions. The phenomenon of photoinduced phase transitions will probably be found in a wide range of organic systems with bistable electronic phases, although it has seldom been extensively investigated.

4.2.3 Photoelectron Spectroscopy of Molecular Solids

4.2.3.1 Principles and Experimental Aspects

The principles and the technical aspects of photoelectron spectroscopy are described briefly, but detailed descriptions can be found in the references [17]. When a solid in a vacuum is irradiated with high-energy monochromatic photons, electrons are emitted by the photoelectric effect. Figure 4.2-11 illustrates the photoelectric emission (or photoemission) process. When monochromatic photons of energy $h\nu$ impinge on the sample, electrons can be emitted from occupied states with various binding energies. For electrons escaping from the solid without inelastic scattering, the binding energy, E_b, relative to the vacuum level (or ionization energy) is related to the kinetic energy of the photoelectron, E_k, by the Einstein relation

$$E_b = h\nu - E_k \qquad (4.2\text{-}2)$$

Thus, the occupied-state structure can be deduced by measuring the E_k distribution of these electrons. The electrons excited deep inside the solid cannot escape without scattering, making this technique surface-sensitive. The probing depth of the electrons spans several to tens of nanometers depending on E_k.

According to the photon energy of the light source, the field of photoelectron spectroscopy can be divided roughly into two types. One is UV photoelectron spectroscopy (UPS) using vacuum ultraviolet light (e.g., HeI (21.2 eV) and HeII (40.8 eV) lines), which can emit only valence electrons. The other is X-ray photoelectron spectroscopy (XPS; also called ESCA) using soft X-rays (e.g. MgK$_\alpha$

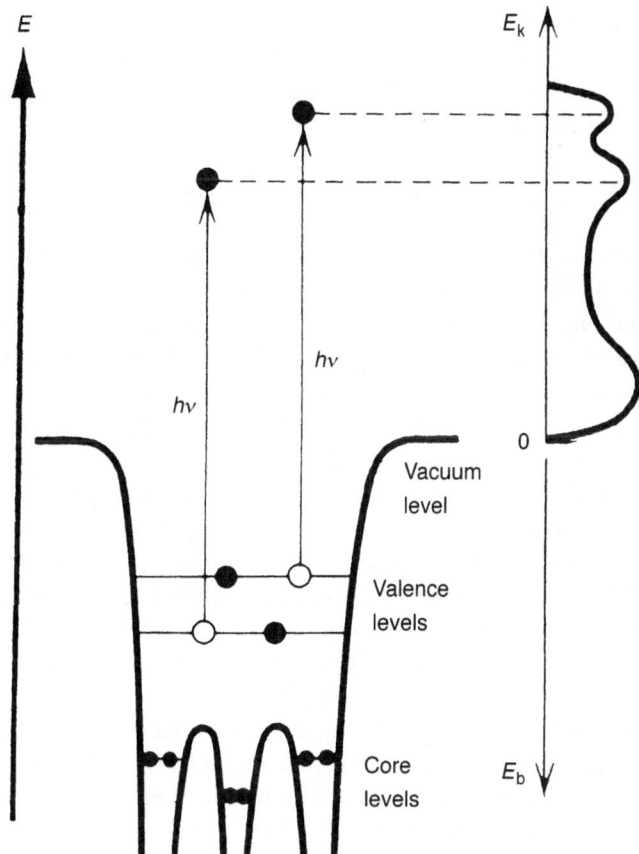

FIG. 4.2-11. Principle of photoelectron spectroscopy. E_b and E_k denote the binding energy and kinetic energy of an electron

(1253.6 eV) and AlK$_\alpha$ (1486.6 eV) lines), which can excite both valence and core electrons. Recently the gap between these two types has been bridged by synchrotron radiation.

The energy resolution in UPS (10–200 meV) is generally higher than that of XPS (0.7–0.8 eV). This makes UPS more suitable for probing valence states, which dominate the electronic and optical properties of solids. This is particularly true for organic solids, mainly consisting of C, N, and O atoms, because the relative photoemission cross sections of the $2p$ orbitals forming the topmost valence states for X-rays are much smaller than those of $2s$ electrons, while they are comparable for UV photons. The following sections will be mainly concerned with UPS.

Energy analysis of photoelectrons can be performed by various types of analyzers, e.g., the electrostatic-deflector type, the retarding-field type, or the time-of-flight type [18]. Angle-resolved measurements (ARUPS), in which the E_k

distribution of electrons emitted into a small steric angle is analyzed, offer detailed information about oriented molecules.

The following section describes how this technique can be used for obtaining information about molecular systems.

4.2.3.2 Comparison with Vapor Phase Spectra and Polarization Energy

Figure 4.2-12a shows the UPS spectra of naphthacene, which is a typical aromatic hydrocarbon [19]. The abscissa is the solid-state ionization energy I_s (binding energy) plotted against vacuum level. For comparison the spectrum of naphthacene vapor is also shown (Fig. 4.2-12b), with the gas phase ionization energy I_g as the abscissa. The peaks in the vapor spectrum can be assigned by comparison with MO calculations, as shown in Fig. 4.2-12 for π-orbitals. The large peaks in the solid state are due to electrons scattered within the solid.

It can be seen that there is good one-to-one correspondence between the spectra of the solid and vapor phases, showing that the electronic structure of a naphthacene molecule is not much perturbed by molecular aggregation caused by weak van der Waals interactions. This indicates that the electronic structure of a molecule can be investigated using the solid-state spectrum when the vapor spectrum cannot be measured. This is particularly useful for biological or polymeric materials which are decomposed by heat.

Two differences between solid- and vapor-phase spectra should be noted. One is a lowering of the ionization energy from a vapor to the solid state. The lowest ionization energy peak, which corresponds to the ionization from the highest occupied molecular orbital (HOMO), is at 7 eV in the vapor phase, while it is at 5.8 eV in the solid state. Thus the solid-state spectrum shows a rigid shift towards lower ionization energy. This shift can be mainly ascribed to the stabilization of the photoionized molecule. The Coulombic field, by a positive charge on the cation, polarizes the electrons in the surrounding molecules, leading to a stabilization of the whole system. This quantity is called polarization energy, P_+. Thus we obtain

$$I_s = I_g - P_+ \tag{4.2-3}$$

Table 4.2-1 shows selected values of I_s, I_g, and P_+ [19 and 20]. The physical nature of P_+ suggests that it depends on the polarizability of surrounding molecules and also on the intermolecular distance, which affects the strength of the Coulombic field. The results in Table 4.2-1 show such a trend. For aromatic hydrocarbons, the value of P_+ is nearly constant. An increase in molecular size leads to an increase in both polarizability and intermolecular distance, leading to some canceling effects for P_+ [20].

The other difference between gas and solid-state spectra is the peak broadening in the solid spectrum. This has been ascribed to several factors, such as (i) the excitation of intermolecular vibrations and phonons at photoionization, and (ii) the variation of P_+ between the molecules in the bulk and at the surface, where about half the polarizable molecules are absent [21].

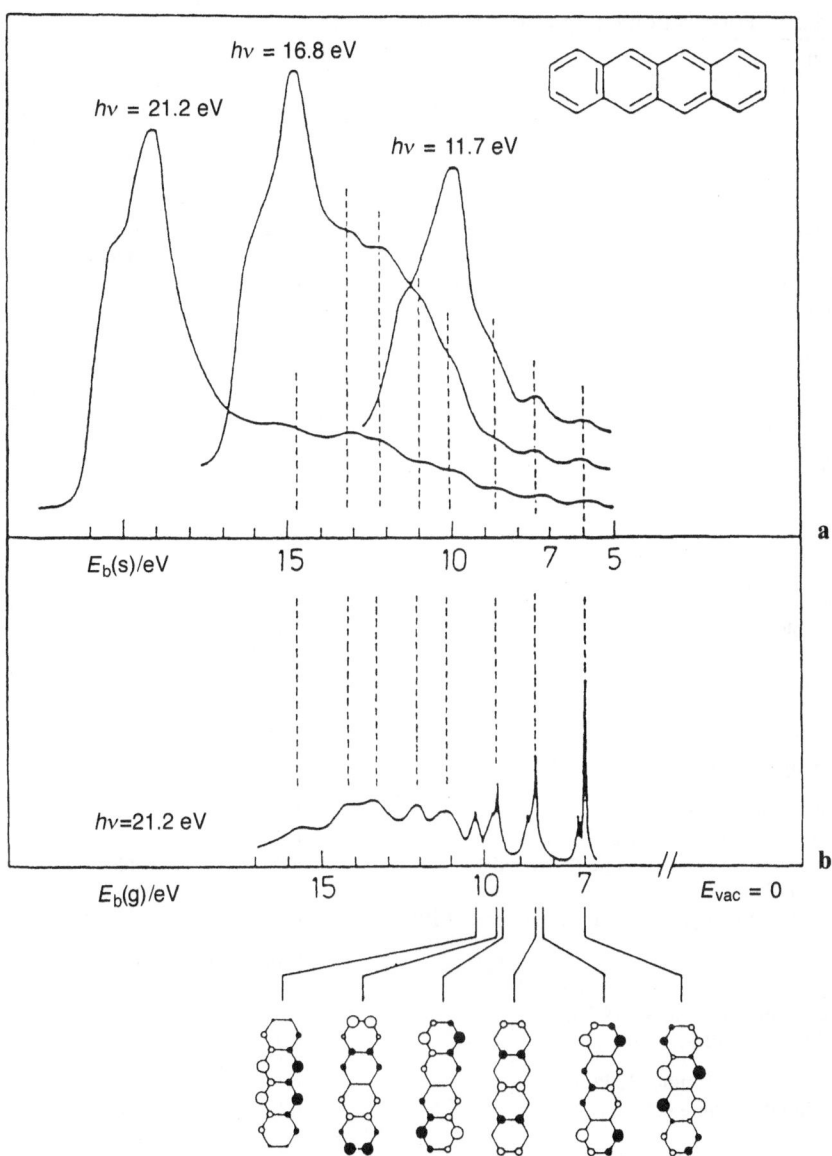

FIG. 4.2-12. UV photoelectron spectra of naphthacene in the solid (**a**) and vapor (**b**) phases. The abscissa is the binding energy relative to the vacuum level [23]

A similar idea of polarization energy can be applied to excess electrons (molecular anions) in a solid, leading to a relation between the electron affinities of the gas and solid states (A_g and A_s)

$$A_s = A_g + P_- \qquad (4.2\text{-}4)$$

TABLE 4.2-1. Polarization energies of molecular solids

Compound	I_g/eV	I_s/eV	P_+/eV
Methane	12.70	11.2	1.2
Hexane	10.25	8.55	1.7
Iodine	9.26	6.34	2.9
Benzene	9.17	7.3	2.1
Naphthalene	8.12	6.4	1.7
Anthracene	7.36	5.67	1.7
Naphthacene	6.89	5.10	1.8
Pentacene	6.58	4.85	1.7
Phenanthrene	7.86	6.08	1.7
Chrysene	7.51	5.8	1.7
Benz[a]anthracene	7.38	5.64	1.7
Triphenylene	7.81	6.2	1.6
Pyrene	7.37	5.58	1.7
Perylene	6.90	5.12	1.8
Coronene	7.25	5.52	1.7
Sexiphenyl	7.2	5.9	1.3
Rubrene	6.41	5.3	1.1
Quaterrylene	6.11	4.76	1.4
9,10-Dichloroanthracene	7.5	5.8	1.7
Pyridine	9.26	7.3	2.0
Phenothiazine	6.82	5.15	1.5
Tetrathianaphthacene (TTT)	6.07	4.4	1.7
Tetrathiafulvalene (TTF)	6.4	5.0	1.4
Bis(ethylenedithiolo)-TTF (BED–TIF)	6.21	4.78	1.4
Tetramethyltetraselena fulvalene (TMSTSF)	6.27	4.84	1.4
p-Phenylenediamine	6.84	5.2	1.7
N,N,N',N'-Tetramethyl-p-phenylenediamine (TMPD)	6.2	4.7	1.5
p-Chloranil	9.74	8.1	1.6
Anthraquinone	9.05	7.6	1.5
Tetracyanoquinodimethane (TCNQ)	9.5	7.4	2.0
Pyromellitic dianhydride (PMDA)	10.9	8.0	2.9
Metal-free phthalocyanine (H$_2$PC)	6.1	5.15	1.0
Copper phthalocyanine (CuPC)	6.1	(α) 4.88	1.2
		(β) 4.62	1.5
Zinc tetraphenylporphyrin (ZnTPP)	6.2	5.0	1.2
Ferrocene	6.72	5.4	1.3
Decamethylferrocene	5.7	4.7	1.0
β-Carotene	6.5	5.4	1.1

I_g, gas phase ionization energy; I_s, solid state ionization energy; P_+, polarization energy.

where P_- is the polarization energy for an electron. Thus the effective bandgap for forming an electron–hole pair becomes

$$I_s - A_s = I_g - A_s - \left(P_+ + P_-\right) \qquad (4.2\text{-}5)$$

Thus quantitative knowledge of the polarization energy is essential for a discussion of the energetics of charge-carrier formation [22].

The concept of polarization energy also applies to polymers, as discussed in Sect. 5.3.1. A comprehensive list of ionization threshold energies can be found in [23]. The intrachain order of regularly repeated units makes the evolution of the electronic structure of polymers particularly interesting. This will be described in Sect. 5.3.1.

4.2.3.3 Intermolecular Bands in Strongly Interacting Systems

In some organic compounds, the intermolecular interaction is significantly stronger than the usual van der Waals interaction, and sometimes even leads to the formation of intermolecular bands of appreciable bandwidth. This band formation causes an additional lowering of the ionization energy from the vapor phase as well as the polarization energy described above. This effect can be deduced by the extraordinarily large shift between the gas- and solid-state ionization energies, as in the case of the iodine compounds listed in Table 4.2-1.

More detailed studies can be performed in favorable cases by angle-resolved UPS. Figure 4.2-13 illustrates such a study [24] for bis(1,2,5-thiaziazolo)-p-quinobis(1,3-dithiole) (BTQBT) (see structure in Fig. 4.2-13b), which is a good organic semiconductor with a conductivity of $10^{-5}\,S\,cm^{-1}$ at room temperature. By careful evaporation of BTQBT on highly oriented pyrolitic graphite, well-ordered films could be prepared in which the molecules are oriented with the molecular plane almost parallel to the substrate surface. This orientation could also be deduced by analyzing the emission-angle dependence of peak intensity in the UPS spectra. Figure 4.2-13a shows the $h\nu$-dependence of UPS spectra for electrons emitted normal to the substrate surface in the energy region of the HOMO and the level just below the HOMO (next-HOMO). There is a large variation in the HOMO peak energy with $h\nu$, while a smaller peak-energy variation is found for the next-HOMO.

According to the theory of ARUPS, measurements at a specific $h\nu$ in such an experimental set-up probe only a selected part of an energy band [24], while the whole set of measurements at various values of $h\nu$ gives the total width of the band formed. From the results in Fig. 4.2-13a we can deduce a HOMO bandwidth of $0.37\,eV$, which corresponds well with a theoretically estimated value of $0.5\,eV$. By further analysis of these data, the energy dispersion relation $E = E(k)$ can be deduced for the HOMO and the next-HOMO, as shown in Fig. 4.2-13b. This is the first direct observation of intermolecular band formation in a single-component organic molecular solid.

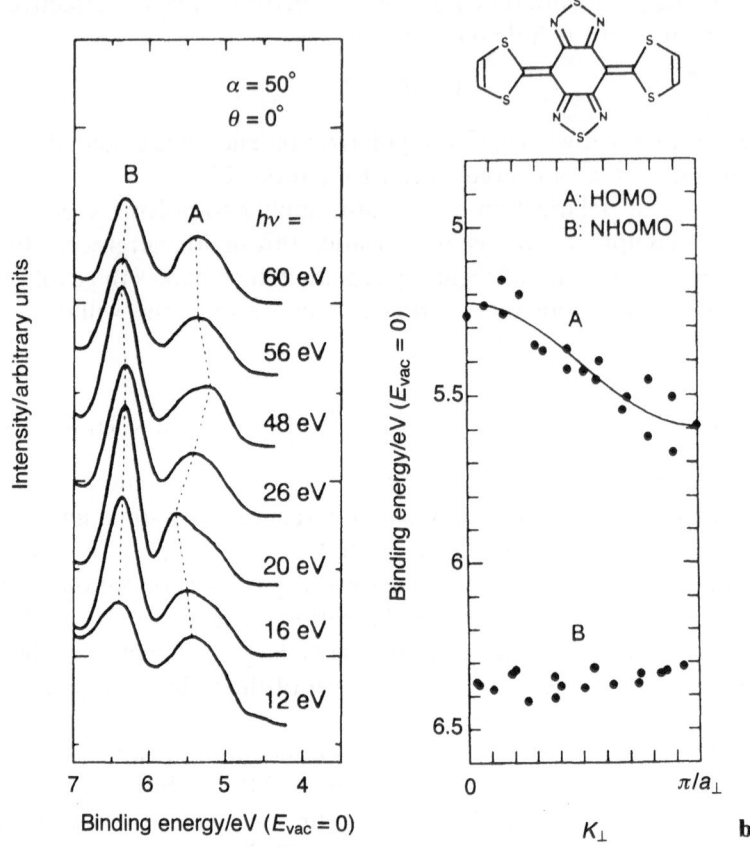

FIG. 4.2-13. **a** Photon energy dependence of the photoelectron spectra of bis(1,2,5-thiaziazolo)-*p*-quinobis(1,3,-dithiole) (BTQBT) at normal emission. **b** Energy band dispersion relation $E = E(k)$ deduced from the spectra in (**a**). *HOMO*, highest occupied molecular orbital; *NHOMO*, next-HOMO. k is the wave vector, α and θ are the incidence angle of photons and the emission angle of photoelectrons relative to the surface normal. (From [24], with permission)

4.2.3.4 Interfaces

In the development of molecular devices, interfaces that include molecular systems play an important role, e.g., in light-emitting diodes, solar cells, electrophotography, and silver halide photography. Thus there is a rapidly growing interest in interfacial electronic structures. UPS can provide direct information about these structures by monitoring the changes in the spectra when one component is deposited onto another.

As an example of such a study, Fig. 4.2-14a shows the changes in the UPS spectra of AgBr during stepwise deposition of a merocyanine dye [25]. The lower abscissa is the retarding voltage applied between the sample and the electron-

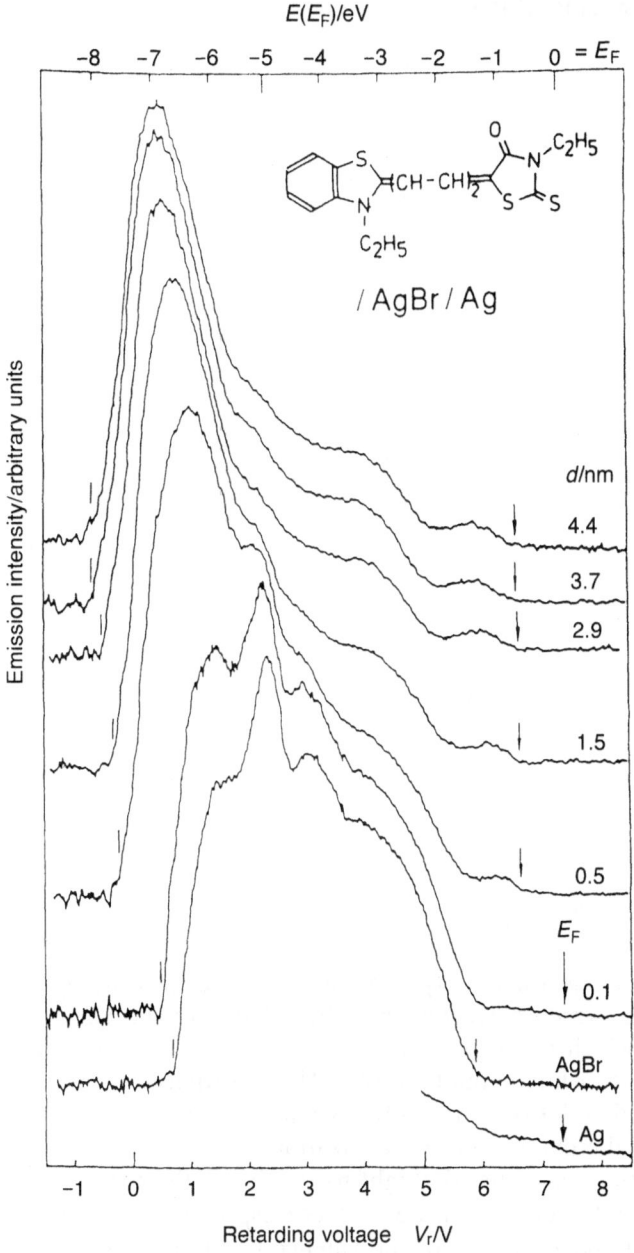

FIG. 4.2-14. **a** Sequential change in the UV photoelectron spectroscopy (UPS) spectra by depositing a merocyanine dye on AgBr [25]. *d* is the thickness of the dye layer. The spectrum of the Ag substrate on which the AgBr film in the Fermi edge region is grown is also shown. The *abscissa* is the retarding voltage applied between the sample and the analyzer, which roughly corresponds to the kinetic energy of the photoelectrons. **b** Energy diagrams at the interfaces formed between the AgBr and merocyanine dyes (A–D) of systematically changing central polymethine chain length, which is denoted by the number of double bonds *n* [25]. The dye in **a** corresponds to dye C. E_F, Fermi level of AgBr; $E(E_F)$, electron energy relative to E_F; *VL*, vacuum level; *CB* and *VB*, the bottom of the conduction band and the top of the valence band of AgBr, respectively; *LUMO* and *HOMO*, the lowest unoccupied and the highest occupied molecular orbital of the dye, respectively

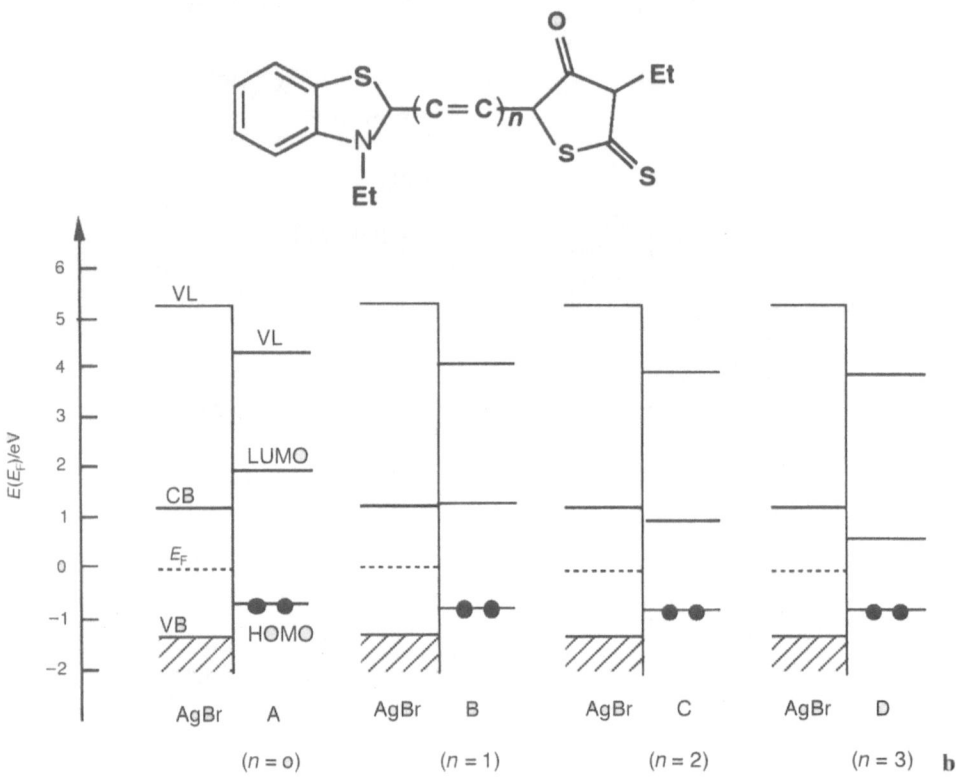

FIG. 4.2-14. *Continued*

collecting electrode for energy analysis. This energy scale is converted to the upper abscissa, which corresponds to E_b relative to the Fermi level of the Ag substrate on which the AgBr film is formed. Such dye/AgBr systems are very important in silver halide photography. The sensitivity of AgBr, which is limited to the blue and violet region, can be extended by dye-adsorption to the absorption region of the dye (spectral sensitization).

Figure 4.2-14a shows that the right-hand edge of the spectrum shifts further to the right due to the growth of the peak corresponding to the HOMO of the dye that is located above the top of the valence band of AgBr. The location of the HOMO does not depend on the thickness of the dye. The left-hand edge of the spectrum, corresponding to the location of the vacuum level, also shows a large shift to the right. This indicates a large lowering of the vacuum level by vacuum deposition. From such data and the optical spectra of AgBr powders with adsorbed dyes, detailed interfacial energy diagrams (Fig. 4.2-14b) can be estimated for a series of dyes with central methine chains of different lengths. Theoretical studies of the electron-transfer mechanism of spectral sensitization, in which electrons excited from the HOMO to the LUMO are injected into the conduction band of AgBr, indicate that this dye is expected to be a poor sensitizer, since the LUMO of the dye is below the conduction band minimum of AgBr. In

UPS experiments on similar dyes, where the number of central repeating double-bonds, n, was varied from $n = 2$ (Fig. 4.2-14) to $n = 0, 1$, or 3, the sensitizing ability of the dye was expected to be good for $n = 0$ and 1, and poor for $n = 3$. These expectations showed excellent agreement with the observed sensitizing abilities of the dyes, strongly supporting the theory of an electron-transfer mechanism.

An important question when considering such interfaces is how the relative electronic structures at both sides are aligned. The simple idea of a common vacuum level (Fig. 4.2-15a) is frequently applied in estimating interfacial energy diagrams, while textbooks of solid-state physics suggest a Fermi-level alignment (Fig. 4.2-15b).

When an organic compound is deposited on a metal of work function Φ_m, we would not expect any change in the energy of the vacuum level for case in Fig. 4.2-15a, while for Fig. 4.2-15b a shift of $\Phi_m - \Phi_s$ would be expected for a thick film with some band bending, where Φ_s is the work function of the organic sample. For the combination of a fixed organic material and metals of various work functions, the gap, Δ, between the Fermi level of the metal and the top of the valence band of the organic solid should be independent of Φ_m for case (b) for a sufficiently thick film, while Δ should change for case (a) following the change in Φ_m.

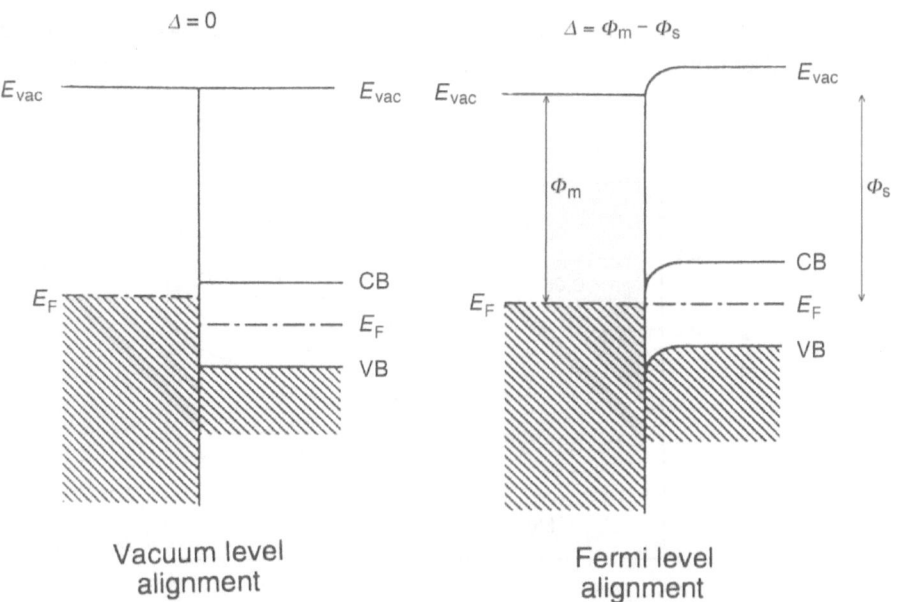

FIG. 4.2-15. **a,b** Two traditional models for estimating the interfacial electronic structure between a metal and an organic solid. A common vacuum level (**a**) and Fermi level (**b**) is assumed, with band bending in **b**. The expected relation between the vacuum level shift Δ and the work functions of the metal and organic samples (Φ_m and Φ_s, respectively) are indicated at the top of each figure. *Shaded areas* represent occupied electronic states

It is only very recently that such fundamental problems about interfaces which include organic molecules have been studied. Figure 4.2-16 shows the observed shift of the vacuum level, Δ, at the deposition of Zn tetraphenylporphyrin (ZnTPP) and metal-free tetra(4-pyridyl)porphyrin (H$_2$TPyP) on various metals plotted against the work function of the metal substrate Φ_m [26]. The results follow the linear relationship $\Phi_m = a\Delta + b$, but the value of a is smaller than 1, and therefore does not agree with that expected in Fig. 4.2-15b ($a = 1$). In particular, the results for ZnTPP indicate that a is nearly equal to 0, which is similar to the value expected from Fig. 4.2-15a ($a = b = 0$). However, the observed b is about 0.7 eV, which is very different from the value of $b = 0$, corresponding to the common vacuum level. This large value for b corresponds to the abrupt change at the interface, indicating the formation of an interfacial dipole layer. The energy of the vacuum level shows little dependence on film thickness, suggesting that the number of carriers in organic molecules needed to attain electrical equilibrium is very small and the thickness of the diffusion layer is large. Although the full elucidation of the implications of these data awaits further experiments, this kind of information will play an important role in clarifying the mechanism of charge carrier injection and separation at organic interfaces.

4.2.4 NEXAFS Spectroscopy of Organic Materials

In the X-ray absorption spectrum of a compound, a sharp increase in absorption is observed at energies characteristic of the constituent elements. Such energy is

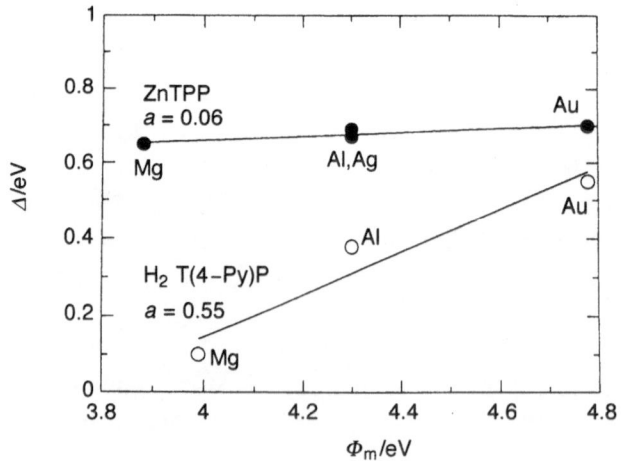

FIG. 4.2-16. Observed relationship between the vacuum level shift Δ and the work functions of the metal and organic samples (Φ_m and Φ_s, respectively). *Solid* and *open circles* represent data for ZnTPP and H$_2$T(4-Py)P, respectively. a is the slope of the best-fit straight line

called the *absorption edge* of the element, and corresponds to the energy of the core electrons. At each edge, the spectrum often shows fine structures. These fine structures are divided into two kinds, i.e., near-edge absorption fine structures (NEXAFS) up to about 30 eV above the edge, and extended X-ray absorption fine structures (EXAFS) beyond this energy.

NEXAFS, also called X-ray absorption near-edge structures (XANES), arise from the excitation of core electrons into various vacant states, as shown in Fig. 4.2-17. The bold lines represent the effective potential well for an electron formed by nuclei and other electrons, and occupied (core and valence) and unoccupied (discrete and continuous) levels are formed by this potential. X-ray irradiation can excite core electrons into both discrete vacant states and a continuum above the vacuum level.

The EXAFS arise from another mechanism. When an electron is excited into a continuum above the vacuum level, it can be emitted from there as a photoelectron. On the way out, the electron wave is scattered by the surrounding atoms, and the reflected wave can interfere with the outgoing wave. This interference is reflected in the absorption spectrum, giving rise to the EXAFS. In organic materials, EXAFS are often weak owing to the small scattering ability of the light elements forming the system, while NEXAFS appear as sharp fine structures, and have recently been developed as a powerful tool for probing molecular orientation and electronic structure in organic ultrathin films and adsorbed states [27]. Therefore, our focus is on this aspect of X-ray absorption.

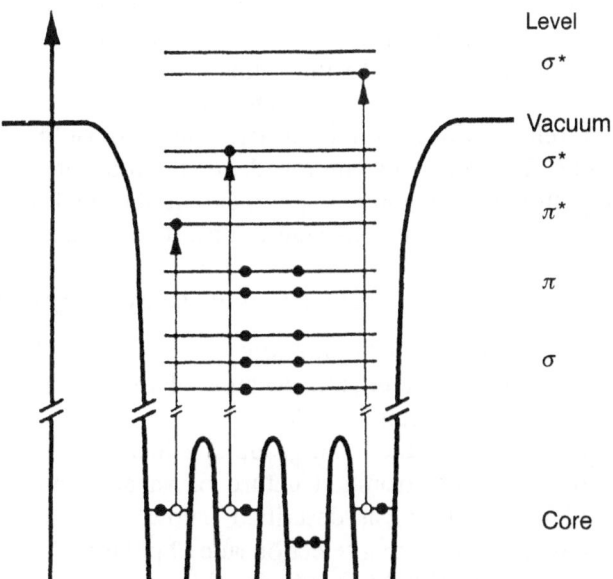

FIG. 4.2-17. Principle of near-edge absorption fine structure (NEXAFS) spectroscopy. Soft X-ray absorption occurs by the excitation of core electrons into various unoccupied states, giving rise to fine structures in the X-ray absorption spectrum

TABLE 4.2-2. Energies of soft X-ray absorption edges for the elements forming organic compounds

Core level	Energy (eV)	Core level	Energy (eV)
C 1s	290	S 1s	2480
N 1s	410	2s	230
O 1s	540	2p	160
F 1s	700	Cl 1s	2820
Si 1s	1840	2s	270
2s	150	2p	200
2p	100		
P 1s	2150		
2s	190		
2p	140		

Table 4.2-2 summarizes the energies of the edges of elements forming organic materials. For first-row elements, they are 1s (K shell) atomic orbitals, and for second-row elements they are 1s, 2s, and 2p orbitals (K, L_I, and $L_{II,III}$ edges).

The NEXAFS spectra are usually measured using synchrotron radiation, which is one of the few practical continuous light sources in the soft X-ray region. Information which is almost equivalent can be obtained by measuring the inelastic scattering of high-energy monochromatic electrons (electron energy-loss spectroscopy, EELS), which is particularly useful for gas-phase studies [28].

Although the measurement of absorption in the transmission mode is simple in principle and gives an absolute value of the absorption coefficient, it needs self-standing thin film samples of appropriate thickness (of the order of a few μm). Also, it is difficult to apply to thin films on substrates, which are often important in studies of molecular systems. Thus it is preferable to measure other quantities which reflect absorption intensity. Electron yield (EY) or X-ray fluorescence yield (FY) are usually employed, i.e., the intensity of photoelectron emission or X-ray fluorescence emission per incident photon. For details of these techniques and other experimental considerations, see the excellent textbook by Stöhr [27].

NEXAFS have the following advantages in the study of organic materials:

1. The ability to probe molecular orientation;
2. Surface sensitivity for EY detection because of the small escape depth of photoelectrons (a few nanometers or less);
3. Element-specificity and/or functional-group-specificity;
4. Information about the electronic structure of vacant states, although this should be used with caution, as described below. The study of molecular orientation is based on the strong selection rule of polarized X-ray absorption:
   a) Excitation from a 1s orbital to a π^* orbital is allowed only for the electric vector of light, E, parallel to the axis of the 2p orbital forming the π^* orbital. For systems containing only double-bonds, this means that E should be perpendicular to the molecular plane.

b) Excitation from a $1s$ orbital to a σ^* orbital localized in a bond is allowed only when $\boldsymbol{E}$ is parallel to the bond. Such polarization-dependence can be determined by measuring the spectra with polarized light at various incidence angles of light θ. Fortunately, synchrotron radiation is an ideal light source for this kind of study, as its electric field vector $\boldsymbol{E}$ is strongly polarized within the horizontal plane including electron orbits in the synchrotron.

Figure 4.2-18 shows an example of such a study for the rubbed surface of an evaporated film of poly(tetrafluoroethylene) (PTFE) [29], which has the molecular structure shown in Fig. 4.2-18a. In Fig. 4.2-18b, the C K-edge spectra are shown for a rubbed PTFE film measured in EY mode for normal incidence ($\boldsymbol{E}$ parallel to the substrate surface, $\theta = 90°$) and glancing incidence ($\boldsymbol{E}$ almost perpendicular to the substrate surface, $\theta = 15°$), with the rubbing direction in the plane of the incident light. The two peaks in the low energy region are already assigned, as indicated in the figure, to transitions from the C $1s$ orbital to the σ^* orbitals localized in the CF and CC bonds. The spectra in Fig. 4.2-18b show clear polarization-dependence, with the $1s \rightarrow \sigma^*$ (CC) peak much stronger at normal incidence. In contrast, the $1s \rightarrow \sigma^*$ (CF) peak is stronger at glancing incidence. According to the criteria described for (b), this polarization-dependence indicates that the molecular chain is lying nearly flat on the substrate surface. Moreover, the two normal incidence spectra in Fig. 4.2-18c, with $\boldsymbol{E}$ parallel and perpendicular to the rubbing direction, show that σ^* (CC) excitation is much stronger for $\boldsymbol{E}$ parallel to the rubbing direction. This indicates that the molecular chain is also oriented in the plane of the sample surface, and parallel to the rubbing direction. Measurements for evaporated films without rubbing show a similar (but less perfect) flat configuration, but without azimuthal order in the substrate surface.

For such studies, the use of surface-sensitive EY detection is essential, since the rubbing process affects only the surface region of the film. This surface sensitivity is an advantage compared with other techniques such as attenuated-total-reflection IR (ATR–IR) spectroscopy, which has a probe depth of the order of a few micrometers.

Figure 4.2-19a shows another study of molecular orientation for Zn tetraphenyl porphyrin (ZnTPP) film evaporated onto Ag substrate kept at 367 K during evaporation [30]. The NEXAFS spectra at the N K-edge show a large dependence on θ, with peaks A–C at low photon energy and the peaks in the higher energy region showing opposite polarization-dependence. The former and the latter are assigned to excitations of the π^* and σ^* orbitals, respectively. According to the criteria mentioned above, the central macrocycle containing the N atoms is lying almost flat on the substrate surface. For a more quantitative analysis, Fig. 4.2-19b shows the intensity of peak A plotted against θ. By comparing this result with a theoretical simulation for various values of the angle α between the plane of the central macrocycle and the substrate surface, a best fit is obtained at $\alpha = 28 \pm 10°$, shown in Fig. 4.2-19b by the curved line.

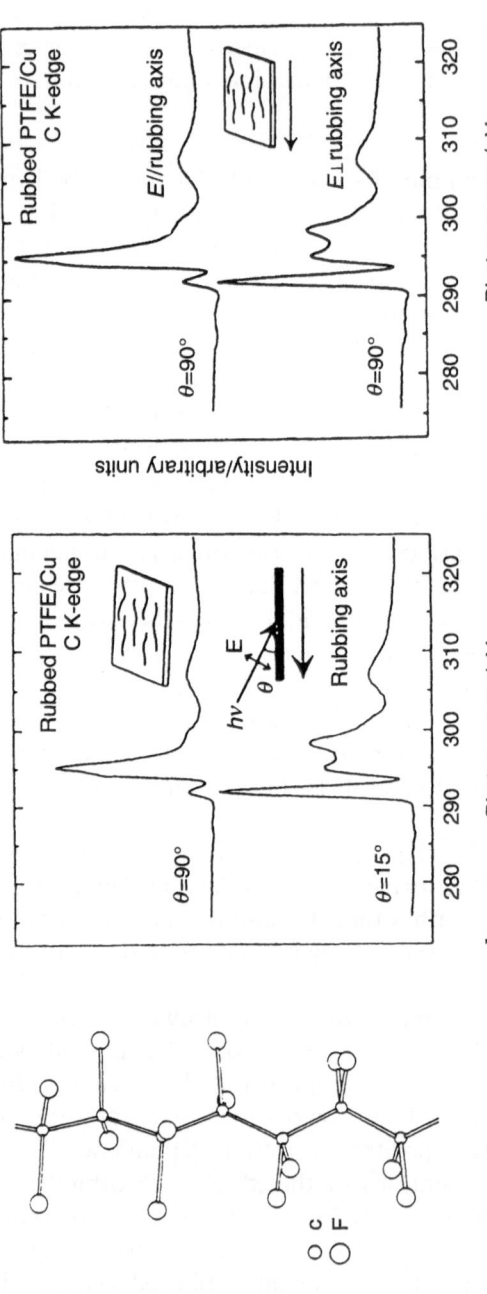

FIG. 4.2-18. Polarized X-ray absorption spectra of the rubbed surface of an evaporated film of poly(tetrafluoroethylene) (PTFE) [29]. **a** Molecular structure of PTFE. **b** Polarized C K-edge NEXAFS spectra at normal ($\theta = 90°$) and grazing ($\theta = 15°$) incidence of light. The rubbing direction is in the plane of incident light. **c** Polarized C K-edge NEXAFS spectra for normal incidence with the electric vector of light, E, parallel and perpendicular to the rubbing direction. (From [29a], with permission)

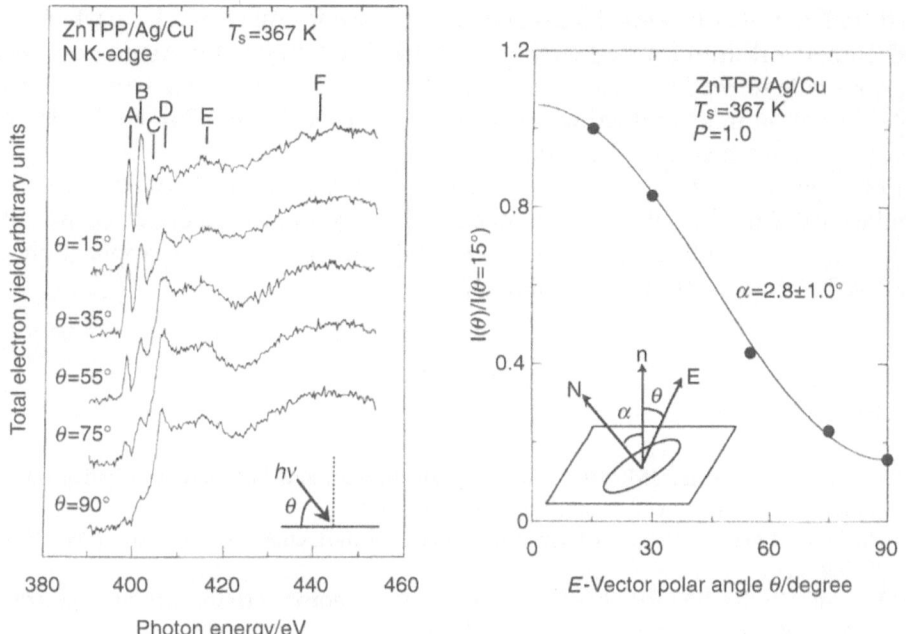

FIG. 4.2-19. **a** Polarized N K-edge absorption spectra of Zn tetraphenylporphyrin (ZnTPP) evaporated on an Ag substrate at 367 K [30]. **b** Intensity of peak A in **a** vs. the angle of incidence of light θ. The curved line indicates the simulated intensity variation for the angle α between the surface normal (n) and the normal of the central macrocycle of ZnTPP (N)

This example also shows the usefulness of the element specificity of NEXAFS. The four peripheral phenyl rings are known to be almost vertical to the central macrocycle, and it is difficult to use C K-edge NEXAFS spectra to study the molecular orientation. By measuring the N K-edge NEXAFS, we can selectively study the orientation of the central macrocycle, which is most important in determining the electronic properties of ZnTPP. These examples show the power of NEXAFS spectroscopy for studying molecular orientation in ultrathin films.

NEXAFS measurements can also be used to deduce the adsorption site by detecting the large change in the absorption of atoms involved in a strong inter-action with the substrate. Examples of this are the N atom in pyridine on Pt [31] and the S atom of a merocyanine dye on AgCl [32]. A similar examination can be performed by X-ray photoelectron spectroscopy (XPS), but NEXAFS is a gen-tler method than XPS in the sense that radiation damage by X-rays is not usually a problem.

Another type of information obtained from NEXAFS is the electronic struc-ture of unoccupied states. Owing to the simple character of the initial state,

NEXAFS spectra are expected, in the first approximation, to reflect the unoccupied state density of states. This is actually the case for fullerenes [33], where the carbon network forms a well-connected delocalized π-electron system to screen the core hole formed by excitation. A drastic change in electronic structure, e.g., hole formation in a conducting polymer by acceptor doping [34], could also be successfully detected by NEXAFS.

For organic compounds in general, however, the spectra obtained deviate significantly from the density of unoccupied states due to the effect of the core hole. This is an interesting subject [27, 33], but it makes the application of NEXAFS for such studies very difficult.

References

1. Philpott MR, Turlet JM (1976) Surface, subsurface, and bulk exciton transitions of crystalline anthracene. J Chem Phys 64:3852–3869
2. Rashba EI (1966) Theory of vibronic spectra of molecular crystals. Sov Phys JETP 23:708–718
3. Broude VL, Rashba EI, Sheka EF (1967) A new approach to the vibronic spectra of molecular crystals. Phys Status Solidi 19:395–406
4. Tokura Y, Koda T, Iyechika Y, Kuroda H (1983) Electro-reflectance spectra of charge-transfer excitations in copper phthalocyanine single crystals. Chem Phys Lett 102:174–177
5. Brillante A, Philpott MR (1980) Reflection and absorption spectra of singlet charge transfer excitons in anthracene–PMDA crystals. J Chem Phys 72:4019–4030
6. Haarer D, Philpot MR, Morawitz H (1975) Field-induced charge-transfer exciton transitions. J Chem Phys 63:5238–5245
7. Tokura Y, Koda T (1981) Experimental determination of the charge-transfer exciton bandwidth in anthracene–PMDA crystals. Solid State Commun 40:299–302
8. Kuwata-Gonogami M, Peyghambarian N, Meissner K, Fluegel B, Sato Y, Ema K, Shimano R, Mazumdar S, Guo F, Tokihiro T, Ezaki H, Hanamura E (1994) Exciton strings in an organic charge-transfer crystal. Nature 367:47–48
9. Ueta M, Kanzaki H, Kobayashi K, Toyozawa Y, Hanamura E (1984) Excitonic process in solids. Springer, Berlin Heidelberg New York Tokyo
10. Torrnace JB, Bazquez JE, Meyerle JJ, Lee VY (1981) Discovery of a neutral-to-ionic phase transition in organic materials. Phys Rev Lett 46:253–256
11. Torrance JB, Girlando A, Meyerle JJ, Crowley JI, Lee VY, Batail P (1981) Anomalous nature of neutral-to-ionic phase transition in tetrathiafulvalene-chloranil. Phys Rev Lett 47:1747–1750
12. Tokura Y, Koda T, Mitani T, Saito G (1982) Neutral-to-ionic transition in tetrathiafulvalene-p-chloranil as investigated by optical reflection spectra. Solid State Commun 43:757–760
13. Tokura Y, Okamoto H, Koda T, Mitani T, Saito G (1986) Pressure-induced neutral-to-ionic phase transition in TTF-p-chloranil studied by infrared vibrational spectroscopy. Solid State Commun 57:607–610
14. Toyozawa Y (1992) Condensation of relaxed excitons in static and dynamic phase transitions. Solid State Commun 84:255–257

15. Koshihara S, Tokura Y, Mitani T, Saito G, Koda T (1990) Photo-induced valence instability in organic molecular compound tetrathiafulvalene (TTF)-chloranil. Phys Rev B 42:6853–6856

16. Koshihara S, Tokura Y, Takeda K, Koda T (1992) Reversibly photo-induced phase-transition in alkyl-urethane substituted polydiacetylene. Phys Rev Lett 68:1148–1151

17. Cardona M, Ley L (eds) (1978) Photoemission in solids vol 1 and 2. Springer, Berlin Heidelberg New York

18. Roy D, Carette JD (1977) Design of electron spectrometers for surface analysis. In: Ibach H (ed) Electron spectroscopy for surface analysis. Springer, Berlin Heidelberg New York

19. Seki K (1989) Ionization energies of free molecules and molecular solids. Mol Cryst Liq Cryst 171:255–270

20. Sato N, Seki K, Inokuchi H (1977) Polarization energies of organic solids determined by ultraviolet photoelectron spectroscopy. J Chem Soc Faraday Trans II 77:1621–1633

21. Salaneck WR (1981) Intermolecular relaxation effects in the ultraviolet photoelectron spectroscopy of molecular solids. In: Dwight DW, Fabish TJ, Thomas HR (eds) Photon, electron, and ion probes of polymer structure and properties. ACS, Washington, DC, pp 121–149

22. Wright JD (1987) Molecular crystals. Cambridge University Press, Cambridge

23. Seki K (1989) Photoelectron spectroscopy of polymers. In: Bässler H (ed) Optical techniques to characterize polymer systems. Elsevier, Amsterdam, pp 115–180

24. Hasegawa S, Mori T, Imaeda K, Tanaka S, Yamashita Y, Inokuchi H, Fujimoto H, Seki K, Ueno N (1994) Intermolecular energy-band dispersion in oriented thin films of bis(1,2,5-thiaziazolo-p-quinobis(1,3-dithiole) by angle-resolved photoemission. J Chem Phys 100:6969–6973

25. Seki K, Yanagi H, Kobayashi Y, Ohta T, Tani T (1994) UV photoemission study of dye/AgBr interfaces in relation to spectral sensitization. Phys Rev B 49:2760–2767

26. Narioka S, Ishii H, Yoshimura D, Sei M, Ouchi Y, Seki K, Hasegawa S, Miyazaki T, Harima Y, Yamashita K (1995) The electronic structure and energy level alignment of porphyrin/metal interfaces studied by UV photoelectron spectroscopy. Appl Phys Lett 67:1899–1901

27. Stöhr J (1992) NEXAFS spectroscopy. Springer, Berlin Heidelberg New York

28. Hitchcock A (1990) Core excitation and ionization of molecules. Phys Scr T31:159–170

29. Nagayama K, Mitsumoto R, Araki T, Ouchi Y, Seki K (1995) Polarized XANES studies on the mechanical rubbing effect of fluorinated polyethylene and its model compounds. Phys B 208/209:407–408

29a. Phys B 208/209:419–420 (1995)

30. Narioka S, Ishii H, Ouchi Y, Yokoyama T, Ohta T, Seki K (1995) The electronic structure and energy level alignment of porphyrin–metal interfaces studied by UV photoelectron spectroscopy. J Phys Chem 99:1332–1337

31. Horseley JA, Stöhr J, Hitchcock AP, Newbury DC, Johnson AL, Sette F (1985) Resonances in the K-shell excitation spectra of benzene and pyridine: gas phase, solid, and chemisorbed states. J Chem Phys 83:6099–6107

32. Araki T, Seki K, Narioka S, Takata Y, Yokoyama T, Ohta T, Watanabe S, Tani T (1992) XANES spectroscopic studies of merocyanine dyes and their adsorbed states on AgCl. Jpn J Appl Phys 32:434–436

33. Seki K, Mitsumoto R, Araki T, Ito E, Ouchi Y, Kikuchi K, Achiba Y (1994) X-ray absorption near-edge structure (XANES) spectroscopy of fullerenes: inner-shell excitonic effects in fullerenes and the XANES spectrum of a higher fullerene C76. Syn Metals 64:353–357
34. Tourillon G, Fontaine A, Jugnet Y, Duc TM, Braun W, Feldhaus J, Holub-Krappe E (1987) Evolution upon doping of the π^* and σ^* bands of poly-(3-methylthiophene) grafted on Pt electrodes as studied by near-edge X-ray-absorption fine-structure spectroscopy. Phys Rev B 36:3483–3486

4.3
Thermal and Mechanical Properties

Toshiaki Enoki

The thermal and mechanical properties of molecular assemblies are very important. Since these assemblies have some degrees of freedom in their electronic and lattice systems, their thermal and mechanical properties give clues to the characteristics of their electronic structure and lattice dynamical behavior. In molecular assemblies, molecules form solid crystals in which intramolecular interactions between atoms are more than one order of magnitude stronger than the intermolecular interactions, whose strength ranges from about 0.1 to 1 eV. An assembly of atoms forms molecular orbitals with energies of eV, and then interactions between the molecular orbitals of adjacent molecules form the electronic structure of a molecular crystal. In terms of lattice dynamical behavior, the energies of intramolecular vibrations are approximately 0.1 eV, which are well separated from the energy range of the intermolecular vibrations with strengths around of 0.01 eV. These unusual features of the electronic structure and lattice dynamics give specific characteristics to molecular assemblies which are different from those of ordinary ionic crystals or metal crystals which do not have molecular units.

This chapter considers how to extract information on the electronic structure and lattice system of molecular assemblies from their fundamental thermal properties (specific heat, thermal expansion coefficient, and thermoelectric power) using graphite intercalation compounds as examples. Graphite intercalation compounds [1] are a type of host–guest molecular assembly system having two-dimensional metallic properties, where guest species are inserted between host graphitic planes.

4.3.1 Specific Heat of Molecular Solids

The specific heat of a molecular assembly system consists of contributions from both the electronic and the lattice systems. In cases where the system has unpaired electrons, the magnetic effect associated with electronic spins also contributes to the specific heat. This section considers only the electronic and lattice contributions.

For the lattice contribution, the specific structure of the molecular assembly affects its specific heat. In molecules, there are three types of degrees of freedom: vibration, rotation, and translation. Molecular vibrations which are governed by strong intramolecular interactions make a negligible contribution to the specific heat below room temperature since the excitation energies of these vibrations are well above the thermal energy at room temperature. The rotational degree of freedom occasionally contributes to the specific heat since the rotational energy is well below the thermal energy at room temperature. The translational degree of freedom for the center of mass of a molecule is related to the generation of lattice vibrations which propagate in a crystal. These are called phonons. Since there are three possible independent directions, x, y, z, there are two transverse phonons and one longitudinal phonon for each phonon branch. In transverse phonons, the displacements of molecules by vibration are perpendicular to the direction of the wave vector, while the displacements are parallel to the wave vector in the longitudinal phonons. For crystals having one molecule in a unit cell, the phonon is characterized as an acoustic phonon whose frequency is proportional to the wave vector q:

$$\omega_q = v_s q \tag{4.3-1}$$

where v_s is the velocity of sound. For acoustic phonons, the specific heat contribution is given as the Debye specific heat C_D:

$$C_D(T) = 9R \left[\frac{4T^3}{\Theta_D} \int_0^{\Theta_D/T} \frac{x^3 dx}{\exp x - 1} - \frac{\Theta_D/T}{\exp(\Theta_D/T) - 1} \right] \tag{4.3-2}$$

where R is the gas constant and Θ_D denotes the Debye temperature ($\Theta_D = \hbar \omega_D / k_B$), which represents the upper limit of the vibrational frequencies. The specific heat is proportional to the cube of the temperature at low temperatures, as expressed in the following equation:

$$C_D(T) = \frac{12\pi^4 R}{5\Theta_D^3} T^3 = \alpha T^3 \tag{4.3-3}$$

while the high temperature limit is $3R$, which corresponds to the classical limit of the specific heat (Dulong Petit law).

In the case of a crystal with more than two molecules in a unit cell, optic phonons, i.e., the molecules vibrating against each other, are generated and also contribute to the specific heat. An optic phonon has a high excitation energy. If the excitation energy, $\hbar \omega_0$, is assumed to be independent of the wave vector as a first approximation, the specific heat of optic phonons can be described in terms of the Einstein specific heat C_E, as in the following equation:

$$C_E(T) = \frac{3R(T_E/T)^2 \exp(T_E/T)}{\left[\exp(T_E/T) - 1 \right]^2} \tag{4.3-4}$$

where T_E is the Einstein temperature, i.e., the characteristic temperature representing the frequency of the vibration ω_0: $T_E = \hbar\omega_0/k_B$). As can be seen from Eq. (4.3-4), the Einstein specific heat decrease exponentially with temperature, as given by Eq. (4.3-5):

$$C_E(T) = 3R(T_E/T)^2 \exp(-T_E/T) \qquad (4.3-5)$$

which is extrapolated to $3R$ at the high-temperature limit.

We now consider the electronic contribution to specific heat. In a molecular assembly system with metallic properties, conduction electrons generated through intermolecular interactions make an electronic contribution to the specific heat. Conduction electrons occupy the conduction band up to the Fermi energy E_F according to Fermi–Dirac statistics. Therefore, the electronic contribution reflects the electronic structure of the conduction band, especially the detailed structure around E_F. Equation (4.3-6) gives the temperature-dependence of the electronic specific heat, C_e, for a metallic system, which is expressed using the electronic density of states $N(E_F)$ at E_F,

$$C_e(T) = \frac{1}{3}\pi^2 k_B^2(1+\lambda)N(E_F)T = \gamma T \qquad (4.3-6)$$

where k_B is the Boltzmann constant and λ is the electron–phonon coupling constant, which represents the strength of the interaction between a conduction electron and the lattice vibration (phonon). γ is the electronic specific heat coefficient. The electronic specific heat, which is proportional to the temperature, is small in comparison with the lattice specific heat contribution, since only the conduction electrons existing around E_F participate in the specific heat. Therefore, the electronic specific heat can be observed at low temperatures when the lattice specific heat is considerably reduced.

4.3.2 Specific Heat of Graphite Intercalation Compounds

This section covers the specific heat of alkali-metal-hydride–graphite intercalation compounds, and shows how to extract information about electronic structure and lattice dynamical behavior on the basis of the fundamental knowledge given above. Alkali-metal–graphite intercalation compounds with composition C_8M (M = K, Rb), where alkali-metal atoms play the role of guests (intercalates) in graphitic galleries, adsorb hydrogen chemisorptively at room temperature, resulting in the formation of alkali-metal-hydride–graphite intercalation compounds C_8MH_x. Therefore these compounds are an interesting type of metal–hydrogen system which can be compared with transition metal hydrides such as PdH_x [2]. In the adsorption of hydrogen in C_8K, adsorbed hydrogen species are stabilized at interstitial sites in the alkali-metal layers at low hydrogen concentrations ($x < 0.1$). The additional introduction of hydrogen generates a structural change to a stage-2 structure when an intercalate layer is inserted in

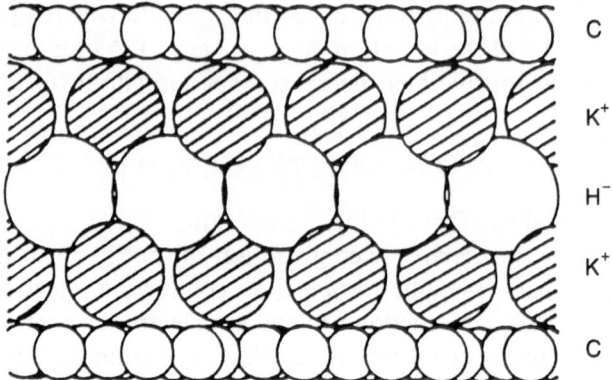

FIG. 4.3-1. Cross-sectional view of a potassium-hydride–graphite intercalation compound parallel to the *c*-axis

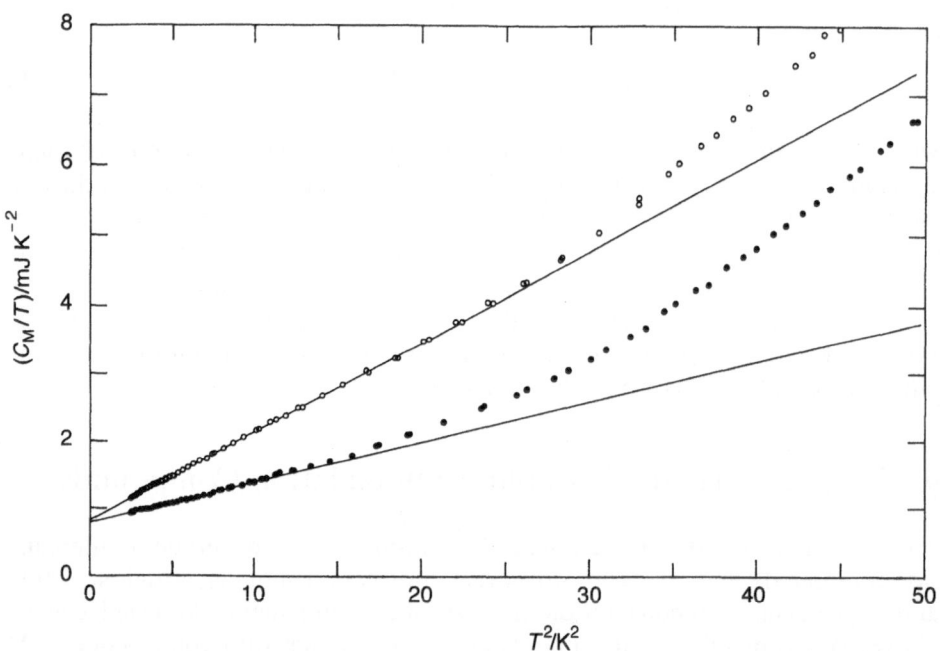

FIG. 4.3-2. Low temperature specific heat (C/T vs. T^2 plots) for C_8Rb and $C_8RbH_{0.055}$. *Open circles*, C_8Rb; *solid circles*, $C_8RbH_{0.055}$. The lines were obtained by least-square fits

every two graphitic galleries. The stage-2 structure is completed around a saturated concentration of $x = 0.6$–0.8. In this structure, at saturated hydrogen concentration, the guest layer comprises ionic triple atomic layers of alkali-metal/hydrogen/alkali-metal in graphitic galleries, as shown in Fig. 4.3-1. The nature of the hydrogen in graphitic galleries varies from ionic to metallic depending on the alkali-metal species.

The temperature-dependence of the heat capacity is shown in Fig. 4.3-2 for alkali-metal-hydride–graphite intercalation compounds in the low temperature region [3]. At low temperatures, this can be expressed in the following equation:

$$C(T) = \gamma T + \alpha T^3 + \Delta C(T) \qquad (4.3\text{-}7)$$

As discussed above, the first term, which has linear temperature-dependence, is associated with the electronic contribution and is related to the electronic density of states at E_F. The second term, which has cubic temperature-dependence, originates for acoustic phonons. The last term, which indicates a deviation from linear behavior in the C/T vs. T^2 plot, comes from the optic phonon mode.

First, we consider lattice vibrational properties. In graphite intercalation compounds, which have intercalate sheets having heavy alkali-metal atoms in graphitic galleries, two-dimensional vibrations of the intercalates with respect to the graphitic sheets behave as low-energy optic phonons [3]. Therefore, the observed optic phonon is considered to be associated with the two-dimensional vibrational degrees of freedom for the intercalates. The Debye contribution is mainly caused by the acoustic phonons propagating in the c-axis direction, which is associated with the force constants of the intercalate–graphitic plane interaction and the intergraphitic plane interaction, since these forces are very small in comparison with the intraplane vibrations of the stiff graphitic skeleton.

Figure 4.3-3 shows the hydrogen concentration-dependence of the Debye temperatures, Θ_D, in acoustic phonon modes and the Einstein characteristic temperature, T_E, in optic phonon modes. In the case of C_8HK_x, Θ_D increases as a function of x in the concentration range $x < 0.1$ where the introduction of hydrogen does not cause a structural change. At the saturated hydrogen concentration, $x = 0.65$, where there is a structural change to stage-2, Θ_D decreases by 11%. T_E increases by 10% as a function of x in the low concentration range, but decreases by 29% at the saturated concentration. For C_8RbH_x, Θ_D increases by 30% and T_E decreased by 7% for $x = 0.055$, which is the maximum hydrogen concentration realized in ambient hydrogen pressure.

In metal–hydrogen systems, the introduction of hydrogen at interstitial sites causes a lattice expansion, which contributes to a reduction in lattice energy. Conversely, the introduction of hydrogen generates a chemical bond between the hydrogen and the host metal, which enhances the lattice energy. In actual cases, the competition between these two effects determines the change in vibrational energy. For alkali-metal-hydride–graphite intercalation compounds at small hydrogen concentration, hydrogen species are stabilized at interstitial sites in the two-dimensional alkali-metal lattice. These interstitial sites are relatively large in comparison with the size of the alkali-metal ions, so the lattice expansion effect is relatively small. Moreover, the chemical bonding between alkali-metal ions and hydrogen ions is also small. The changes in the Debye and Einstein temperatures are governed by a subtle balance between the two effects. At the saturated hydrogen concentration, the introduction of hydrogen induces structural change to form a stage-2 structure where triple atomic layers of alkali-metal/hydrogen/ alkali-metal are inserted between graphitic layers. Because in the stage-2 struc-

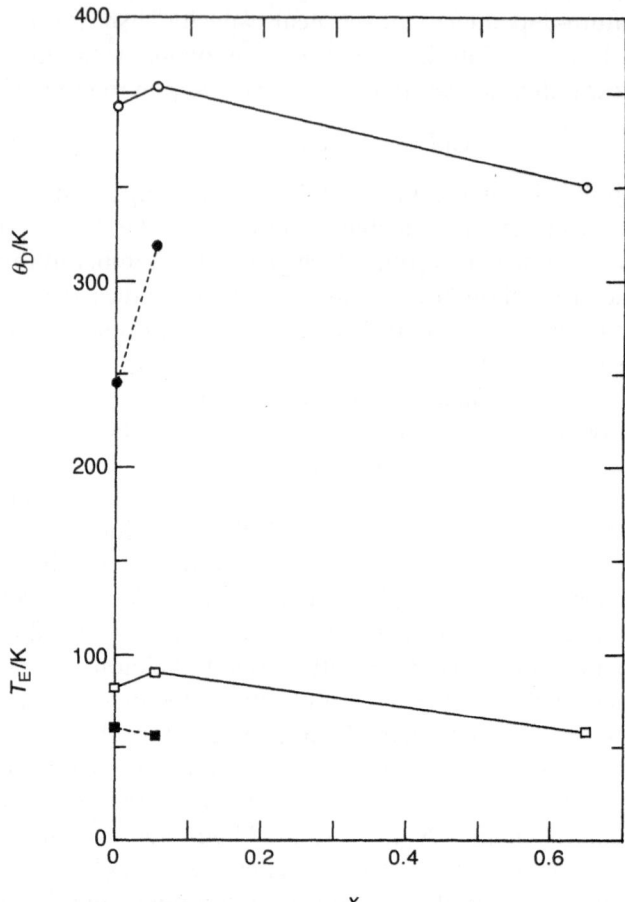

FIG. 4.3-3. The hydrogen concentration-dependence of Θ_D and T_E for C_8KH_x and C_8RbH_x. *Open* and *solid circles* denote Θ_D for C_8KH_x and C_8RbH_x, respectively; *open* and *solid squares* denote T_E for C_8KH_x and C_8RbH_x, respectively

ture the triple atomic layer has a larger mass than a pure single alkali-metal layer, the reductions in the Debye and Einstein temperatures in a potassium-hydride–graphite intercalation compound at the saturated hydrogen concentration can be explained in terms of the increase in mass of the intercalates.

Next, we deal with the electronic specific heat, and consider the change in electronic structure caused by the introduction of hydrogen. From an analysis of the electronic specific heat, the hydrogen concentration-dependence of γ or the density of states $N(F_E)$ can be extracted, as shown in Fig. 4.3-4, for the two kinds of alkali-metal-hydride–graphite intercalation compounds C_8KH_x and C_8RbH_x. From this figure we can see that the density of states $N(E_F)$ decreases linearly as the hydrogen concentration x increases for both C_8KH_x and C_8RbH_x.

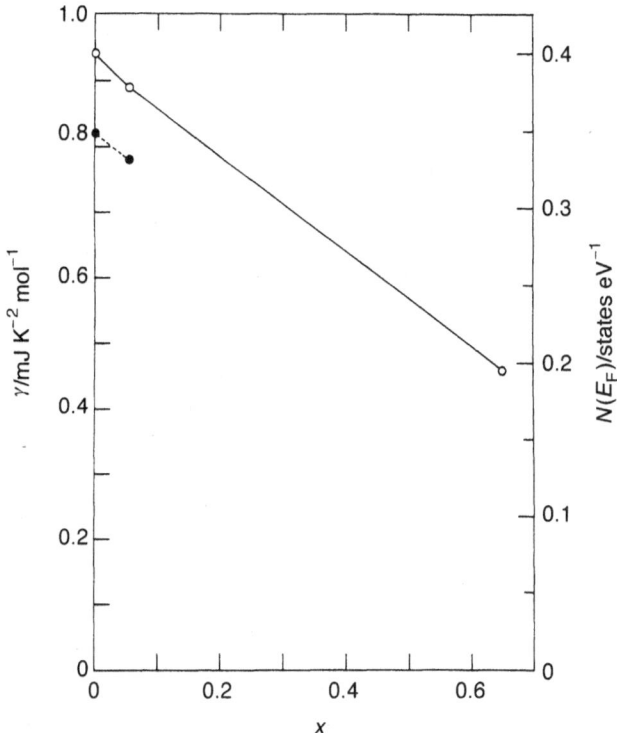

FIG. 4.3-4. The hydrogen concentration dependence of γ and $N(E_F)$ for C_8KH_x and C_8RbH_x. *Open circles*, C_8HK_x; *solid circles*, C_8RbH_x. $N(E_F)$ is not corrected for λ

Since hydrogen has a strong electron affinity in metallic media, it is believed that the introduction of hydrogen induces charge transfer from the host alkali-metal–graphite intercalation compounds to the adsorbed hydrogen species, leading to the formation of H^- ions in the intercalate space [2]. Therefore, the change in $N(E_F)$ can be understood in terms of the charge transfer. We now discuss the effect of the chemisorbed hydrogen on the electronic structure on the basis of the rigid-band electronic structure model. According to this model, two bands coexist: two-dimensional graphitic π-bands and bands originating from alkali-metal s-electrons, as shown in Fig. 4.3-5a. The Fermi energy, E_F, and the location of the bands are determined by a balance between electronic and lattice energies. The ratio of occupation between π- and alkali-metal s-bands gives the charge transfer rate from the alkali-metal to graphite. In the case of host C_8K, the charge transfer rate is estimated to be $f = 0.6$ ($C_8^{-f/8}K^{+f}$) [1]. The introduction of hydrogen generates an additional band associated with hydrogen $1s$-electrons, which is expected to be located below E_F, as shown in Fig. 4.3-5b, since hydrogen has a strong electron affinity. From the band structure model, the decrease in $N(E_F)$ is considered to be caused by a transfer of donor electrons from the graphitic π^*- and alkali-metal s-bands around E_F to the hydrogen $1s$-band.

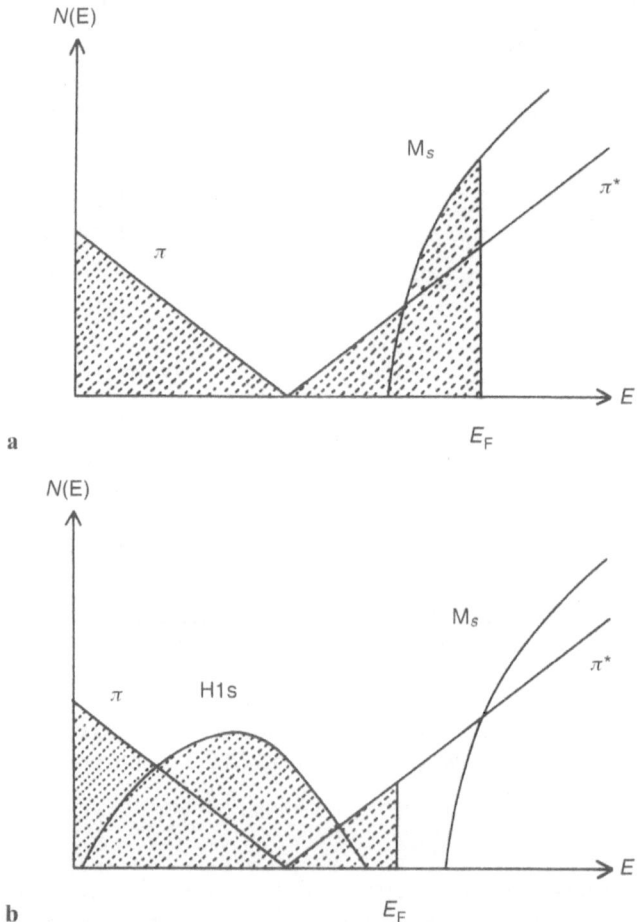

Fig. 4.3-5. The electronic structure model of **a** alkali-metal–graphite intercalation compounds, and **b** alkali-metal-hydride–graphite intercalation compounds. M_s denotes the alkali-metal s-band; π and π^* denote the graphitic bands; $H1s$ is the hydrogen $1s$-band.

We now consider the change in $N(E_F)$ in relation to the charge transfer to hydrogen on the basis of the rigid-band model in a potassium-hydride–graphite intercalation compound. Taking into account that hydrogen atoms are stabilized as hydride ions H^-, we can use the formula. $C^{-f(1-x)/8}{}_8K^{+f}{}_{(1-x)}(K^+H^-)_x$ for C_8KH_x, where f is the charge transfer rate per K atom. After detailed calculations involving the potassium 4s- and graphitic π-bands that give conduction electrons, the total density of states is expressed by the following equation:

$$N\left(E_F\right) = uf^{1/2}\left(1-x\right)^{1/2} + v\left(1-f\right)^{1/3}\left(1-x\right)^{1/3} \qquad (4.3\text{-}8)$$

where the first and second terms represent the partial densities of states at E_F for the graphitic π- and potassium 4s-bands, respectively. The parameters u and v

provide detailed information on the electronic structures of both energy bands and are estimated at 0.324 and 0.199, respectively. Using Eq.(4.3-8) for the change in $N(E_F)$, it is suggested that the charge transfer to hydrogen causes change in the valence of K from a partial positive of +0.6 to +1 when x is increased from 0 to the saturation value 0.65. Thus, from the rigid-band model, the linear decrease in $N(E_F)$ observed in the specific heat experiment can be understood in terms of charge transfer from host alkali-metal–graphite intercalation compound to adsorbed hydrogen with a strong electron affinity. The important information obtained from the experimental results is that the charge transfer to hydrogen enhances the ionicity of the intercalates, leading to the formation of a K^+H^- ionic lattice at the saturated concentration $C_8KH_{0.65}$. The slope of the $N(E_F)$ vs. x curve, $dN(E_F)/dx$, is estimated from Fig. 4.3-4 to be –0.31 states per eV.mol atom of H for both C_8KH_x and C_8RbH_x, so that the system becomes less metallic with increasing x, and $N(E_F)$ can be extrapolated to zero at $x \sim 1.3$. This suggests that the formation of ionic K^+–H^- intercalates in the intercalate space tends to reduce the charge transfer between graphite and the intercalate.

4.3.3 Thermal Expansion of Molecular Solids

Thermal expansion occurs due to the anharmonicity of the potential acting between molecules in solid crystals [4]. Under intermolecular interaction, the force between two molecules is attractive when the intermolecular distance is larger than its equilibrium value, and is repulsive when the intermolecular distance is smaller than the equilibrium value. In general, the shape of the potential curve of the attractive force is gentler than that of the repulsive force since repulsive is an extremely short-range force associated with electron repulsion. Therefore, the potential has an anharmonic nature. The equilibrium intermolecular distance for molecules vibrating in such an anharmonic potential tends to increase with an increase in temperature since the amplitude of the vibration is enhanced.

We now examine the thermal expansion of alkali-metal-hydride–graphite intercalation compounds. From the considerations given above, the linear thermal expansion coefficient α_c parallel to the c-axis can be described on the basis of the lattice degree of freedom using the following equation [4, 5]:

$$\alpha_c = \frac{1}{I_c}\frac{dI_c}{dT} = \frac{1}{C_cV_0}\sum_{q,p}\frac{dE_{q,p}}{dT}\Gamma_{q,p} \qquad (4.3-9)$$

where I_c and V_0 are the c-axis lattice constant and the unit cell volume, respectively. C_c is an effective elastic constant for the expansion of the crystal length, L_c, parallel to the c-axis, and is C_{33} by first-order approximation. The Grüneisen parameter $\Gamma_{q,p} = -(d\ln\omega_{q,p}/d\ln L_c)$ indicates the magnitude of the anharmonicity of the phonon mode $\omega_{q,p}$ with polarization p, and is probably largest for modes involving an interatomic restoring force for c-axis compression and expansion, so that we restrict the sum Σ to phonon branches having c-axis polarization. The

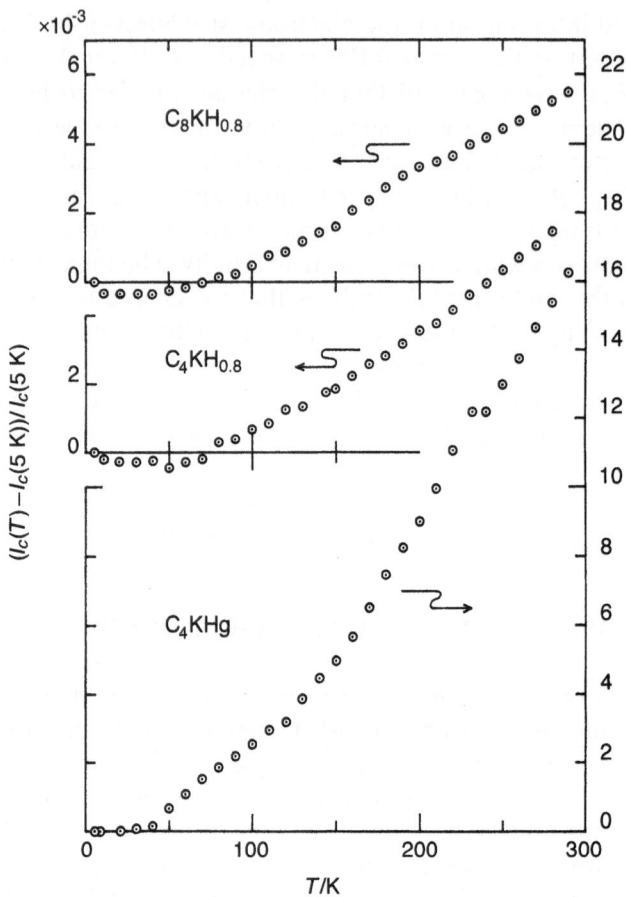

Fig. 4.3-6. Temperature-dependence of the c-axis lattice constants I_c for $C_8KH_{0.8}$, $C_4KH_{0.8}$, and C_4KHg

term $dE_{q,p}/dT$, which depends on temperature, gives the specific heat contribution associated with the $\omega_{q,p}$ phonon mode, where $E_{q,p}$ is the expected value of the energy of the $\omega_{q,p}$ phonon mode.

The temperature-dependence of the c-axis lattice constant I_c is shown in Fig. 4.3-6 for $C_8KH_{0.8}$, $C_4KH_{0.8}$, and C_4KHg, where C_4KHg is isostructural to $C_4KH_{0.8}$. Table 4.3-1 gives the linear thermal expansion coefficients α_c obtained from Fig. 4.3-6, as well as those of the host alkali-metal–graphite intercalation compound C_8K and graphite (highly oriented pyrolytic graphite; HOPG), and the thermal expansion coefficient parallel to the a-axis for pure KH. C_4KHg has the largest α_c among the materials investigated here, while the small thermal expansion coefficients of $C_8KH_{0.8}$ and $C_4KH_{0.8}$ are comparable to the thermal expansion coefficient of pure KH, which plays the role of guest to the graphite.

As mentioned in Sect. 4.3.2, the low-temperature heat capacity of C_8K and $C_8KH_{0.8}$ suggests the presence of acoustic and low-energy optic phonon modes.

TABLE 4.3-1. Elastic constant C_{33}, Debye temperature Θ_D, and c-axis thermal expansion coefficient α_c (at 250 K)

	$C_{33}(\times10^{11})$ (dyn cm^{-2})	Θ_D (K)	$\alpha_c(\times10^6)$ (K^{-1})
HOPG	3.65 ± 0.1	427	2.63
KH			2.97
C_8K	4.85 ± 0.14	393.5	4.67 ± 0.04
C_4KHg	2.3 ± 0.1	269	7.31 ± 0.55
C_4KH_x	3.4		3.73 ± 0.11
C_8KH_x		350.3	2.36 ± 0.18

According to theoretical studies of the lattice dynamics of graphite intercalation compounds, low-energy optic phonon modes are considered to be associated with two-dimensional vibrations of the intercalate with polarization parallel to the c-plane [3]. This means that the optic phonon is less important for linear thermal expansion parallel to the c-axis. Figure 4.3-7 shows the thermal expansion coefficient α_c of $C_8KH_{0.8}$ derived from the results in Fig. 4.3-6. The temperature-dependence of α_c is in good agreement with the calculated depen-

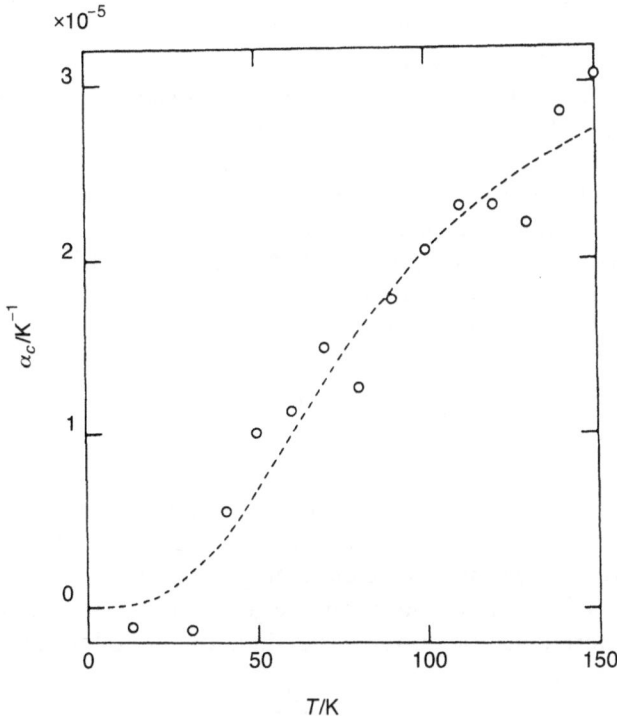

FIG. 4.3-7. The c-axis linear thermal expansion coefficient α_c for $C_8KH_{0.8}$. The line denotes the expected behavior of α_c for the longitudinal acoustic phonon (Debye temperature $\Theta_D = 350$ K), where α_c at 100 K is fitted to the experimental results

dence based on the acoustic phonon mode which has Debye temperature of Θ_D = 350 K from heat capacity measurements (see Sect. 4.3.1). Therefore, we can say that the longitudinal acoustic phonon mode with polarization parallel to the c-axis dominates the thermal expansion coefficient α_c.

Table 4.3-1 summarizes values for the elastic constant C_{33}, the Debye temperature Θ_D, and the linear thermal coefficient α_c for the binary and ternary graphite intercalation compounds, KH and graphite (HOPG). The elastic constant C_{33}, which represents the magnitude of the lattice stiffness, is obtained from the c-axis compressibility:

$$\kappa_c = -\frac{1}{I_c}\frac{dI_c}{dP} = \frac{1}{C_{33}} \tag{4.3-10}$$

From the microscopic point of view, C_{33} is related to the velocity of sound for the longitudinal acoustic phonons propagating in the direction of the c-axis, as given below:

$$v_{sz}^2 = \frac{C_{33}}{d} \tag{4.3-11}$$

where d is the crystal density. Figure 4.3-8 shows the pressure dependence of the c-axis lattice constant at room temperature for HOPG and C_8K [6], which gives the c-axis compressibility according to Eq. (4.3-10). Judging from the magnitude of C_{33}, the lattice stiffness parallel to the c-axis increases in the order C_4KHg < $C_4KH_{0.8}$ < HOPG < C_8K, while α_c decreases in the order HOPG < KH < $C_4KH_{0.8}$ < C_8K < C_4KHg. From Eq. (4.3-8) the Grüneisen parameter Γ, which indicates the magnitude of the anharmonicity of the vibration, is equal to the product $\alpha_c V_0 C_{33}/C_V$. Taking into account the Debye temperatures, the specific heat C/V_0 associated with the longitudinal acoustic phonon varies by less than 10% at about room temperature among the compounds shown in Table 4.3-1. Therefore, the Grüneisen parameter Γ for anharmonicity is found to increase in the order HOPG < $C_4KH_{0.8}$ < C_4KHg < C_8K.

4.3.4 Thermoelectric Power

When one end of a long crystal is heated, an electromotive force is generated between the two ends of the crystal. In a crystal which has conduction electrons, the temperature gradient makes the electron gas particles move toward the cold end by a process of diffusion, resulting in the generation of an electromotive force [7]. Thus, the thermoelectric power S is defined by the following equation:

$$S = \frac{\Delta V}{\Delta T} \tag{4.3-12}$$

where ΔV and ΔT are the differences in the electromotive force and the temperature, respectively, between the two ends. The value of the thermoelectric power has a negative sign in cases where the conduction is governed by electron carriers

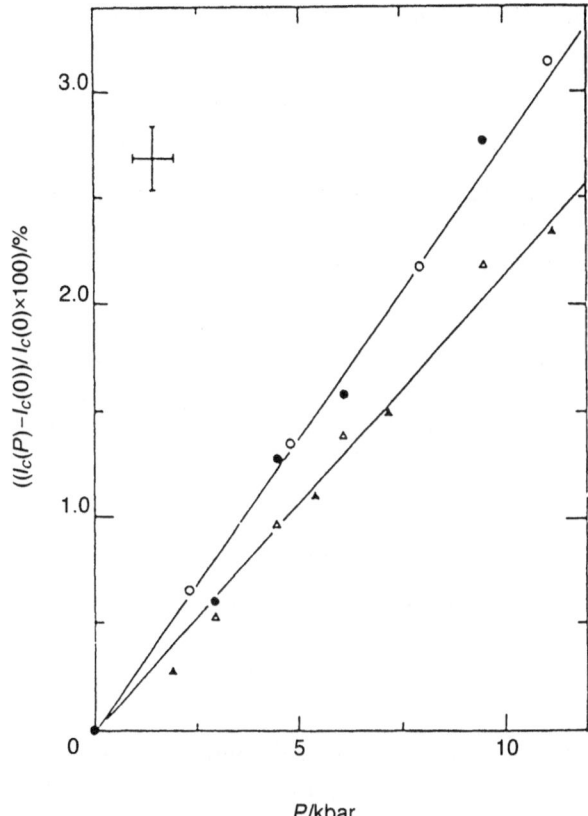

FIG. 4.3-8. Percentage of lattice constant contraction *vs.* hydrostatic pressure. *Open circles*, HOPG; *solid circles*, pure single crystal graphite; *open triangles*, C_8K prepared from a single crystal; *solid triangles*, C_8K prepared from HOPG. The estimated error bars are also shown. (From [6], with permission)

with negative charges, while hole carriers with positive charges give positive thermoelectric power. Microscopic examination shows that electron diffusion is affected by the presence of positive ions in the crystal. Since an electron with its negative charge can deform a lattice consisting of positively charged ions, the motion of the electron generates phonons by an electron–phonon interaction, resulting in participation of the phonons in thermoelectric power, i.e., the phonon-drag effect. Therefore, thermoelectric power can be explained in terms of the electron diffusion process (S_d) and the phonon-drag process (S_p) [7]. The electron diffusion contribution gives a linear temperature-dependence for thermoelectric power in a metallic material, which is expressed by the following equation:

$$S_d = \frac{\pi^2 k_B^2}{2e} T \left(\frac{\partial \ln \sigma}{\partial E} \right)_{E_F}$$

(4.3-13)

where σ is conductivity. The conductivity is expressed by the following equation:

$$\sigma = Ne\mu = \frac{Ne^2\tau}{m^*} \tag{4.3-14}$$

where N, μ, τ, and m^* are the conduction electron density, mobility, relaxation time, and effective mass of the conduction carriers, respectively. The relaxation time, τ, has the following energy dependence:

$$\tau(E) = \tau_0 E^P \tag{4.3-15}$$

where the value of index P depends on the carrier scattering process (impurity scattering, phonon scattering, etc.) and the dimensionality of the electronic structure. Therefore, using Eqs. (4.3-13)–(4.3-15), the electron diffusion term can be re-expressed in the following equation:

$$S_d = \frac{\pi^2 k_B^2}{3e} \frac{1+P}{E_F} T \tag{4.3-16}$$

The phonon drag contribution [8] is given by

$$S_p = \frac{\langle C_D r \rangle}{2eN} \tag{4.3-17}$$

where C_D is the specific heat of the acoustic phonon system and $\langle \rangle$ indicates the statistical average according to the phonon distribution. r is the momentum transfer ratio through the electron–phonon interaction from the phonon system to the carriers, which is defined by the ratio of the relaxation rate in the electron–phonon scattering process $1/t_c$ to the total phonon relaxation rate $1/t$. In the phonon relaxation process, the momentum of the phonon is transferred to the environment by the following scattering processes: electron–phonon scattering, domain boundary scattering, Rayleigh scattering associated with point defects and vacancies, and phonon–phonon scattering. Consequently, the momentum transfer ratio r which depends on the phonon momentum q is given by

$$r(q) = \frac{t}{t_c} = \frac{aq}{B + aq + fq^3 + AqT^3} \tag{4.3-18}$$

where B, aq, fq^3, and AqT^3 are the relaxation rates for domain boundary scattering, electron–phonon scattering, Rayleigh scattering, and phonon–phonon scattering, respectively. The crystal domain boundary scattering is related to the domain size L by $B = v_s/L$, where v_s is the sound velocity of the acoustic phonons. Thus, the phonon-drag term is derived by taking into account the specific heat of acoustic phonons given in Eq. (4.3-2).

We now consider the thermoelectric power of a sodium-hydride–graphite intercalation compound as an example of molecular assemblies. The temperature-dependence of thermoelectric power parallel to the c-plane, S, is shown in Fig. 4.3-9 for the stage-3 compound $C_{12}NaH$, where the triple atomic layer unit Na^+–

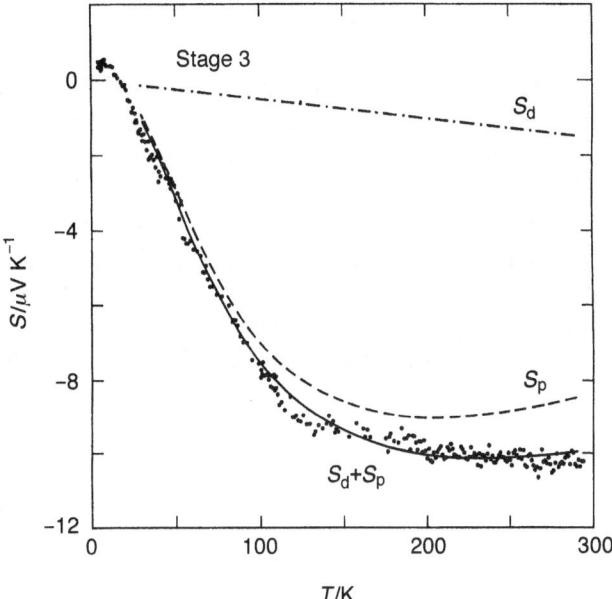

FIG. 4.3-9. Temperature-dependence of the thermoelectric power, S, for a stage-3 NaH–graphite intercalation compound. S_d, S_p, and $S_d + S_p$ are the calculated electron diffusion term (*chained line*), the phonon-drag term (*dashed line*), and the total thermoelectric power (*solid line*), respectively, obtained by theoretical fits

$H^- – Na^+$ is inserted in every third graphitic gallery [9]. The thermoelectric power has negative values over the whole temperature range. This suggests that the majority of carriers originate from the graphitic π-electrons with an electronic nature and are generated by charge transfer from NaH ionic intercalate to graphitic conduction π-bands. An investigation of the electronic structure shows that there is an absence of carriers on the NaH intercalate layers. The thermoelectric power increases slowly with a decrease in temperature below room temperature. Below about 100 K the thermoelectric power increases steeply and approaches zero. The thermoelectric power observed can be shown to be the sum of the electron diffusion and phonon-drag contributions ($S = S_d + S_p$) of the graphitic conduction π-electrons, as was explained above. In a stage-3 sodium-hydride–graphite intercalation compound, the conduction graphitic π-band is split into three sub-bands. Therefore, it is necessary to calculate the electron diffusion and phonon-drag contributions for each sub-band, and then the total thermoelectric power is given by

$$S = \frac{\sum_{i=1}^{3} \sigma_i S_i}{\sum_{i=1}^{3} \sigma_i}$$ (4.3-19)

where i is the index of the sub-band. The best fit between experimental and calculated results is shown in Fig. 4.3-9, where the sound velocity of the in-plane acoustic phonon of graphite, $v_s = 2.1 \times 10^6 \, \text{cm s}^{-1}$ was used. The calculation gives an estimate of the Fermi energy ($E_F = 0.45 \, \text{eV}$) and the domain size ($10\,000\,\text{Å}$). From the information obtained above, the charge transfer rate from the ionic intercalate NaH to graphite is estimated to be $f_c = 0.004$, suggesting that the NaH intercalate does not transfer charge to the graphitic π-system to the same extent as occurs for the intercalates in ordinary graphite intercalation compounds, which indicates the strongly ionic nature of the triple atomic layer intercalate Na$^+$–H$^-$–Na$^+$ in graphitic galleries. The domain size, estimated at $L = 10\,000\,\text{Å}$, is consistent with the typical size of domains for intercalate in graphite intercalation compounds.

References

1. Dresselhaus MS, Dresselhaus G (1981) Intercalation compounds of graphite. Adv Phys 30:139–326
2. Enoki T, Miyajima S, Sano M, Inokuchi H (1990) Hydrogen–alkali-meatal–graphite intercalation compounds. J Mater Res 5:435–466
3. Enoki T, Sano M, Inokuchi H (1985) Low-temperature specific heat of hydrogen-chemisorbed graphite–alkali-metal intercalation compounds. Phys Rev B 32:2497–2505
4. Grimvall G (1986) Thermophysical properties of materials. Selected topics in solid state physics, vol XVIII. North Holland, Amsterdam
5. Enoki T, Jeszka JK, Inokuchi H (1990) C-axis thermal expansion of donor graphite intercalation compounds. Synth Met 38:235–241
6. Wada N, Clarke R, Solin S (1980) X-ray compressibility measurements of the graphite intercalates KC$_8$ and KC$_{24}$. Solid State Commun 35:675–679
7. Ziman JM (1963) Principles of the theory of solids. Cambridge University Press, London
8. Sugihara K (1983) Thermoelectric power of graphite intercalation compounds. Phys Rev B 28:2157–2165
9. Enoki T, Sakamoto N, Nakazawa K, Suzuki K, Sugihara K, Kobayashi K (1993) C-axis electrical conductivity and thermoelectric power of sodium-hydride graphite intercalation compounds. Phys Rev B 47:10662–10670

Part V
Environmental and External Effects

5.1
Solvent Effects

Tadashi Okada and Hiroshi Miyasaka

Peaks in absorption and fluorescence spectra of molecules in solution are generally broader than those in the gas phase because the energy of solute molecules in a solvent is fluctuating due to the orientational fluctuations of the surrounding solvent molecules. The fluctuating motion of the solvent molecules plays an important role in chemical reactions in the solution phase. For example, in an electron transfer reaction the energy fluctuation of the reactants caused by the fluctuating solute–solvent interaction is believed to be essential to reach the transition state of that reaction. One of the most fundamental and challenging problems regarding chemical reactivity in solution is to explain how the microscopic structure and dynamics of the solute–solvent interaction can assist or impede a chemical reaction. The dynamic aspects of solvation, however, are very complex, since the orientational and translational motions of the surrounding solvent molecules occur on the same time-scale as the reactant dynamics. The most important and characteristic features of the solution phase are the types of motional degrees of freedom which are possible for both solute and solvent molecules, and which include intra- and intermolecular rotational, vibrational, or librational motion, as well as translational diffusive motion under multidimensional interaction potentials.

The first part of this chapter considers a theoretical model of solvent effects on electronic spectra based on the theory developed by McRae and recent developments in the study of the microscopic dynamics of solvation, while the second part discusses the effects of solvent viscosity on chemical processes in solution. The energy of the solute molecule interacting with the surrounding solvent molecules is determined mainly by the dielectric properties of the solvent, while the speed of the chemical processes is controlled by the friction of motion, which is believed to be linearly related to the viscosity of the solvent.

5.1.1 Solvation and Solvent Cage

5.1.1.1 Solvent Effects on Electronic Spectra

The method of treatment developed by McRae [1] to study electronic spectra in the steady state in solution is also suitable for studying the origins of solute–

solvent as well as solvent–solvent interactions that affect the energy of the solute molecules. The method consists of the application of second-order perturbation theory to the calculation of the electronic state energies of a solution containing N identical solvent molecules and one solute molecule.

In the zero order of approximation, the molecules are considered not to interact with each other. The zero-order electronic state functions of the solution are then made up of products of state functions for the unperturbed component molecules. Neglecting the nonorthogonality of the latter,

$$\Phi^0_{a(p)b(q)j} = \phi_0^{V(1)} \cdots \phi_a^{V(p)} \cdots \phi_b^{V(q)} \cdots \phi_0^{V(N)} \phi_j^u \tag{5.1-1}$$

$$\Phi^0_{a(p)j} = \phi_0^{V(1)} \cdots \phi_a^{V(p)} \cdots \phi_0^{V(q)} \cdots \phi_0^{V(N)} \phi_j^u \tag{5.1-2}$$

and

$$\Phi^0_{a(p)} = \phi_0^{V(1)} \cdots \phi_a^{V(p)} \cdots \phi_0^{V(q)} \phi_0^{V(N)} \phi_0^u \tag{5.1-3}$$

where, for example $\Phi^0_{a(p)b(q)j}$ denotes the zero-order function representing the state of the solution in which the solute molecule is in its excited electronic state j ϕ_j^u, and solvent molecules p and q are in their excited states a and b $\Phi_a^{V(p)}$, $\Phi_b^{V(q)}$, respectively. The superscript notations u and $V(p)$ refer to the solute and solvent molecule p, respectively, and the subscript zero indicates the ground electronic state of a single molecule.

The zero-order electronic state energies are sums of the electronic state energies of the unperturbed component molecules. The energy corresponding to $\Phi^0_{a(p)j}$ is given by

$$W^0_{aj} = (N-1)w_0^V + w_a^V + w_j^u \tag{5.1-4}$$

where w denotes the electronic state energy of an unperturbed molecule. Then the energy difference between states i and j of the solute molecule will be expressed by

$$v_{ji}^u = (w_j^u - w_i^u)/hc = -v_{ij}^u \tag{5.1-5}$$

The interaction energy of the molecules in solution, using the classical dipole interaction under the point–dipole approximation, is given by the perturbation Hamiltonian,

$$\mathcal{H}' = -\sum_{p=1}^{N} \theta^{uV(p)} m^u m^{V(p)} - \frac{1}{2} \sum_{p=1}^{N} \sum_{q=1}^{N} \theta^{V(p)V(q)} m^{V(p)} m^{V(q)} \tag{5.1-6}$$

where m denotes the instantaneous magnitude of the dipole moment of a molecule and θ is a geometrical factor dependent on the mutual orientation and separation of two dipoles.

The energy of the state Φ_i, corresponding in the zero-order approximation to Φ_i^0, is governed principally by the matrix elements

$$H_{j,i} = W_i^0 \delta_{ji} + \left\langle \Phi_{a(p)i}^0 \middle| \mathcal{H}' \middle| \phi_j^0 \right\rangle \tag{5.1-7}$$

$$H_{a(p)j,i} = \left\langle \Phi_{a(p)i}^0 \middle| \mathcal{H}' \middle| \phi_j^0 \right\rangle \tag{5.1-8}$$

and

$$H_{a(p)b(p)i,i} = \left\langle \Phi_{a(p)b(p)i}^0 \middle| \mathcal{H}' \middle| \phi_j^0 \right\rangle \tag{5.1-9}$$

The matrix elements may by be written

$$H_{jj} = W_i^0 \delta_{ji} - \sum_{p=1}^{N} \theta_{ji,00}^{uV(p)} M_{ji}^u M_{00}^V - \frac{1}{2} \sum_{p=1}^{N} \sum_{q=1}^{N} \theta_{00,00}^{V(p)V(q)} \left(M_{00}^V \right)^2 \delta_{ji} \tag{5.1-10}$$

$$H_{a(p)j,i} = -\theta_{ji,a0}^{uV(p)} M_{ji}^u M_{a0}^V - \frac{1}{2} \sum_{q=1,\ne p}^{N} \theta_{a0,00}^{V(p)V(q)} M_{a0}^V M_{00}^V \tag{5.1-11}$$

and

$$H_{a(p)b(p)i,i} = -\frac{1}{2} \sum_{q=1,\ne p}^{N} \theta_{a0,b0}^{V(p)V(p)} M_{a0}^V M_{b0}^V \tag{5.1-12}$$

Here $M_{ji}^u = <\phi_j^u | m^u | \phi_i^u>$, the matrix element of the dipole moment.

Confining our attention to nondegenerate states of the solute molecules, the energy of state Φ_i is given by

$$W_i = H_{i,i} + \sum_{j\ne i} \frac{\left(H_{j,i}\right)^2}{W_i^0 - W_j^0} + \sum_{p=1}^{N} \sum_{a\ne 0} \frac{\left(H_{a(p)i,i}\right)^2}{W_i^0 - W_{ai}^0}$$

$$+ \sum_{p=1}^{N} \sum_{a\ne 0} \sum_{j\ne i} \frac{\left(H_{a(p)j,i}\right)^2}{W_i^0 - W_{ai}^0} + \frac{1}{2} \sum_{p=1}^{N} \sum_{q=1,\ne p}^{N} \sum_{a\ne 0} \sum_{b\ne 0} \frac{\left(H_{a(p)b(p)i,i}\right)^2}{W_i^0 - W_{abi}^0} \tag{5.1-13}$$

The frequency shift, Δv, is given by

$$hc\Delta v = \overline{\left(W_i - W_0\right)} - \left(W_i^0 - W_0^0\right) \tag{5.1-14}$$

where the bar indicates a time-average value.

Since numerical calculation using these equations is difficult, McRae made some assumptions to obtain a simple expression for the frequency shift. First, the term which describes the interaction between solvent molecules should be neglected. Second, all the point dipoles associated with any one molecule may be considered to lie at the same point in that molecule, and the molecules are considered to be optically isotropic. He then applied Onsager's reaction field model, where the interaction of the solute–solvent system is described as a function of the dielectric constant and the refractive index of the solvent, and also as a function of the dipole moment and polarizability of solute molecules. The final expression for the frequency difference between the 0–0 bands in the absorption and emission transitions between the ground and excited states of an isotropic solute molecule is predicted to be

$$\Delta v_{absorption} - \Delta v_{emission} = \frac{2}{hc} \frac{\left(M_{00}^u - M_{ii}^u\right)^2}{a^3} \left[\frac{\varepsilon-1}{\varepsilon+2} - \frac{n_0^2-1}{n_0^2+1}\right]$$

$$+ \frac{2}{hc} \frac{\left(\alpha_0^u - \alpha_i^u\right)\left[3\left(M_{00}^u\right)^2 - 5\left(M_{ii}^u\right)^2 + 2M_{00}^u M_{00}^u\right]}{a^6} \left[\frac{\varepsilon-1}{\varepsilon+2} - \frac{n_0^2-1}{n_0^2+2}\right]^2 \quad (5.1\text{-}15)$$

where ε and n_0 are the static dielectric constant and refractive index of the solvent, respectively, α_0^u and α_i^u are the polarizabilities in the ground and excited states, respectively, of the solute molecules, and a is the cavity radius in Onsager's model.

Detailed description of solvent effects has been made in the book written by Mataga and Kubota [2]. The results are widely used to obtain reliable information about the electronic state of molecules in the excited state as well as in the ground state, and to estimate the dipole moment of solute molecules in the excited state.

5.1.1.2 Molecular Aspects of Solvation Dynamics

During the last decade, there have been many attempts, both experimental and theoretical, to understand the solvation dynamics of solvent fluctuation. One of the most convenient ways to observe solvent reorganization and relaxation processes in excited-state molecules is time-resolved fluorescence spectroscopy. By using time-resolved techniques, a time-dependent fluorescence peak shift, known as the dynamic Stokes shift, has been detected down to femtosecond time regions. Another method of observing solvent relaxation processes is time-resolved absorption spectroscopy. This method is suitable for observing the ground-state recovery of the orientational distribution of the surrounding a solute molecule.

The time-evolution of the ground-state hole as well as of fluorescence spectra initiated by short-pulse laser irradiation of a solution may be understood in terms of the configuration coordinate model shown in Fig. 5.1-1. Two adiabatic potentials, corresponding to the ground and excited states, are shown as functions of the configuration coordinate, which includes each atomic position of both solute and solvent molecules.

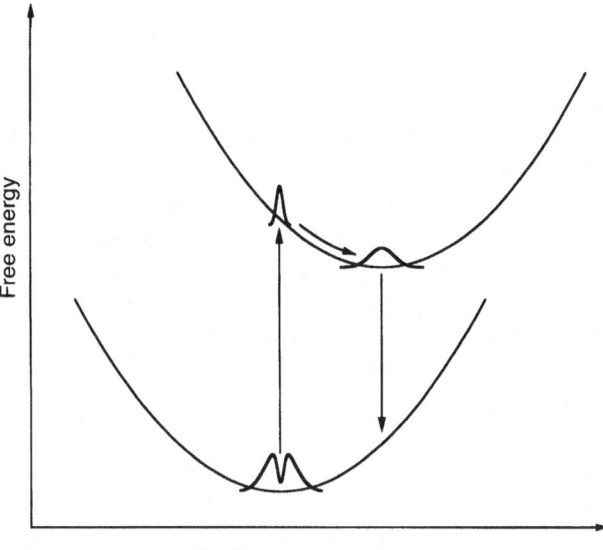

Fig. 5.1-1. The relaxation process in the excited state in solution

Free energy

Configuration coordinate

The time-dependence of the dynamic Stokes shift due to the fluctuational behavior of the solvent molecules is characterized by the normalized time-correlation function of the fluctuation, $\rho_m(t)$, which is given by the following equation using experimentally observed time-resolved fluorescence spectra:

$$\rho_m(t) = \frac{\tilde{v}_f(t) - \tilde{v}_f(\infty)}{\tilde{v}_f(0) - \tilde{v}_f(\infty)} \tag{5.1-16}$$

where $\tilde{v}_f(0)$, $\tilde{v}_f(t)$, and $\tilde{v}_f(\infty)$ denote the peak energy of the fluorescence spectrum at $t = 0$, t, and ∞, respectively. The fluorescence Stokes shift, i.e., $\tilde{v}_f(0) - \tilde{v}_f(t)$, is usually explained as the dielectric dispersion of the surrounding medium, and the interaction between the solute dipole and the reaction field, as described in the first part of McRae's theory, which states that the energy shift in the fluorescence spectrum is roughly proportional to the square of the difference of the dipole moments between the ground and excited states of the solute molecules.

Early theoretical treatments of solvation dynamics used a time-dependent Onsager dielectric continuum model of the solvent. This treatment uses macroscopic solvent response data obtained by microwave loss and dispesion measurements to predict the microscopic solvation behavior about the excited-state probe molecule. Approximation of the solvent dielectric response by a single Debye relaxation time leads to a result for the energy relaxation of the solute dipole in the excited state, i.e., that the initial non-equilibrium reaction field should decay exponentially with a time-constant of

$$\tau_1 = \frac{\left(2\varepsilon_\infty + \varepsilon_c\right)}{\left(2\varepsilon_0 + \varepsilon_c\right)}\tau_D \cong \frac{\varepsilon_\infty}{\varepsilon_0}\tau_D \qquad (5.1\text{-}17)$$

where ε_∞ is the infinite frequency dielectric constant, which is approximately equal to the square of the refractive index of the solvent, ε_0 is the static dielectric constant, and τ_1 and τ_D are the longitudinal and Debye relaxation times of the solvent, respectively. ε_c is the dielectric constant of the cavity containing the point dipole of the excited probe molecule, which is often set equal to one. According to this model, the longitudinal relaxation time, τ_1, is considered to be equal to the time-constant of the correlation function of the fluctuation, $\rho_m(t)$. It should be noted, however, that this expression may only be true when the dielectric response of the solvent can be modeled with a single relaxation time, which is not the case for most solvents.

The observed time-correlation function obtained using Eq. (5.1-16) has been compared with the dielectric continuum model for each solvent, and much shorter time components have been discovered in the solvation relaxation functions for many solvents [3, 4]. For example, about 80% of the time-correlation function in an acetonitrile solution decayed in 100 fs, although the longitudinal relaxation time of the solvent was found to be 200 fs. The ultrafast relaxations observed were explained by the vibrational characteristics of liquid dynamics compared with the optical Kerr effect measurements on acetonitrile [3].

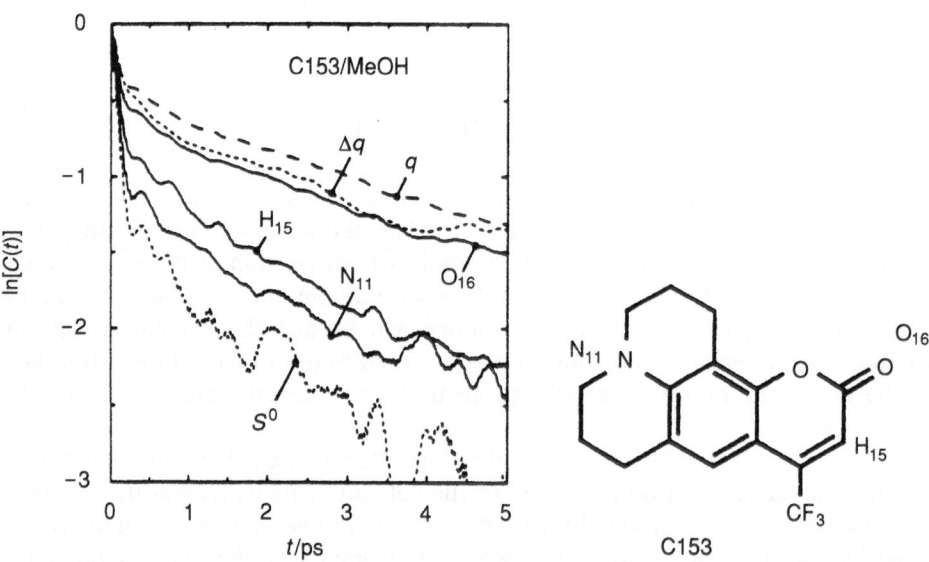

FIG. 5.1-2. Time-correlation functions obtained from simulations of the coumarin 153 (C153) in methanol system. *Solid curves*, time-correlation functions of the electrical potential at the atomic sites N_{11}, H_{15}, and O_{16}. The Δq function is the charge-difference time-correlation function $C(t)$, which is the function being compared with experimental dynamic Stokes shift measurements. (From [5], with permission)

Recent studies on polar solvation using molecular dynamics simulations [5], as well as the statistical theory of molecular liquids, employed the interaction–site model [6] and provide a relatively unified molecular picture of the basic dynamics involved in energy relaxation processes in solution.

Time-correlation functions, $C(t)$, obtained from computer simulations of coumarin 153 (C153) in methanol [5] are shown in Fig. 5.1-2. The simulated time-correlation function, $C(t)$, corresponds to the experimentally observed $\rho_m(t)$ in Eq. (5.1-16). The function $C(t)$ can be written

$$C(t) = \frac{\sum\limits_{i,j} \Delta q_i \Delta q_j \left\langle \delta v_i(0) \delta v_j(t) \right\rangle^{(0)}}{\sum\limits_{i,j} \Delta q_i \Delta q_j \left\langle \delta v_i \delta v_j \right\rangle^{(0)}} \tag{5.1-18}$$

where v_i is the electrical potential, Δ_{qi} is the $(S_1 - S_0)$ charge difference at site i, q_i is the discrete atomic charge, $<x>^{(0)}$ denotes the average of the quantity x throughout a system in equilibrium with the solute in state S_0, and δx is a fluctuation ($\delta x = x - <x>$). The total response may be considered as being composed of contributions from individual atomic sites:

$$c_i(t) = \frac{\left\langle \delta v_i(0) \delta v_i(t) \right\rangle^{(0)}}{\left\langle \delta v_i^2 \right\rangle^{(0)}} \tag{5.1-19}$$

The solid curves in Fig. 5.1-2 are time-correlation functions of the electrical potential at atomic sites N_{11}, H_{15}, and O_{16} in which the atoms have charges of -0.63, $+0.21$, and -0.62, respectively, in the ground state of C153. S^0 indicates the case of a monatomic solute subject to a charge jump, which is identical to the single-site correlation function calculated using Eq. (5.1-19). In the case of methanol, the molecular solvation response appears to be very similar to that of the carbonyl oxygen. This site differs from the others in showing a much slower site time-correlation function owing to the fact that it forms strong, long-lived hydrogen bonds with methanol. However, the authors noted that owing to the small charge change on this atom, its contribution to $C(t)$ is negligible.

Forms of the site-dependent as well as solvent-dependent time-correlation functions can also be predicted by statistical theory for the interaction-site dynamic structure factor of a liquid. This theory has been developed for solvent dynamics based on the generalized Langevin equation combined with the reference interaction site model (RISM) equation that describes the static structure of molecular liquids in terms of the positional correlation between a pair of interaction sites [6]. Figure 5.1-3 shows the time-correlation function of solvation for the Cl → Cl⁻ process in water, methanol, methyl chloride, and acetonitrile. The time-correlation function for methanol is characterized by a tri-exponential decay. Some authors have suggested a possible mechanism for

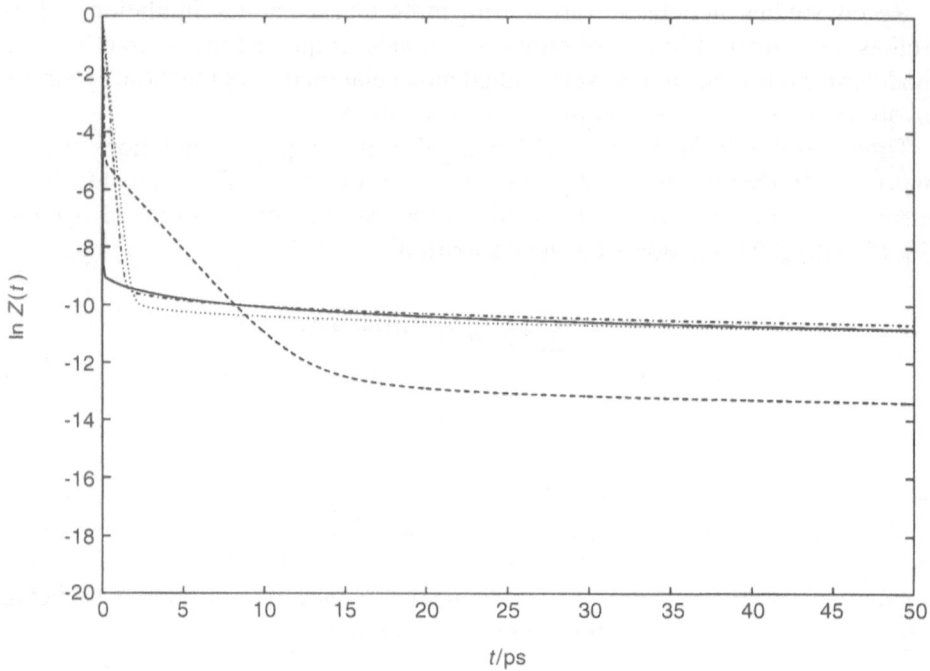

FIG. 5.1-3. Time-correlation functions for the Cl → Cl⁻ process in water (*solid line*), methanol (*dashed line*), methyl chloride (*dotted line*), and acetonitrile (*chained line*). (From [6], with permission)

this result, but further study is required, including a molecular dynamics simulation. From calculation of the radial distributions of solvent atoms around chloride in equilibrium before and after the abrupt production of a charge on the solute atom, the solute is primarily coordinated before the change either by the methyl group or the oxygen site. After the change, the solute is predominantly coordinated by the hydrogen site. The first decay, characterized by a time-constant of about 30 fs, is attributed to the initial rotation of the methyl group in the first coordination shell, which has a positive partial charge. This process is quick because the solute is already well coordinated by the methyl group before the charge change. The solvent molecules in the first coordination shell further rotate to bring the hydrogen atom into the final equilibrium position in which the hydrogen atom and the ion are "bonded" by a strong electrostatic interaction. This gives the second decay, with a time-constant of about 1.7 ps in the plot. After the solvent molecules in the first shell attain their final orientation, the molecules in the second and further shells must be reorganized cooperatively to adapt to the new equilibrium state. This process will be very slow, since it involves the relaxation of many molecules.

5.1.2 The Role of Viscosity in Solution Processes

5.1.2.1 Solvent Viscosity Effects on Diffusion and Reaction Processes

Electrostatic interactions play an important part in the effects of solvent viscosity on a molecular level, and a detailed description of the origin of the friction experienced by the solute in solutions is rather complicated. However, solvent viscosity, in its original meaning, is based on the macroscopic properties of the solvent, and most viscosity effects on solute reactivity can be interpreted without a precise treatment of the specific microscopic interactions between the solute and the solvent.

In general, the viscosity of the solvent affects intermolecular as well as intramolecular motions in solutions, e.g., translational and rotational diffusion processes. Hence, the rate constant of an intermolecular reaction which requires translational diffusion to produce the resultant complex is strongly dependent on the viscosity of the solvent. The orientation of the polar solvent molecules, which was described in the previous section, is partly regulated by the viscosity of the solvent. In addition, the intermolecular reaction processes that accompany any large structural change in the solute such as the *cis–trans* isomerization reaction are also affected by the solvent viscosity. Not only the intramolecular processes but also the rearrangement of the mutual intermolecular geometry of the reacting molecules such as distance and orientation is sensitive to the solvent viscosity.

To summarize, the viscosity of a solvent is closely related to the dynamic aspects of molecular diffusion, rotational relaxation, and intramolecular as well as intermolecular geometrical rearrangement processes. By a combination of these dynamic processes with competitive reactions, the viscosity of a solvent produces a variety of reaction profiles in a solution.

Translational Diffusion Processes and Bimolecular Reaction Rate Constants. The translational diffusion constant of a solute molecule is connected with the solvent viscosity, η, by the Stokes–Einstein equation

$$D = k_B T / 6\pi\eta r \tag{5.1-20}$$

Here, r, k_B, and T are the radius of the solute molecule, the Boltzmann constant, and the temperature, respectively. In this equation, the solute molecule is treated as a sphere. In the case where a bimolecular reaction between A and B takes place in a solution, the number of encounters, n_c, can be written as

$$n_c = 2\pi(D_A + D_B)(r_A + r_B)N_A N_B \tag{5.1-21}$$

Here $D_{A,B}$, $r_{A,B}$, and $N_{A,B}$ are the diffusion coefficient, the radius, and the number of molecules per unit volume of A and B, respectively. Substitution of Eq. (5.1-20) for $D_{A,B}$ in Eq. (5.1-21) gives

$$n_C = \frac{2k_B T(r_A + r_B)}{3\eta r_A r_B} \qquad (5.1\text{-}22)$$

By assuming that A and B are of equal radius, one can obtain the following equation:

$$n_C = \frac{8k_B T}{3\eta} N_A N_B \qquad (5.1\text{-}23)$$

where N_A and N_B are given in M. This expression corresponds to the maximum rate of reaction that could take place in a liquid solution. In the case where the intrinsic reaction of the encounted complex is fast, the reaction becomes diffusion-limited. The diffusion-limited bimolecular rate constant is given by

$$k = \frac{8RT}{3\eta} \qquad (5.1\text{-}24)$$

where R is the gas constant.

Although this model is rather crude, it has been confirmed that the upper limit of the rate constant of very fast bimolecular reactions in solutions, such as the fluorescence quenching reaction of aromatic compounds, is quantitatively in reasonable agreement with that predicated by Eq. (5.1-24). Typical values for the diffusion-limited bimolecular rate constant in various solutions, estimated by Eq. (5.1-24), are of the order of 10^9–$10^{10}\,s^{-1}M^{-1}$.

Translational Diffusion Reactions in the Coulombic Potential. In the treatment described above external fields were not taken into account. The diffusion process of charged particles is affected by the electric field. As an example of a translational diffusion process in an electric field, a geminate recombination between two oppositely charged particles has been widely investigated. This problem was originally treated by Onsager using the Smoluchowski equation, and theoretical improvements have been reported by many researchers [7].

First, consider a charged particle produced at a distance r_0 from an oppositely charged particle in solution. The motion of the charged particles is treated as translational diffusion in a Coulombic field, which is given by the Smoluchowski equation

$$\frac{\partial n}{\partial t} = \frac{D}{r^2} \frac{\partial}{\partial r}\left(r^2 \frac{\partial n}{\partial r} + \frac{e^2}{\varepsilon k_B T} n \right) \qquad (5.1\text{-}25)$$

Here, n is the number of charged particles at r and t, e is the elementary electric charge and ε is the dielectric constant of the medium. The second term of the equation relates to the Coulombic interaction, and the first is the usual diffusion equation. The boundary condition provides complete recombination for all particles at $r = 0$. The fraction of remaining particles at t (survival probability) is given by the following equation:

$$F(t) = \int_0^\infty 4\pi r^2 n(r,t) dr \qquad (5.1\text{-}26)$$

The time profile of $F(t)$ is dependent on the initial distribution of the charged particles, and rather complicated behaviors were obtained in the early stage of the time span. However, the decay profile, $t^{-1/2}$, frequently has been observed at the nanosecond scale in fluid solution.

The time-dependent Smoluchowski equation has mainly been used to analyze the dynamic behavior of electron–hole recombination processes in solution. In polar solutions, the electron ejected by photoionization or radiolysis is usually solvated, and the mobility of the solvated electron is almost of the same order as that of normal molecules. Moreover, the rather high dielectric constant does not seriously affect the diffusion process of charged particles in polar solutions. However, in nonpolar solutions such as alkanes, either the electron is not strongly solvated or the solvent does not accept the electron in its molecular orbital because the energy level of the solvent alkane anion is usually beyond the conduction level of an electron in an alkane solution. Hence, the diffusion coefficient of the electron, which is proportional to the mobility, m, by the Nernst–Einstein equation ($D = mk_BT/e$), is much larger (10^2–10^4 times) than that of a normal molecule in an alkane solution. In addition, the low dielectric constant strongly affects the diffusion process of the charged particles. For this reason, electron–hole recombination processes in nonpolar solutions have been studied as typical examples of the diffusion process in an electric field.

Rotational Diffusion Processes. As with the translational process, the solvent viscosity also affects the rotational diffusion process. According to the Stokes–Einstein–Debye relation, the reorientation time of a solute molecule in a solution with viscosity η is given by

$$\tau_R = \frac{VC\eta}{k_BT}f \qquad (5.1\text{-}27)$$

where V, C, and f are the volume of the solute, the hydrodynamic boundary condition, and the Perrin parameter for molecular shape, respectively. Hence, the rotational relaxation time is proportional to the viscosity. In the case of $C = 1$, the boundary condition is "stick". The stick boundary condition implies that the velocity of the fluid (the solvent) relative to the solute vanishes on the surface of the solute. The motion of the solute is perfectly coherent with that of the neighboring fluid. On the other hand, the slip boundary condition requires that the fluid exerts no tangential force on the rotating body. The parameter C is dependent on the molecular structure, e.g., prolate or oblate spheroids. The rotational relaxation time in the slip boundary condition is in general shorter than that predicted in the stick boundary condition.

The parameter f is also dependent on the molecular shape. The rotational relaxation process and its relation to solvent viscosity have been extensively studied, and the detailed results of these studies appear in several books [8, 9].

5.1.2.2 Solvent Viscosity Effects on Isomerization

The *trans–cis* isomerization process of solute molecules such as stilbene and its derivatives has been investigated from various viewpoints [10]. The isomerization processes of *trans* → *cis* and *cis* → *trans* involve an activated structure where two phenyl rings are perpendicular to each other. Since this activated structure has a much larger volume than those of the *cis*- and *trans*-forms, the solvent molecules must be pushed out of the way during the activation process of the solute. Hence, the viscosity of the solvent strongly affects the isomerization rate constants. In direct investigations of the relation between the isomerization rate constant of stilbene and the solvent viscosity, it was observed that the isomerization rate constant decreases with an increase in solvent viscosity [10].

The theoretical work of Kramers [11] has frequently been applied in attempts to explain the effect of viscosity on the *trans–cis* isomerization process in solutions. In the Kramers theory, it is assumed that the reacting particle, propelled by the random forces of the solute, passes the activation barrier, as shown in Fig. 5.1-4. The motion of the particle is given by the following Langevin equation:

$$m\frac{du}{dt} = -\frac{dv(x)}{dx} - m\gamma u + X(t)$$

(5.1-28)

Here, u is the velocity of the particle, $m\gamma u$ is the friction force produced by the solvent, and $v(x)$ is the potential function. $X(t)$ is the Langevin force, or random force produced by the solvent, and is given by $<X> = 0$ and $<X(0)X(T)> = 2\gamma m k_B T \delta(t)$. $\delta(t)$ is the delta function and γ is the friction coefficient, which is proportional to the solvent viscosity in the first approximation. In this equation, the solvent has two roles: providing the friction to reduce the velocity of the particle, and the Brownian motion, or collision with the particles (the solute), to increase the velocity. Both roles are related to the same friction coefficient, γ.

Kramers calculated the rate constant of the passage of the potential barrier using the solution of Eq. (5.1-28). For the case $\gamma \to 0$, the rate constant is given by

$$k = \gamma \frac{E_A}{k_B T} \exp\left(-\frac{E_A}{k_B T}\right)$$

(5.1-29)

where E_A is the height of the activation barrier. The rate constant for $\gamma \to \infty$ is given by

$$k = \frac{\omega_R \omega_A}{2\pi\gamma} \exp\left(-\frac{E_A}{k_B T}\right)$$

(5.1-30)

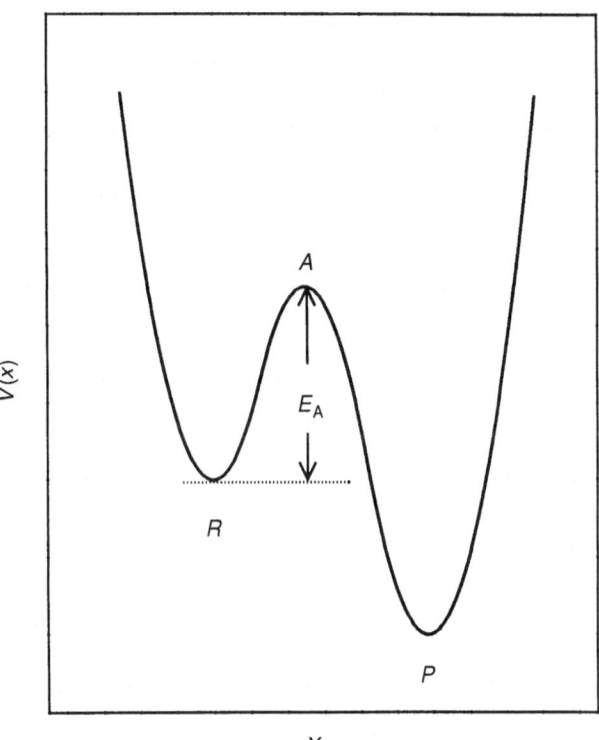

FIG. 5.1-4. Reaction with a potential barrier in the course of the reaction

where ω_R and ω_A are the frequencies obtained by the curvatures of the potential at R and A, respectively, in Fig. 5.1-4. As is clear in these equations, the rate constant increases with an increase in γ when γ is small, indicating that the friction coefficient in the acceleration ($X(t)$) regulates the reaction. However, the rate constant decreases with increasing γ in large γ regions where the friction term, $m\gamma u$, in Eq. (5.1-28) is effective in slowing the velocity. Hence, the rate constant is maximum at a specific value of γ, and the bell-shaped γ-dependence of the rate constant is confirmed.

Much research has been carried out on the relation between Kramers' predictions and the *trans–cis* isomerization processes of stilbenes in solution [10]. Although a change in the rate constant with an increase in solvent viscosity was observed in some cases, no quantitative agreement was confirmed. In the original theory the molecular friction was assumed to be constant. This implies that the solvent motions are much faster than the potential motions of the particles. In the case where both motions are comparable, the frequency-dependence of γ should be taken into account, and the following generalized Langevin equation is required for the analysis:

$$m\frac{du}{dt} = -\frac{dv(x)}{dx} - m\int_0^t \gamma(t)u(t-\tau)d\tau + X(t) \tag{5.1-31}$$

Although better agreement was reached using this type of equation, a clear conclusion on the reaction profiles has not yet been obtained. Part of the disagreement between the theories and the experimental results may be attributed to the difference between the viscosity on the macroscopic scale and friction at the molecular level, and many investigators are currently attempting to find a description of viscosity at the microscopic level.

5.1.2.3 Solvent Viscosity Effects on Torsional Relaxation Dynamics in the Excited State of 9,9′-Bianthryl

As a typical example of the relaxation of a large amplitude motion (LAM) combined with solvent friction, and one of the simplest, the dynamics of intramolecular torsional relaxation of 9,9′-bianthryl in solution are discussed below. For molecules with one LAM degree of freedom (e.g. the torsional angle of biaryl-type molecules), separation of spectral structures arising from vibronic progression from the LAM potentials can be achieved by applying high-resolution spectroscopy to Franck–Condon (FC) active vibrations as obtained from molecular beam experiments [12]. The distribution along the LAM coordinate often shows as a characteristic band shape, which reflects the broadening of the individual vibronic transitions. By means of simultaneous FC and band-shape analyses of the temperature-dependent steady-state fluorescence spectra, the effective S_1 torsional potential (Fig. 5.1-5) can be determined for bianthryl (BA) in a nonpolar solvent [12].

Irradiation at the red edge of the absorption spectra results in selective excitation of the molecules belonging to a narrow torsional distribution that may be exploited to probe torsional relaxation. Since solute–solvent interactions are of minor significance in nonpolar solvents, torsional relaxation is expected to be the predominant cause of the variation in the fluorescence spectra. The time-resolved fluorescence spectrum can be represented by

$$\frac{\Phi_{\tilde{v}}(\tilde{v},t)}{\tilde{v}^3} = \frac{1}{3} C^{fl} N_E(t) \left| \mu_{EG}^0 \right|^2 \sum_e \sum_g W_{Ee}(e|g)^2 S_{EG}^{fl}(\tilde{v} - \tilde{v}_{eg}^{0,fl}, t) \qquad (5.1\text{-}32)$$

where Φ_v is the emitted spectral photon current density related to the wave number $\tilde{v}$, C^{fl} is a constant, and N_E is the total number of molecules in any excited vibronic state related to the probe volume. μ_{EG}^0 is the electronic transition dipole at reference configuration between the electronic ground and the lowest excited state |G> and |E>, respectively, and W_{Ee} is the Boltzmann factor describing the population of excited vibronic states. (e|g) is the Franck–Condon overlap integral between the vibrational states |g) and |e) belonging to the ground and the excited electronic state, respectively. The semiclassical band shape S_{EG}^{fl} describes the spectral broadening of each vibronic transition |G>|g) ← |E>|e) caused by intramolecular as well as intermolecular effects, and is given by

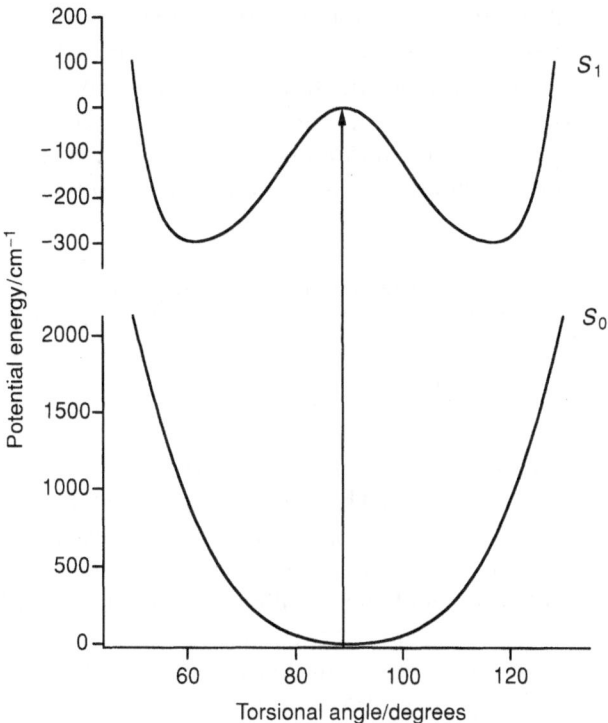

FIG. 5.1-5. Potential surfaces of the ground and excited states of bianthryl as a function of the torsional angle

$$S_{EG}^{fl}\left(\tilde{v}-\tilde{v}_{eg}^{0,fl},t\right)=\int_0^\pi d\varphi w_E^L\left(\varphi,t\right)p_{EG}^{fl}\left(\tilde{v}-\tilde{v}_{eg}^{fl}\left(\varphi,t\right),t\right) \qquad (5.1\text{-}33)$$

$$\tilde{v}_{eg}^{fl}\left(\varphi,t\right)=\tilde{v}_{eg}^0+\Delta\tilde{v}_{eg}^{0,fl}\left(t\right)+V_E^L\left(\varphi\right)-V_G^L\left(\varphi\right) \qquad (5.1\text{-}34)$$

The superscript "L" indicates quantities depending on the torsional (LAM) degree of freedom. w_E^L is the normalized probability distribution function with respect to the torsional distribution of angle φ in the excited electronic state. p_{EG}^{fl} is the spectral shift and broadening effects caused by librations and intermolecular interactions. It may be empirically described by a Gaussian distribution with width σ^{fl}. $\tilde{v}_{eg}^0$ is the transition frequency at reference configuration with respect to the torsional angle immediately after excitation. $\Delta\tilde{v}_{eg}^{0,fl}$ is the solvent-induced shift of $\tilde{v}_{eg}^0$. V_G^L and V_E^L are the torsional potentials in the electronic states $|G\rangle$ and $|E\rangle$, respectively, which may be represented by Fourier series of the form

$$V_I^L\left(\varphi\right)=\sum_n V_{2nI}\left(\left(-1\right)^n-\cos 2n\varphi\right) \qquad I=G,\ E \qquad (5.1\text{-}35)$$

where V_{2nI} are expansion coefficients.

On condition that the torsional motion is treated as an overdampled motion (high friction regime), the time evolution of the torsional distribution can be described by a Smoluchowski-type equation. Assuming the population decay rate to be independent of the torsional angle and neglecting any coupling between the torsional and the overall rotational motion, the following equation holds for the excited-state torsional distribution of a solute molecule consisting of two equal parts:

$$\frac{\partial w_E^{\mathrm{L}}(\varphi,t)}{\partial t} = \frac{2k_{\mathrm{B}}T}{\varsigma}\frac{\partial^2 w_E^{\mathrm{L}}(\varphi,t)}{\partial\varphi^2} + \frac{2}{\varsigma}\frac{\partial}{\partial\varphi}\left(\frac{\partial V_E^{\mathrm{L}}(\varphi)}{\partial\varphi}w_E^{\mathrm{L}}(\varphi,t)\right)$$
$$+ k_{\mathrm{abs}}(t)\left(w_E^{\mathrm{L,FC}}(\varphi,\tilde{\nu}_{\mathrm{ex}}) - w_E^{\mathrm{L}}(\varphi,t)\right) \tag{5.1-36}$$

where ς is the friction coefficient of one anthracene moiety, k_{B} is the Boltzmann constant, and T is the temperature. The quantity k_{abs} is related to the time profile of the exciting laser pulse. $w_E^{\mathrm{L,FC}}$ is the Franck–Condon excited-state torsional distribution created by the excitation process, which depends mainly on the excitation wave number $\tilde{\nu}_{\mathrm{ex}}$ and the equilibrium torsional distribution in the electronic ground state $w_G^{\mathrm{L,eq}}$. In case of excitation near the 0_0^0 transition, $w_E^{\mathrm{L,FC}}$ can be approximated by the following equation provided that the 0_0^0 transition is sufficiently separated from other vibronic transitions:

$$w_E^{\mathrm{L,FC}}(\varphi,\tilde{\nu}_{\mathrm{ex}}) = K(\tilde{\nu}_{\mathrm{ex}})w_G^{\mathrm{L,eq}}(\varphi)\int d\tilde{\nu}A(\tilde{\nu}-\tilde{\nu}_{\mathrm{ex}})p_{EG}^{\mathrm{abs}}(\tilde{\nu}-\tilde{\nu}_{00}^{\mathrm{abs}}(\varphi)) \tag{5.1-37}$$

where K is a normalization constant, A is the spectral band shape of the exciting light, and p_{EG}^{abs} is intermolecular broadening related to the absorption process. As in the case of emission, this may be represented by a Gaussian distribution with width σ_{abs}. On condition that the duration of the exciting light pulse is short compared with the time-scale characterizing the torsional relaxation, k_{abs} can be approximated by the Dirac delta function. Equation (5.1-36) then reduces to an initial-value problem including the first two terms with the initial distribution $W_E^{\mathrm{L,FC}}$.

The time-resolved fluorescence spectra of 9,9′-bianthryl in methylcyclohexane (MCH) at 127 K are shown in Fig. 5.1-6 [13]. Irradiation at the maximum of the first vibronic band gives rise to a narrow initial torsional distribution centered around $\varphi = 90°$. A common feature of all spectra was a continuous loss of structure and a red shift of the band centers with increasing time.

The results of the band-shape analysis are shown by dotted curves in Fig. 5.1-6. Good correspondence between experimental and simulated fluorescence spectra was achieved for all temperatures. In particular, the features of the first vibronic band are well reproduced by the simulated spectra. The spectra exhibit a shoulder at the blue edge that arises from transitions of solute molecules with perpendicular conformation. The decrease in size of this shoulder, which represents a characteristic attribute of the time-dependent band shape, reflects the diminution of the S_1 torsional distribution within the range of the potential barrier at $\varphi = 90°$.

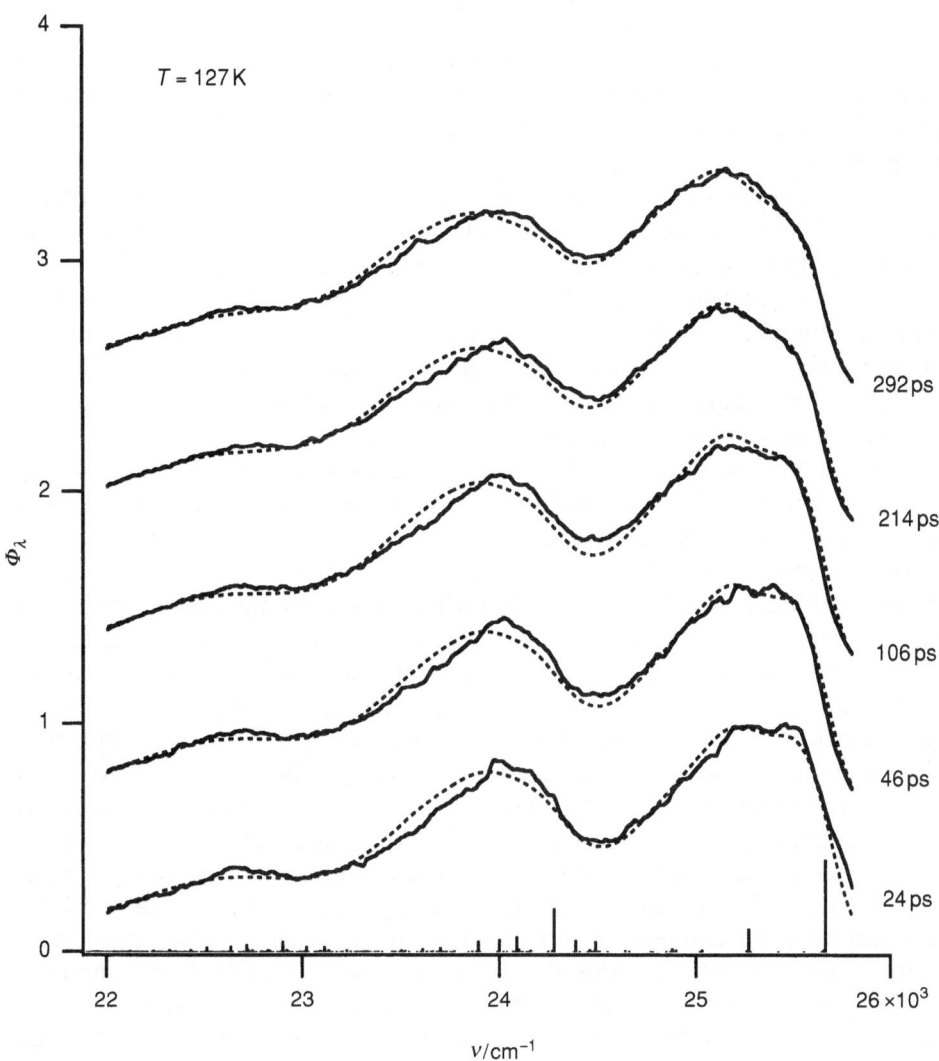

FIG. 5.1-6. Time-resolved fluorescence spectra of bianthryl at 127 K (*solid lines*) and simulated spectra (*dashed lines*)

The spectra simulations were performed using Eqs. (5.1-32) and (5.1-33). The initial torsional distribution was calculated using Eq. (5.1-37) assuming a Gaussian profile for the exciting light pulse with width $\sigma_{ex} = 50\,\text{cm}^{-1}$. The time-dependent band shape itself is characterized by the parameters representing the S_0 and the S_1 torsional potentials V_{2nG} and V_{2nE}, respectively, the friction parameter ζ, and the parameter δ representing the extent of the initial fast torsional relaxation. The potential parameters V_{2G} and V_{4G}, as obtained from analysis of the free jet spectra, were used assuming the S_0 potential to be unaffected by solvent interaction. S_1 torsional potentials were determined for bianthryl in methylcyclohexane. The height of the local barrier at perpendicular conformation and the minimum angle were found to be $\Delta V \approx 300\,\mathrm{cm}^{-1}$ and $\varphi_{\min} \approx 62°$, respectively. The resulting values for the friction coefficient ζ are represented as a function of the viscosity η. A good data fit was achieved by the phenomenological power law of the form $\zeta = B\eta^a$. The fit yields an exponent $a = 0.62 \pm 0.05$ for BA in MCH.

The analysis shows that the dynamics disclosed by the fluorescence spectra can be predominately ascribed to torsional relaxation. In particular, the characteristic variation of the first vibronic band directly indicates the temporal change of the band shape which reflects the motion of the S_1 torsional distribution towards the potential minimum. The band shape itself shows a red shift and a broadening with increasing time due to the smaller transition wave number and the larger slope of the S_0 torsional potential at more planar conformations. Solvent relaxation, which contributes to only a minor extent, causes a shift of the whole spectrum without affecting its shape. The magnitude of the shift ($\sim 100\,\mathrm{cm}^{-1}$) suggests that it is mainly the dispersion interaction which gives rise to the solvent dynamics, as was expected for nonpolar solvents.

Another important outcome of this study concerns the relation between the friction coefficient and the viscosity. In a wide range of isomerization studies, the barrier crossing-rate coefficient k_{iso} was found to be weakly dependent on the viscosity according to the relation $k_{\mathrm{iso}} \sim \eta^{-\alpha}$, with $\alpha < 1$. Three alternative approaches have frequently been discussed in the literature to account for the deviation from linear dependency predicted by Kramers' model in conjunction with the hydrodynamic approximation: (1) coupling between the reactive mode and other modes which requires a multidimensional treatment, (2) the presence of a frequency-dependent friction, and (3) divergence between the microviscosity, affecting the isomerization reaction, and the medium viscosity (breakdown of the Stokes–Einstein hydrodynamic approximation). For bianthryl, the one-dimensional treatment of the torsional motion seems to be appropriate. There is no indication that this should be coupled with other large amplitude motion modes from the high resolution spectra. It has been suggested that frequency-dependent friction has an effect on reactions distinguished by a high barrier with a strong curvature. Such effects should be of minor importance for bianthryl, which shows a comparatively shallow potential barrier. However, the presence of an initial fast component indicates that the hydrodynamic approximation does not give an adequate description of the torsional relaxation of bianthryl. Therefore, small systematic deviations of the simulated spectra from the experimental ones might be explained by a continuous increase in the friction coefficient instead of the simple two-step approximation used in the simulation.

References

1. McRae EG (1957) Theory of solvent effects on molecular electronic spectra. Frequency shifts. J Chem Phys 61:562–572
2. Mataga N, Kubota T (1970) Molecular interactions and electronic specrtra. Marcel Dekker, New York

3. Cho M, Rosenthal SJ, Scherer NF, Ziegler LD, Fleming GR (1992) Ultrafast solvent dynamics: connection between time-resolved fluorescence and optical Kerr measurements. J Chem Phys 96:5033–5038
4. Kahlow MA, Jarzeba W, Kang TJ, Barbara PF (1989) Femtosecond resolved solvation dynamics in polar solvents. J Chem Phys 90:151–158
5. Maroncelli M, Kumar PV, Papazyan A, Horng ML, Rosenthal SJ, Fleming GR (1994) Studies of the inertial component of polar solvation dynamics. In: Gauduel Y, Rossky PJ (eds) Ultrafast reaction dynamics and solvent effects. AIP, New York, pp 310–333
6. Hirata F, Munakata T, Raineri F, Friedman HL (1995) An interaction-site representation of the dynamic structure factor of liquid and the solvation dynamics. J Mol Liq 65/66:15–22
7. Hong KM, Noolandi J (1978) Solution of the Smoluchowski equation with a Coulomb potential. 1. General results. J Chem Phys 68:5163–5171
8. Berne BJ, Pecora R (1976) Dynamic light scattering. Wiley, New York
9. Fleming GR (1986) Chemical applications of ultrafast spectroscopy. Oxford University Press, New York
10. Waldeck DH (1991) Photoisomerization dynamics of stilbenes. Chem Rev 91:415–436
11. Kramers HA (1940) Brownian motion in a field of force and the diffusion model of chemical reactions. Physica 7:284–304
12. Wortmann R, Elich K, Lebus S, Liptay W (1991) Experimental determination of the S_1 torsional potential of 9,9′-bianthryl in 2-methyl-butane by simultaneous Franck–Condon and band shape analysis of temperature-dependent optical fluorescence spectra. J Chem Phys 95:6371–6381
13. Elich K, Kitazawa M, Okada T (in press) The effect of S_1 torsional dynamics on the time-resolved fluorescence spectra of 9,9′-bianthryl in solution. J Phys Chem

5.2
Size and Structure Effects

Nobuyuki Nishi

The functionality of molecules cannot be discussed without considering the role of environmental molecules. Since 1980 the role of environmental molecules has been studied in terms of molecular clusters. This chapter considers some typical examples that demonstrate the size-dependence of proton transfer reactions, the size-dependence of the location of two positive charges in molecular clusters, and structure generation of environmental molecules around solute species. The last topic is particularly important for analyzing molecular functionality in an aqueous environment.

5.2.1 Size Effects on Photo-Induced Chemical Reactions of Molecular Clusters

5.2.1.1 Proton Transfer Reaction

Proton transfer is an important and fundamental process in chemical reactions in the liquid phase. It is an acid–base reaction strongly influenced by the properties of the solvent. According to the Brønsted–Lowry theory [1], "an acid is a substance that is capable of giving up protons, while a base is a substance that is capable of accepting protons." Proton transfer reaction in an acid–base complex (HAB) is expressed as

$$HA \cdot B \rightleftharpoons A^- (HB)^+ \qquad (5.2\text{-}1)$$

A conventional measure of the acidity or basicity of a substance is an acid dissociation constant defined as $pK_a = -\log K_a$ for acid–base equilibrium:

$$HA \xrightleftharpoons{K_a} A^- + H^+ \qquad (5.2\text{-}2)$$

where K_a is an equilibrium constant. This type of dissociation in an excited state was investigated for 1-naphthol [2, 3]. 1-Naphthol (ROH) lives long enough to produce excited 1-naphtholate (RO$^-$*) but does not fluoresce in water. On the basis of the changes of dipole moment and dissociation constants on hydrogen-bonding, Nagakura [4] pointed out that 1-naphthol is a much stronger proton donor than phenol. 1-Naphthol and phenol are known to show large changes in acidity in their excited states. For example, the pK_a value of 1-naphthol is 9.4 in its ground state, while in the excited state it becomes a strong acid with $pK_a = 0.5 \pm 0.2$ in water [5]. Hara and Baba [6] found that the photoreaction of 1-naphthol in the excited triplet state is related to a specific hydrogen-bonding interaction within the solvent rather than to the general polarity. Harris and Selinger [5] reported the pH dependence of the fluorescence from ROH* and RO⁻* in aqueous solution. At pH larger than 10, only the emission from RO⁻* was observed, while at low pH this emission decreased dramatically. The expected proton transfer is strongly related to the environmental change that promotes reaction (5.2-2) in 1-naphthol. From this point of view, the solvent-number-dependence of the proton transfer reaction in naphthol-in-water and napthhol-in-ammonia binary cluster systems is very useful for extracting the fundamental factors in an acid–base reaction. In particular, the following questions need to be considered: (1) What is the critical number of solvent molecules in the cluster which induces excited state proton transfer? (2) What determines this critical number?

5.2.1.2 Proton Transfer in Excited States of Naphthol

Electronic spectra of size-selected 1-naphthol·$(NH_3)_n$ clusters were observed using supersonic-molecular-beam and resonance-two-photon-ionization (R2PI) techniques coupled with time-of-flight mass spectrometry [7, 8]. The proton affinity of gas-phase ammonia is $207\,kcal\,mol^{-1}$ and is 1.26 times that of water [9, 10]. Since the hydrogen-bond formation induces charge transfer from the hydrogen acceptor to the hydrogen donor, the basicity of the ammonia coordinated with the OH of 1-naphthol is expected to increase with increasing cluster size n. R2PI spectra of 1-naphthol·$(NH_3)_n$ clusters, with $n = 1$ and $n = 2$ (middle), showed sharp lines with red shifts in the origin bands with an increase of NH_3. The $n = 2$ spectrum is composed of at least two different isomers. Apart from the size of the spectral shifts, the spectral features of these clusters indicate little difference in the nature of the excited state. For $n = 3$, the low-frequency intermolecular vibrational band structure becomes a dominant feature, suggesting a complex change in the solvent cluster structure on electronic excitation. The $n = 4$ and $n = 5$ spectra are very different; they are red-shifted, featureless, and very broad. Fluorescence emission spectra observed $40\,\mu s$ after the excitation of the $n = 1$–3 clusters [11] were essentially the same as that of the bare 1-naphthol, as shown in Fig. 5.2-1. With a delay time for the observation window longer than $80\,\mu s$, a relatively strong non-structured broad emission band was observed for the excitation of 1-naphthol·$(NH_3)_{n \geq 4}$ clusters, as shown in Fig. 5.2-1(d). The spectral features of this band were similar to the emission of the 1-naphtholate (RO⁻*) anion in aqueous solution at pH > 8.5. This broad band was assigned to RO⁻* emission following intracluster proton transfer [7, 8]. This blue–violet emission showed little size-dependence for $n \geq 4$, indicating that the microscopic

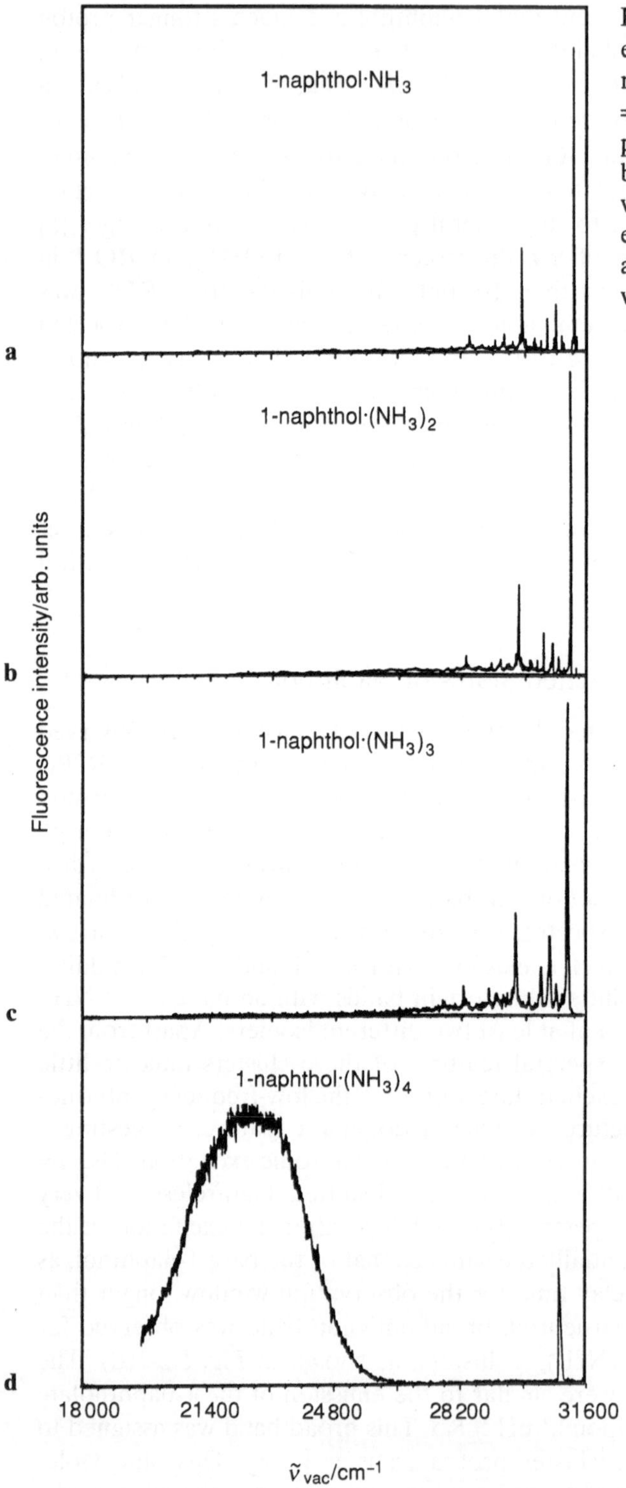

FIG. 5.2-1. Fluorescence emission spectra of 1-naphthol·$(NH_3)_n$ clusters with n = 1–4. The excitation was performed at the respective 0_0^0 bands. The observation window was opened at $40\,\mu s$ after the excitation for n = 1–3 (**a–c**) and at $80\,\mu s$ for n = 4 (**d**). (From [8], with permission)

1-naphthol·NH_3

1-naphthol·$(NH_3)_2$

1-naphthol·$(NH_3)_3$

1-naphthol·$(NH_3)_4$

Fluorescence intensity/arb. units

$\tilde{\nu}_{vac}/cm^{-1}$

18 000   21 400   24 800   28 200   31 600

solvent environment which is necessary for the appearance of RO$^-$* emission is reached at, or just above, $n = 4$.

The time-dependent spectral change invokes real-time probing of the proton transfer reaction in size-selected clusters. The lifetime of the bare 1-naphthol was measured as 60 ± 2 ns, while the $n = 1$ and two isomers of the $n = 2$ clusters of 1-naphthol $\cdot$ (NH$_3$)$_n$ had lifetimes of 38 ± 1 ns, 43 ± 2 ns (blue isomer), and 39 ± 1 ns, respectively [12]. These values approximately correspond to the radiative lifetimes of naphthols. The fast decay values (τ_1) of the $n = 3$ and $n = 4$ clusters at the origins are 60 ps and 70 ps, respectively, and the long decay values (τ_2) are 500 ps and 800 ps, respectively [13]. The fast decay time, τ_1, was attributed to the proton transfer time, and the long decay time, τ_2, to either cluster reorganization subsequent to the proton transfer event or a proton transfer associated with a cluster of different geometry. The proton transfer time for $n = 3$ is a factor of 2–3 slower than the time observed for the argon matrix-isolated cluster (20–25 ps). Other convincing evidence for this assignment is the effect of deuteration. At the origin, excitation τ_1 became longer than 1000 ps for the fully deuterated $n = 3$ cluster. However, the observed difference in the critical ammonia number, n, for proton transfer should be noted: spectroscopic observation showed a critical number of $n = 4$ [7, 8], while dynamic results gave $n = 3$ [10].

Most of the $n = 3$ intensity disappeared and a "new feature" emerged at sufficiently low ionization energy [11]. This fact was attributed to the presence of many (perhaps five or more) $n = 3$ cluster conformations. It was concluded that excited-state proton transfer occurs for $n = 3$ and $n = 4$, while for $n \geq 5$ proton transfer occurs in their ground states [13]. Figure 5.2-2 summarizes these results represented as changes in the potential curve in the excited and ground states. Proton-transfered excited states have minima lower than those of optically pumped states for $n \geq 3$, while in the ground states proton transfer does not occur for $n < 5$.

When we use water as the proton acceptor, the situation changes even for 1-naphthol. Water is not as strong a base as ammonia. As stated above, its proton affinity is 164 kcal mol^{-1} and is smaller than that of ammonia by 43 kcal mol^{-1} [9]. Figure 5.2-3 shows the fluorescence spectra of 1-naphthol $\cdot$ (H$_2$O)$_n$ in different size ranges of $n = 1$–40 [14]. The inserts show time-of-flight mass spectra under the same experimental conditions as the emission spectra. A broad emission in the longer wavelength region is prominent in the top spectrum (d), a large part of which is composed of the emission from the clusters with $n > 20$. In 1-naphthol $\cdot$ (H$_2$O)$_n$ the proton transfer reaction does not occur for $n \leq 20$ because the proton affinity of water is low and because the naphtholate anion is not well stabilized or solvated by water [11]. R2PI and emission spectra of the medium-size clusters ($n = 8$~20) showed incremental shifts approaching that of 1-naphthol in bulk ice at n~20, while the fluorescence spectra of the clusters with $n > 20$ are characteristic of the excited-state 1-naphtholate anion (RO$^-$*). Excited-state proton transfer occurs in clusters with $n \geq 30$ [14]. The temperature of such large clusters would be expected to be higher than that of small clusters. The effect of motional activity on proton transfer should also be taken into account.

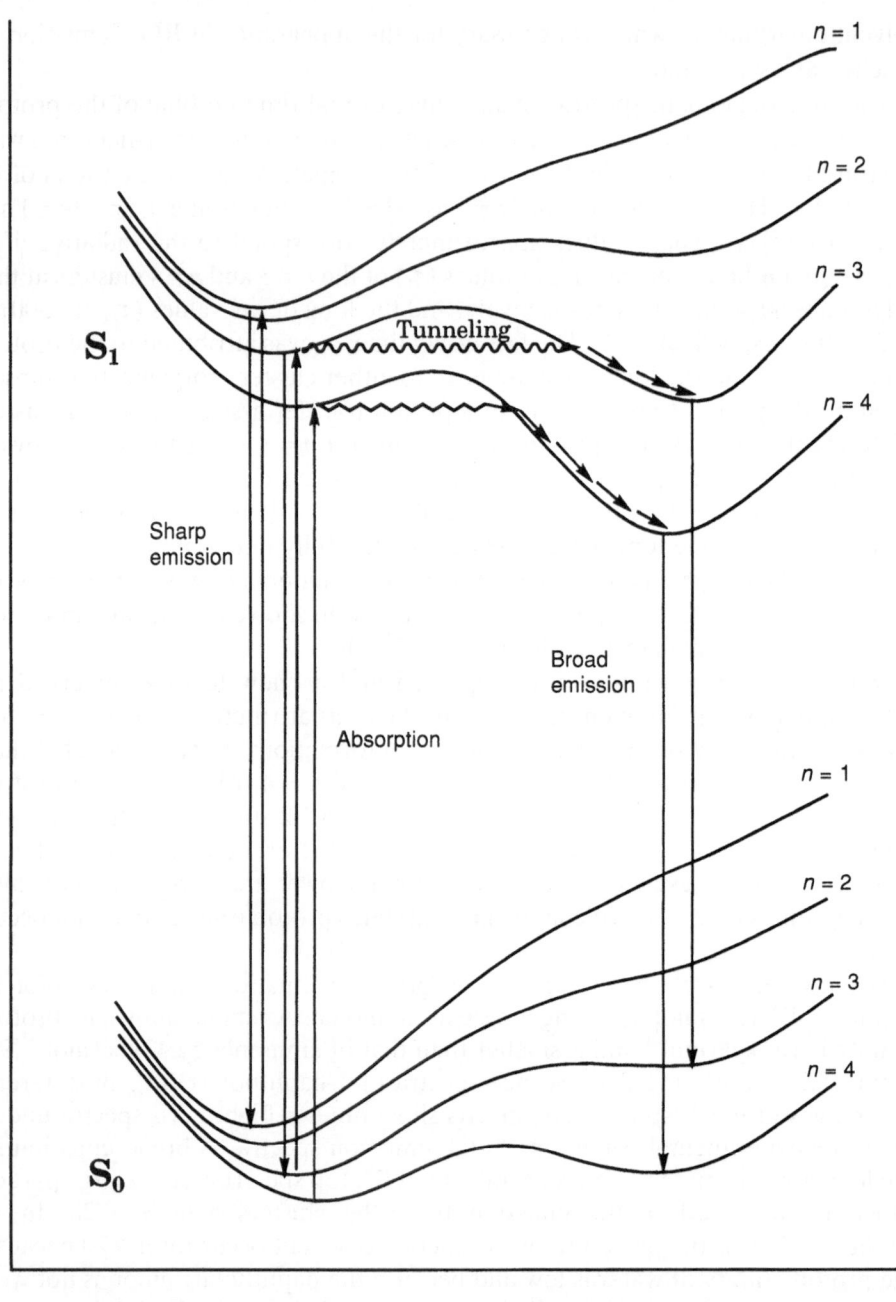

FIG. 5.2-2. A schematic representation of the size-dependent change of the potential curves of 1-naphthol·$(NH_3)_n$ clusters. The minima on the right-hand side correspond to the proton transferred states

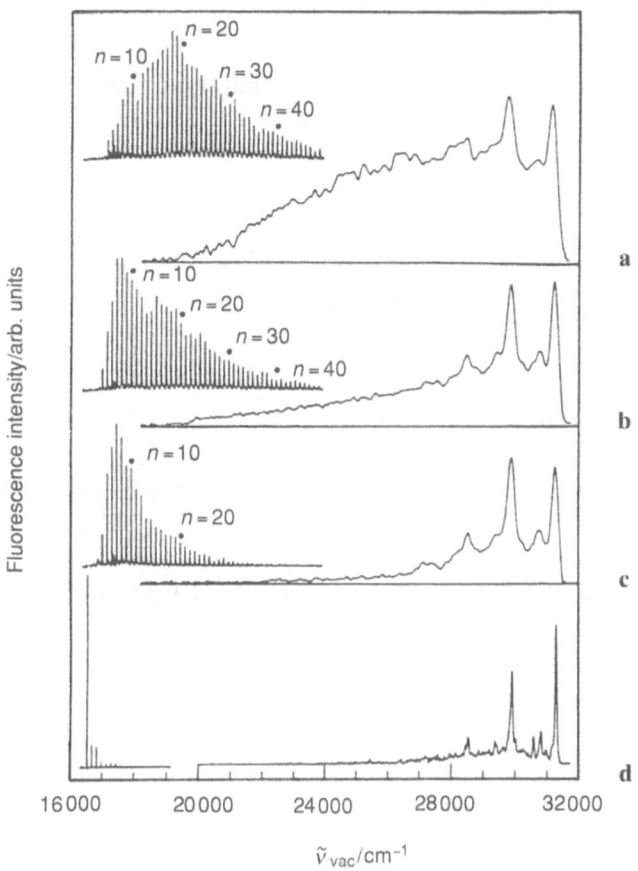

FIG. 5.2-3. Fluorescence spectra of 1-naphthol·$(H_2O)_n$, $n = 1$–40 in different size ranges. The inserts show time-of-flight mass spectra taken under the same experimental conditions as the emission spectra, with dots above the mass spectra indicating the mass peaks of the $n = 10$, 20, 30, and 40 clusters. The excitation wavelengths were: **a** $31153\,cm^{-1}$ (the origin band of the $n = 1$ cluster), **b,c** $31280\,cm^{-1}$, **d** $31311.5\,cm^{-1}$. (From [14], with permission)

5.2.1.3 Excited-State Proton Transfer in Solvated Phenol Clusters

The excited-state proton transfer dynamics of phenol·$(NH_3)_n$ clusters was first studied by Steadman and Syage [15–18]. They found the critical solvent size to be $n = 5$ for proton transfer in the phenol system. The pK_a^* value of excited-state phenol at 298 K is reported to be 4.1 [16]. This value amounts to a free-energy difference of 0.21 eV. In order to gain this energy for proton transfer, the proton affinities of ammonia clusters should be larger than this value. Thus, the addition of two more ammonia molecules forms $(NH_3)_5$, with a structure revealing the highest proton affinity. Hineman et al. [19] criticized the experimental conditions used by Steadman and Syage [16]. The experiments by Hineman et al. indicated that a significant amount of cluster ion fragmentation occurs for phenol $\cdot (NH_3)_{n\geq4}$ cluster systems, and they are doubtful about the determination of a critical size for this system.

5.2.2 Size Effect on the Exact Location of Two Positive Charges in Small Clusters

5.2.2.1 Doubly Charged Benzene Clusters

Formation of multiply charged metal clusters, or fullerenes, is a well-known phenomenon in mass spectrometry with high-energy electron impact ionization [20, 21]. These clusters are mostly bound with valence bonds and in a sense they are molecules containing many conjugated electrons. Thus, two charges in a "molecule" can be located as far as possible from each other.

In benzene cation clusters, a charge is shared with two molecules through resonance interaction, forming a dimer cation core [22, 23]. The electronic

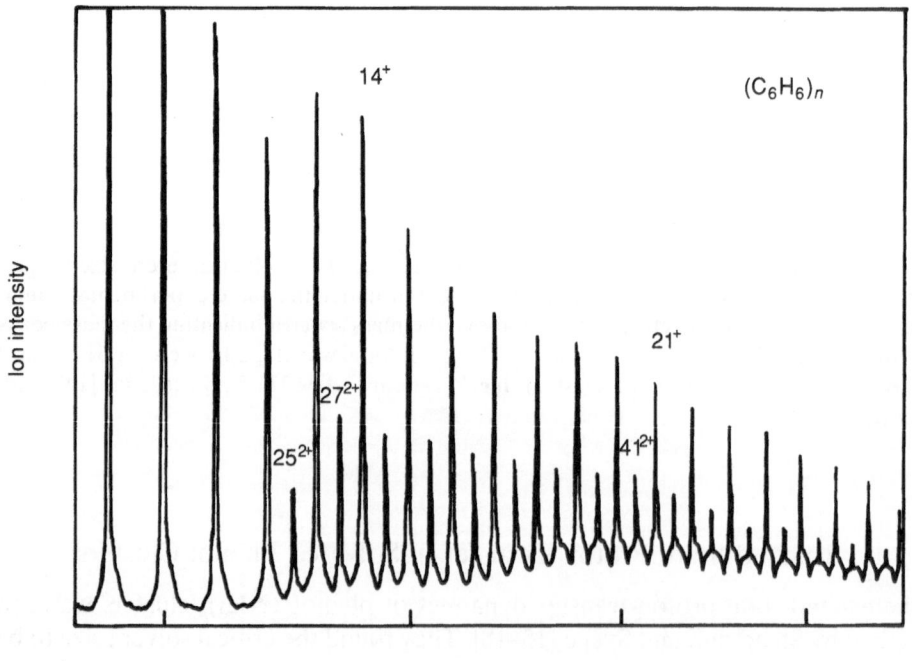

FIG. 5.2-4. Time-of-flight mass spectrum of benzene cluster cations ionized by 6.41 eV laser pulses. The *figures* are the numbers of benzene molecules in the clusters. +, singly charged cluster; 2+, doubly charged cluster

spectra of the benzene trimer, tetramer, pentamer, and hexamer cations are similar to that of the dimer cation, although they show individual site shifts of the order of one hundred wavenumbers. The electronic transitions of the dimer cation itself are not very different from those seen in the monomer cation except for the charge resonance transition peculiar to the dimer cation unit. The charge resonance interaction in a benzene dimer cation is 0.66 eV, which is considerably smaller than the delocalization energy of 1.69 eV for the π orbitals in benzene.

Hopping of a charge occurs in a time that can be predicted from the resonance interaction energy. The resonance interaction takes a maximum value for a pair with identical configurations. However, the molecules surrounding the charged core are oriented to have maximum Coulombic interaction with the ionic core. In many cases, this causes a configuration of neutral molecules which are not equivalent to the ion core molecule(s), resulting in localization of the ion core in the cluster. In hot clusters, where the geometrical structure fluctuates due to the excitation of intermolecular vibration and rotational motion, hopping of the

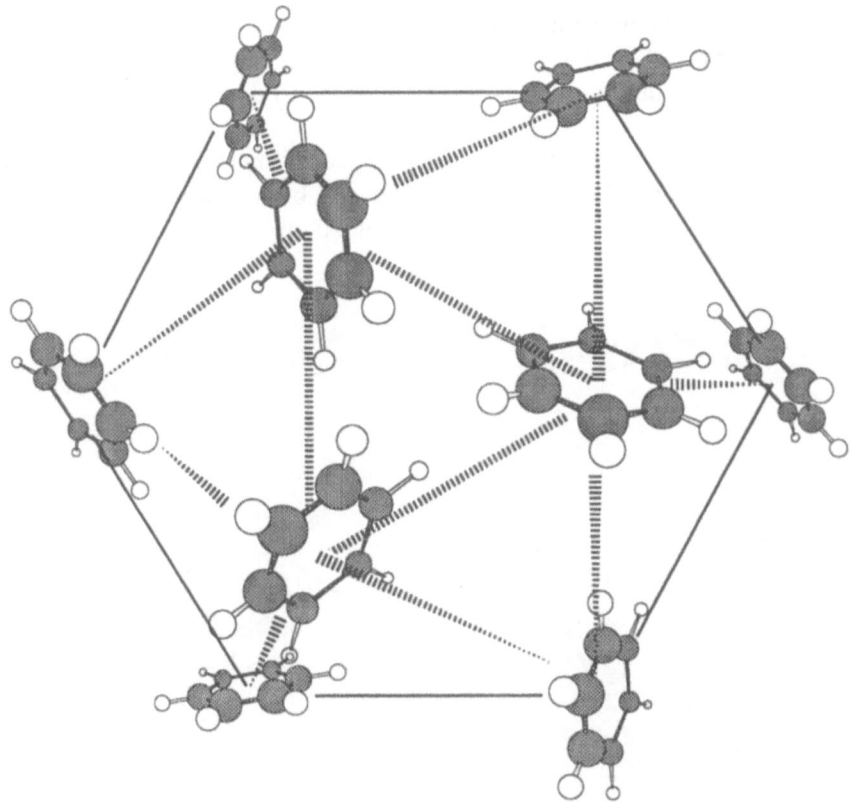

Fig. 5.2-5. A model of the $(C_6H_6)_{12}$ icosahedral cage that may include a benzene dimer cation in the core position

charge is very likely because the temporal configuration is better for resonance interaction.

Whetten and co-workers [24] reported the formation of doubly charged benzene clusters due to the ultrafast fusion of excitons. Figure 5.2-4 shows a time-of-flight mass spectrum of benzene clusters ionized by an ArF (193 nm) excimer laser. The smallest observed doubly charged benzene cluster is the 25-mer; the smaller doubly charged clusters immediately break up because of Coulomb explosion of the two positive charges. As an example of a very stable structure of benzene cation (singly charged) clusters, Whetten and co-workers [25] proposed the (poly)icosahedral packing model composed of a sandwich dimer ion core with 12 nearly equivalent nearest neighbors (total number of molecules, 14), and a pentagonal cap (the double icosahedron, 20), and so on [25]. This is based on the magic numbers of 14, 20, 24, and 27, each of which is one above the magic numbers for Ar cation clusters (13, 19, 23, and 26). The 12 benzene molecules surround a dimer cation forming an icosahedral cage (20 condensed spherically closed triangles). Figure 5.2-5 shows a model of the $(C_6H_6)_{12}$ icosahedral cage that might include a benzene dimer cation in the core position. The orientation of the benzene molecules has not been determined experimentally, but it has been estimated based on the neutral dimer structure where a hydrogen sits perpendicularly in the center of the π ring of the other benzene. This intermolecular binding is a type of "hydrogen-bonding" due to the Coulombic attraction.

One possible structure for the doubly charged 25-mer is a peanut-shell-like cage with two condensed icosahedrons separated by a shared triangular plane (Fig. 5.2-6). Each of the two cavities contains one benzene dimer cation, and thus

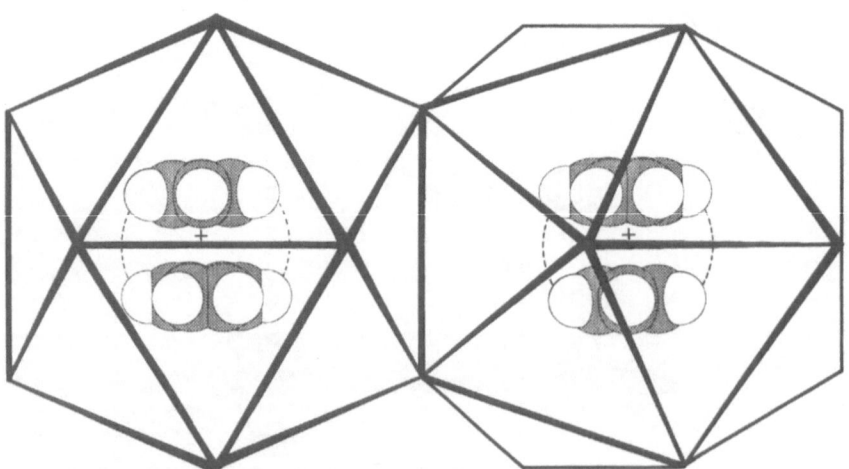

FIG. 5.2-6. A possible structure for the doubly charged benzene 25-mer. Two icosahedrons (shown in Fig. 5.2-5) are joined with a common triangular plane in the middle. Two benzene dimer cations are situated in the two cavities

the 25-mer is composed of 21 neutral benzene and two dimer cations $(2 \times 12 - 3 + 2 \times 2 = 25)$. Completion of the two cavities is essential to the coexistence of the two charges keeping the spatial charge balance. All neutral molecules in the cage sites must participate in the stabilization of the cores, preventing Coulomb explosion of the two positive charges.

5.2.2.2 Doubly Charged Hydrogen-Bonding Clusters

Stace [26, 27] found that the minimum stable size of doubly charged water clusters is $(H_2O)_{35}H_2^{2+}$. He also observed strong signals of $(H_2O)_n^{2+}$ $(n \geq 36)$ with intensities stronger than half of the doubly protonated ones for $n \geq 39$. He used a double-focusing, reverse-geometry mass spectrometer with an ion source voltage of 8 kV that carries metastable species to the detector [28]. Figure 5.2-7 shows a mass spectrum taken by a quadrupole mass spectrometer with an ion source energy of 2 eV. Because of its very low velocity, the metastable species dissociates into two protonated ions during the very slow flight down to the detector. Evidently the intensities of the non-protonated clusters $(H_2O)_n^{2+}$ are less than 20% of the protonated doubly charged clusters. More importantly, even the intensities of the protonated doubly charged clusters are very much weaker than those observed by Stace compared with the intensities of the protonated water cluster with an ^{18}O atom $(H_2O)_{n-1}(H_2^{18}O)H^+$ relative to the doubly charged signals. This means that many of the doubly charged clusters dissociate in a few microseconds leaving very stable doubly charged clusters. The strong sequence of the doubly charged protonated clusters starts from $n = 37$, although the signal of $(H_2O)_{35}(H^+)_2$ can also be seen at $m/z = 316$ with an intensity as weak as the background oil signals.

Water cation cluster spectra taken by electron impact as well as by direct photo-ionization are known to produce a famous magic number at $(H_2O)_{21}H^+$. This cluster, $(H_2O)_{21}H^+$, is expected to have a central H_3O^+ ion in a deformed pentagonal dodecahedron cage of $(H_2O)_{20}$ [29]. Condensation of the two $(H_2O)_{21}H^+$ clusters with a shared pentagonal ring produces the smallest stable doubly protonated water cluster $(H_2O)_{37}(H^+)_2$. A structural model is shown in Fig. 5.2-8. This is a peanut-shell-like structure similar to that of the smallest doubly charged benzene cluster model shown in Fig. 5.2-6. In order to show the pentagonal dodecahedron unit clearly, the core H_3O^+ ions are simply located in the cavities. Since the hydrogen-bonding energy of the hydroxonium ion with water molecules is much larger than that of two water molecules, the core ions interact strongly with cage water molecules so that the hydrogen-bonding network of the cage must be subjected to serious deformation from the basic structure given in Fig. 5.2-6. The signal of the doubly protonated cluster $(H_2O)_{35}(H^+)_2$ is quite weak in our mass spectrum compared with that of $(H_2O)_{37}(H^+)_2$. The absence of $(H_2O)_{35}^{2+}$ indicates the fatal instability of the doubly charged cage. The location of two hydroxonium ions in the cavities of the completed cage may be responsible for the stability of the doubly charged clusters. Although this smallest unit is reasonably stable, attachment of some more water molecules to this cage

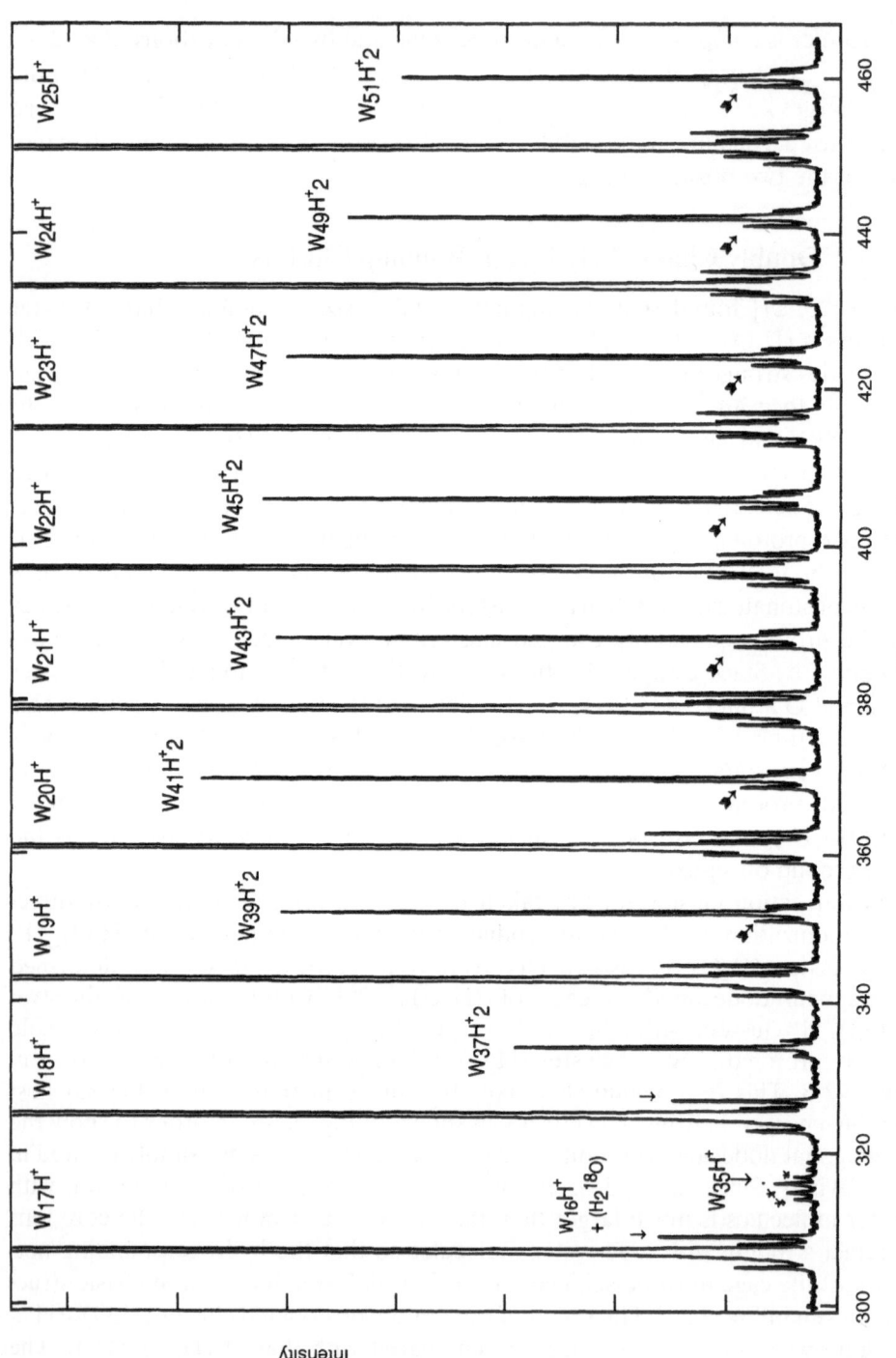

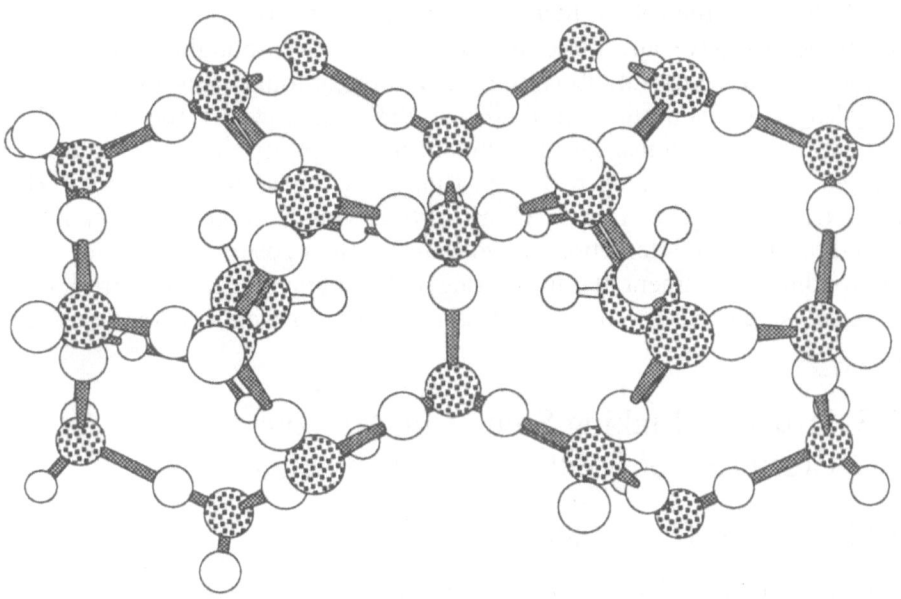

FIG. 5.2-8. A structural model of a doubly charge water cluster, $(H_2O)_{37}(H^+)_2$. In order to show the pentagonal dodecahedron unit clearly, the core H_3O^+ ions are located in the cavities. The actual cage must be deformed because of the electrostatic interaction between the core cations and the cage molecules

could reinforce the doubly protonated cluster because of the network formation of the second layer of the cage. Whatever the actual network structure of the peanut-shell-like cage turns out to be, the minimum number of water molecules surrounding one hydroxonium ion is 20, and for two hydroxonium ions the minimum number of water molecules in the two cavities is 35. The caging of each charge is a necessary condition to account for the two charges being very close. Theoretical calculations may prove to be important in providing a more convincing explanation for the experimental results.

In contrast to water clusters, ammonia cluster ions do not show any cage-type magic number. The smallest doubly charged ammonia cluster is $(NH_3)_{51}^{2+}$ [30]. Interestingly, these doubly charged cluster signals are followed by protonated and doubly protonated species, although the non-protonated signal becomes relatively stronger for larger clusters. Since an ammonia molecule has one lone-pair

◄ ───

FIG. 5.2-7. Mass spectrum of pure water clusters ionized at 70 eV. The ion energy was set to 2 eV and the ions were mass-analyzed by a quadrupole mass spectrometer. *Thin arrows*, protonated water clusters with an ^{18}O atom; *thick arrows*, doubly charged nonprotonated water clusters; *asterisks*, impurity signals

orbital that takes the role of hydrogen acceptor in hydrogen bonding, ammonia clusters do not have any three-dimensional hydrogen-bonding network. One possible structure is essentially composed of three or four chains originating from the hydrogen atoms of the central NH_3^+ or NH_4^+ ion. However, the chains always end with hydrogen atoms. The attachment of an NH_3^+ or NH_4^+ ion would not be expected without assuming hydrogen transfer in a pair unit of NH_3—NH_3^+ $\rightleftarrows$ NH_2—NH_4^+ or NH_3—NH_4^+ $\rightleftarrows$ —HNH_2—NH_4^+. This hydrogen transfer may be induced through the interaction of two chains with opposite directions of donor–acceptor dipole alignment. It is interesting that ammonia clusters require 1.4 times more neutral molecules than water clusters in order to avoid Coulomb explosion.

5.2.3 Structure-Making Solutes in Aqueous Environments: Clustering of Solute and Solvent Molecules

5.2.3.1 Hydrophobic Hydration

Aqueous solutions exhibit unique structural features for solute–solvent associates owing to the fact that they form three-dimensional hydrogen-bonding networks. In 1957, Frank and Wen [31] found that the heat capacity of an aqueous solution of alkyl ammonium increases with increasing solute concentration, in contrast to its dramatic decrease in alkaline halide aqueous solutions. In particular, the tetraalkylammonium ion showed a pronounced increase in heat capacity. They suggested that this was the result of increased order in the environmental water molecules around the alkyl group. The ionic region of tetraalkylammonium is surrounded by hydrophobic groups that may make the water molecules highly ordered through hydrogen bonding.

This phenomenon has been called "hydrophobic hydration," and results in the solvation of apolar residues of amphiphilic molecules in water [31, 32]. Computer experiments by Nakanishi *et al.* [33] gave a model for the hydration of tertiary-butanol (TBA) in an infinitely dilute aqueous solution on the basis of Monte Carlo calculations. They showed that the potential energy and structure of water tend to be stabilized by the introduction of one TBA molecule, and two strong hydrogen bonds between TBA and the surrounding water molecules favor the formation of a bulky, stable hydration shell. The hydrophobic hydration structure of these water molecules is shown in the density diagrams in Fig. 5.2-9, that clearly demonstrate the remarkable structures generated in the water. Figure 5.2-9(a) shows the oxygen atom distribution along the axis perpendicular to the C—O—H plane of a TBA molecule. The distance of each slice from the plane is indicated at the bottom of the diagrams. The structure generated in water around a TBA molecule is more pronounced than in pure water, and also more pronounced than that of methanol. It can also be seen that the water molecule hydrogen-bonded to the TBA molecule is in a key position for the formation of a cooperative hydration structure. The degree of structure promotion is expected to be highly dependent on the size and shape of hydrophobic groups of the solute

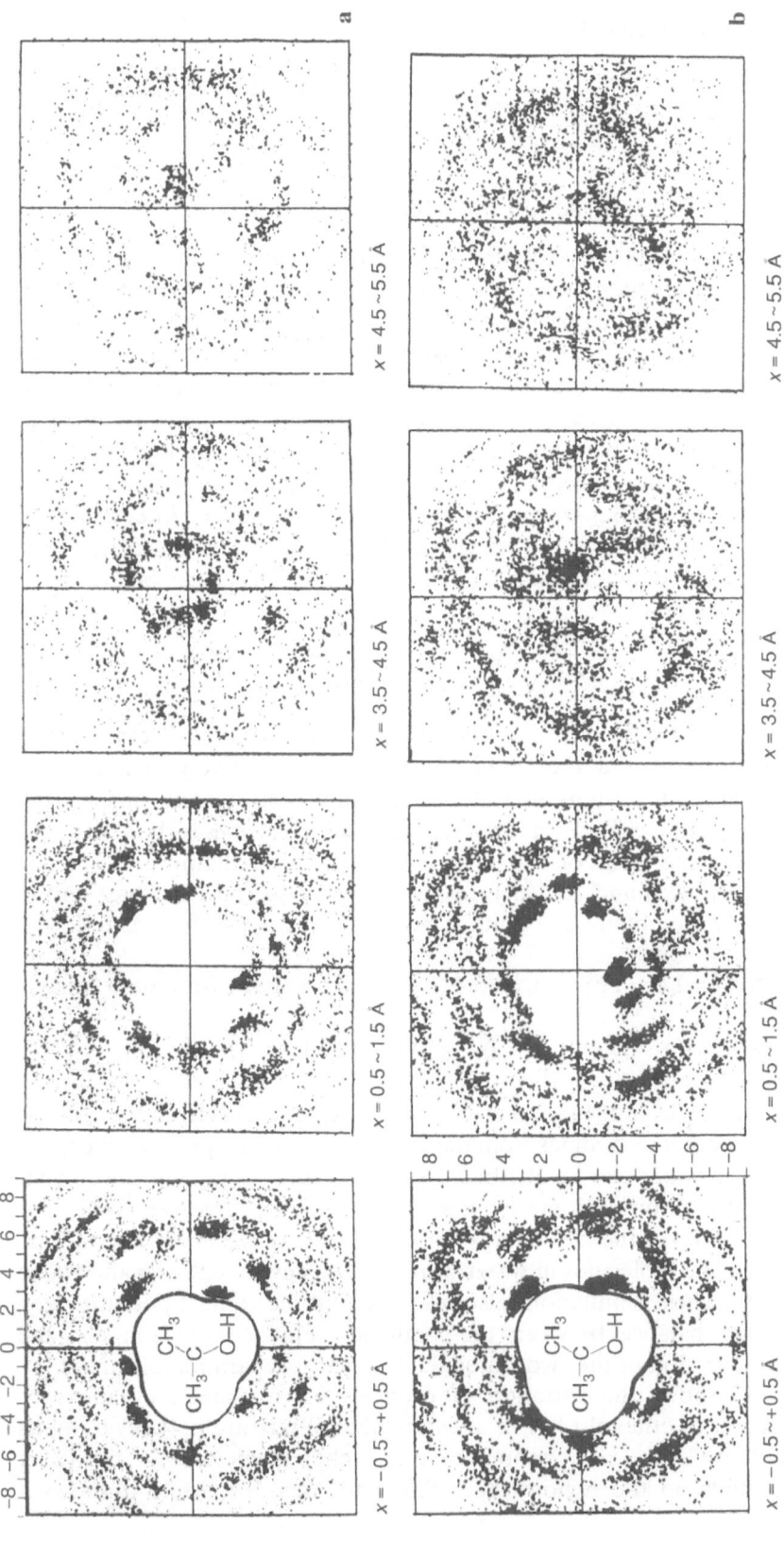

FIG. 5.2-9. Density distribution diagrams for **a** oxygen atoms and **b** hydrogen atoms of water around a tertiary-butanal (TBA) molecule. Each diagram is a sliced plane along the axis perpendicular to the C—O—H plane of the solute molecules. The distance from the C—O—H plane to the slice is indicated at the bottom of each diagram

275

species, but the most important factor is thought to be the adaptability of the hydrophobic group(s) to the various water cages.

5.2.3.2 Hydrophobic Interaction in Aqueous Solution

In the case of structure-making solutes, the interaction between the alkyl groups, the hydrophobic interaction, leads to a fairly large decrease in the partial molar volume of the solute with increasing solute concentration at low initial concentrations [33]. The term "hydrophobic interaction" (HI) stems from "the picturesque view that simple nonpolar solutes hate or have a phobia for water [34]," and hence they tend to avoid exposure to an aqueous environment as much as possible [34–36]. HI has been extensively studied (see the well-known review articles by Eisenberg and Kauzmann [37], Tanford [38], and Ben-Naim [34, 35]). As discussed in detail by Ben-Naim [36], HI is a "solvent-induced interaction" in an aqueous environment. The free energy change caused by this interaction for a pair of solute molecules at a distance σ_1 in a solution $[\delta G^{HI}(\sigma_1)]$ can be evaluated using the following relation:

$$\delta G^{HI}(\sigma_1) = \Delta\mu_D^0 - 2\Delta\mu_M^0 \tag{5.2-3}$$

where $\Delta\mu_D^0$ and $\Delta\mu_M^0$ are the standard free energy changes in solution of the dimer in a sticking position and the monomer species, respectively. To evaluate methane–methane interactions in water, Ben-Naim replaced the $\Delta\mu_D^0$ by the standard free energy change in ethane, $\Delta\mu_E^0$, at a carbon–carbon distance $\sigma_1 = 1.533\,\text{Å}$. On the basis of this approximation, the enthalpy and entropy contributions $[\delta H^{HI}(\sigma_1)$ and $\delta S^{HI}(\sigma_1)$, respectively] to $\delta G^{HI}(\sigma_1)$ in water and in ethanol at 10°C were obtained in units of kcal mol^{-1} (1 cal = 4.184 J) as follows:

$$\delta H^{HI}(\sigma_1) = 1.6, \quad T\delta S^{HI}(\sigma_1) = 3.4, \quad \delta G^{HI}(\sigma_1) = -1.99 \quad (\text{in water}) \tag{5.2-4}$$

$$\delta H^{HI}(\sigma_1) = -1.2, \quad T\delta S^{HI}(\sigma_1) \approx 0, \quad \delta G^{HI}(\sigma_1) = -1.34 \quad (\text{in ethanol}) \tag{5.2-5}$$

It is clear that the main contribution to $\delta G^{HI}(\sigma_1)$ in water comes from the entropy term, whereas in ethanol it comes from the enthalpy term. The positive energy of the HI process in water can be attributed to a structural change in the solvent such that the average binding energy of a water molecule decreases. Molecular dynamics simulations of the free energy, entropy, and internal energy of association of two methane molecules in water were in good agreement with the above estimations [39]. These simulations also revealed that the contact potential-well results from the balance between the methane–methane repulsion and the cavity–cavity attraction of the two water shells at short separation distances.

Figure 5.2-10 shows the variation of $\delta G^{HI}(\sigma_1)$ for methane-methane with the mole fraction of ethanol (X_E) in a water–ethanol mixture [34]. Two points should be noted: (1) the HI is more active at higher temperatures, and (2) it increases gradually in the regions $X_E < 0.03$ and $X_E > 0.2$, while it shows a

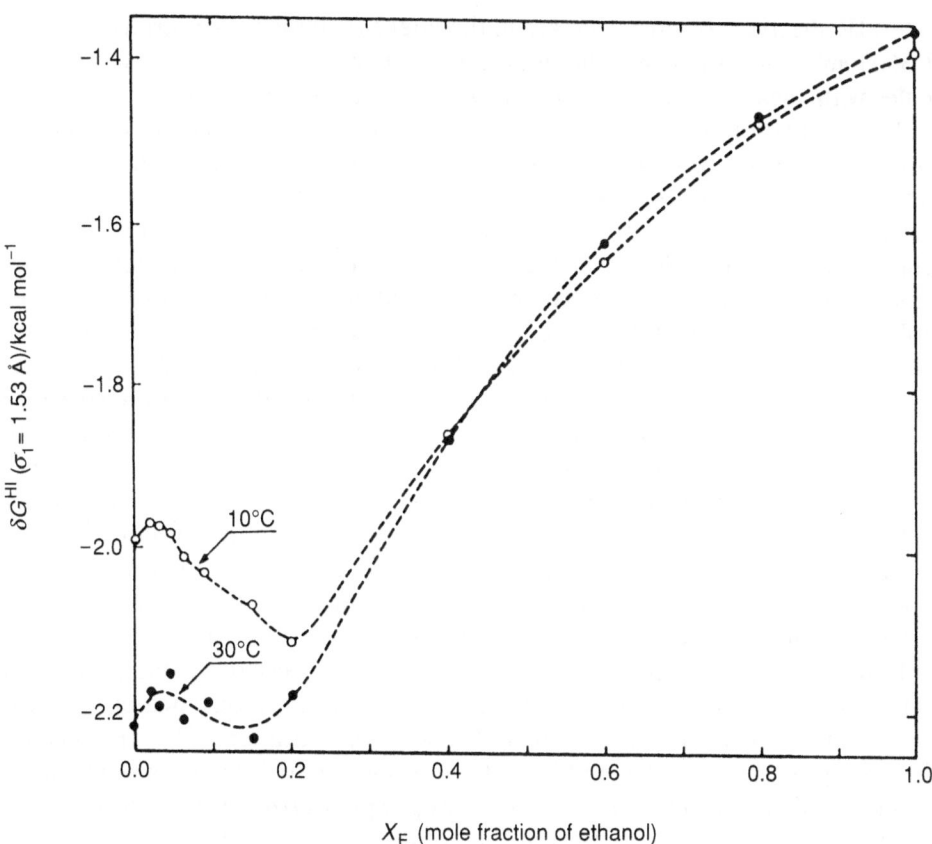

FIG. 5.2-10. Variation of $\delta G^{HI}(\sigma_1)$ (change of the free energy due to the hydrophobic interaction of two methane molecules) with the mole fraction of ethanol in a water–ethanol system. (From [34], with permission)

decrease when $0.03 \leq X_E \leq 0.2$. It appears that there are three concentration regions that provide different environments for methane–methane pairs. The temperature effect (1) indicates that the HI is driven by the positive entropy and enthalpy changes of that interaction [40, 41]. Because an increase in the strength of the HI is accompanied by a decrease in the structure of the solvent, it would be expected that the degree of structure in the water decreases with increasing ethanol in the region $X_E < 0.03$. This has recently been confirmed by Nishi *et al.* [42].

5.2.3.3 Structure Destruction of Bulk Water in the Presence of Stacked Ethanols

The structure of aqueous solutions of ethanol with $X_E < 0.03$ has been studied by three different methods: infrared (IR) absorption spectroscopy, X-ray diffraction, and mass spectrometry of the clusters isolated from liquid droplets [42]. For water, IR spectra showed a decrease in the intensity of the O—H stretching band at $3600\,cm^{-1}$, indicating that the addition of ethanol reduces "free" water molecules with "gas-like" O—H stretching vibrational frequencies. This change may lead to speculation that the structure of environmental water becomes ice-like in nature. However, the absorption of water at $3250\,cm^{-1}$, where low-temperature water or ice shows a prominent peak due to stable hydrogen-bonding network formation, also decreased on the addition of a small amount of ethanol. Thus the relative intensity at $\sim 3400\,cm^{-1}$ (the peak position of water absorption at room temperature) became much stronger. This simultaneous decrease of "free water molecules" and "icy hydrogen bonds" is expected for highly pressurized water. X-ray diffraction measurements provided results which conformed this expectation, not only for the local structure around the solutes but also for the structure of bulk water. The observed structure function $s \cdot I(s)$ of an aqueous solution with $X_E = 0.02$ showed a striking resemblance to the pure water curve at 1 kbar (987 atm). In the radial distribution function shown in Fig. 5.2-11, the linear hydrogen-bond peak at $2.8\,Å$ became notably weaker than that of pure water at 1 atm. New peaks appeared at 3.2 and $3.8\,Å$. The former was attributed to the O—O distance of angular hydrogen-bonding water pairs, and the latter to the distance from a ethanol carbon atom to an oxygen atom of a "cage" water molecule. Soper and Finney [43] reported that the existence of hydration shells of water around the methanol molecule at a carbon-to-water distance of $\sim 3.7\,Å$ is confirmed by neutron diffraction and isotope substitution. The IR spectral changes (loss of free water and reduction of linearly hydrogen-bonding water) are in accordance with those expected for highly pressurized water. Thus one can say that the bulk water structure in this solution is considerably different from that of pure water at 1 atm.

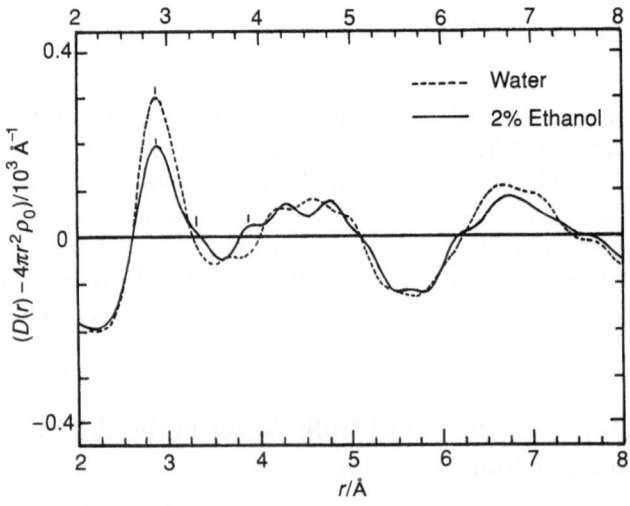

FIG. 5.2-11. Radial distribution function of the form $D(r) - 4\pi r^2 \rho_0$ for pure water (*broken line*) and an aqueous solution of ethanol with $X_E = 0.02$ (*solid line*)

This situation can be understood when we consider the mass spectral changes of clusters isolated from binary solutions. The mass spectrum of a 2% ethanol solution showed that the water-to-ethanol ratios of the prominent peaks are approximately 2($\pm$1) for both the low mass ($m/z = 320$~500) and high mass ($m/z = 320$~500) clusters. The larger clusters can be regarded as simply adducts of the small clusters. The dominant clusters were ethanol tetramer- and pentamer-hydrates in the low mass region, and hexamer-, heptamer-, and octamer-hydrates in the high-mass region. However, the ratio gradually increased with decreasing solute concentration. At 0.2% ethanol, the average water-to-ethanol ratio of dimer hydrates is approximately 20($\pm$5), although ethanol monomer hydrates and pure water clusters are dominant. The change in the water-to-ethanol ratio indicates that the thickness of the hydration shells become very thin with increasing solute concentration. Since pure ethanol dimer, trimer, and even tetramer signals appeared mainly in the low mass region of $m/z = 80$~200 [44], ethanol molecules are expected to associate with each other forming the core of the hydrated clusters. Cohesion of hydrophobic groups is a characteristic of HI that is caused by the "caging" of the hydrophobic groups by hydrogen-bonding networks. At the edge of the hydrophobic area, interfacial tension may arise and reinforce the hydrogen-bonding networks of the water molecules. The ratio of 2($\pm$1) observed in the 2% solution suggests that the number of water molecules is just sufficient to make a monolayer cage around the ethanol polymer core. Since the volume of the polymer core is quite large compared with the volume of water, the fluctuating motion of the core may destroy the thick water shell, leaving just the interfacial skin-like mantle for the core. This fluctuating motion in the hydrate clusters may also affect the structure of bulk water, producing a similar condition to that of highly pressurized pure water. A model of ethanol-hydrate clusters produced at very low ethanol concentrations is shown in Fig. 5.2-12. The thin water shell (or mantle) may be composed of linear hydrogen bonds (LHB) forming a two-dimensional network around the stacked ethanol polymer core.

As explained above, hydrophobic interaction is known to be driven by the entropy term of the change in the free energy of the association. Both theoretical and experimental evidence indicates that direct interaction between the hydrophobic groups is quite weak. Thus the cohesion of the hydrophobic groups is regarded as microscopic phase separation in an aqueous environment. It was believed that the solutes that show hydrophobic hydration make the structure of water molecules stronger. This is true as far as the interface area is concerned. However, the structure of the bulk water itself could be highly perturbed by the cohesively associated solute clusters, resulting in an unexpectedly large destruction of the linear hydrogen-bond networks.

5.2.3.4 Higher-Order Structure Generation in Ethanol–Water Mixtures

Ethanol–water mixtures have been used for extraction of specific constituents from fresh or dried plants, and also for sterilization. Interestingly, the ratio of the mixture is different for different purposes. The chemicals extracted from tea

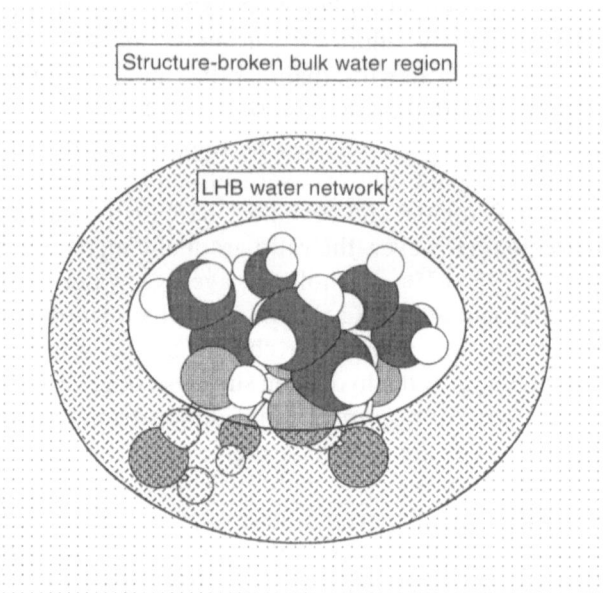

FIG. 5.2-12. Model of ethanol polymer-hydrates with a stacked hydrophobic core of ethyl groups and an interfacing water shell layer coupled with the OH ethanol groups. The structure of the outer "bulk" water area has been destroyed owing to the fluctuating motion of the large polymer core. *LHB*, linear hydrogen bond

leaves are very sensitive to the mixing ratio. A solution with 70% ethanol (by volume) can extract proteins from the membrane of plant cells by destroying the membrane structure. This dependence of the extraction on the mixing ratio could suggest that the hydrophobicity of the mixtures changes depending on the local structure.

On the basis of the hydrogen–hydrogen pair correlation function for a $1:19$ methanol–water ($X_E = 0.05$) mixture obtained from neutron diffraction experiments, Soper and Finney [43] stated that they found no evidence of a more ordered hydration structure than that in pure water. In the case of an ethanol–water solution, the concentration of ethanol has a pronounced effect on the structure of the mixture. As shown in Fig. 5.2-10, $\delta G^{HI}(\sigma_1)$ for methane–methane interaction becomes larger in the negative direction with an increasing mole fraction of ethanol (X_E) in the region of $0.03 \leq X_E \leq 0.2$ [34]. This means that the solvent structure in this concentration region is entirely different from that in dilute solutions with $X_E < 0.03$, where $\delta G^{HI}(\sigma_1)$ becomes smaller. Therefore a solution with an ethanol–water concentration of $1:19$ already possesses different properties from those which are characteristic of water. As shown by mass spectrometry of the clusters, the solution is already so concentrated that most water molecules are dispersed around the ethanol polymer clusters and no water-cluster structures were observed [44, 45]. The dominant clusters were pure etha-

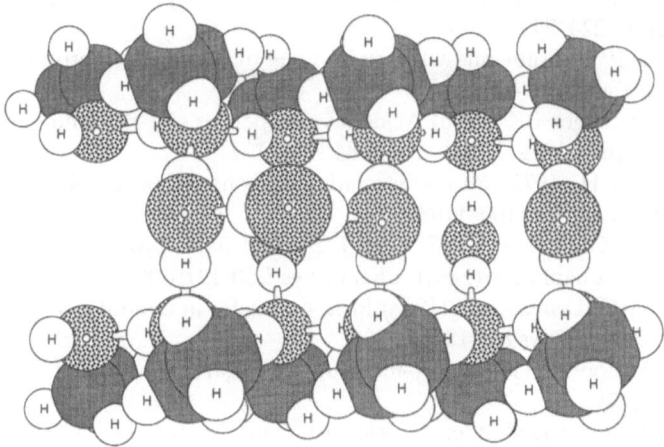

FIG. 5.2-13. Model of ethanol–water binary clusters seen around $X_E = 0.2$. The top and bottom layers are composed of hydrogen-bonded ethanol polymers. Water molecules bridge the two layers with hydrogen-bonding ice-like networks

nol clusters, and the average numbers of water molecules in the major ethanol polymer-hydrates are similar to (or less than) the ethanol numbers. As expected from the results for very dilute solutions, ethanol–ethanol association predominates over water–water association in this system.

X-ray diffraction and mass spectrometer studies of the mixtures in $0.2 \leq X_E < 1.0$ revealed that the structure of the clusters does not change very much with increasing ethanol, although the number of hydrogen bonds does decrease. The radial distribution function of pure ethanol showed C—C peaks characteristic of stacked ethyl groups, while the hydrogen-bond peak at 2.8 Å was very weak. On the basis of the coordination numbers obtained from X-ray analysis and the mass spectral changes of the fragmented clusters, a model of ethanol–water binary clusters was produced, as shown in Fig. 5.2-13 [46, 47]. This model is composed of two hydrophobic layers with a bridging water layer in between. With decreasing water content, the lengths of the ethanol polymer chains becomes shorter and shorter. At the limit, where there is no water, the hydrogen bond between ethanols is expected to switch from one to the other very frequently. This activity must be considered when structure generation in solution systems is investigated. This is important in an understanding of the functionality of biological molecules in an aqueous or a biological environment.

References

1. Rose J (1961) Dynamic physical chemistry. Pitman, London
2. Caldin EF, Gold V (eds) (1975) Proton transfer reactions. Chapman and Hall, London

3. Weller A (1958) Protolytische reaktionen angeregter oxyverbindungen. Z Phys Chem Neue Folge 17:224–245

4. Nagakura S (1954) Study on hydrogen bonding by near UV absorption spectroscopy. J Chem Soc Jpn 75:734–737

5. Harris CM, Selinger BK (1980) Acid–base properties of 1-naphthol. Proton-induced fluorescence quenching. J Phys Chem 84:1366–1371

6. Hara K, Baba H (1975) Photodissociation of α-naphthol in solution: Influence of hydrogen bonding. 71:1100–1108

7. Cheshnovsky O, Leutwyler S (1985) Excited-state proton transfer in neutral microclusters: α-naphthol·(NH_3). Chem Phys Lett 121:1–8

8. Cheshnovsky O, Leutwyler S (1988) Proton transfer in neutral gas-phase clusters: α-naphthol. J Chem Phys 88:4127

9. Lias SG, Ausloos P (1975) Ion–molecule reactions. American Chemical Society, Washington, DC

10. Jouvet C, Lardeux-Dedonder C, Richard-Viard M, Solgadi D, Tramer A (1990) Reactivity of molecular clusters in the gas phase. Proton transfer reaction in neutral phenol–$(C_2H_5NH_2)_n$. J Phys Chem 94:5041–5048

11. Kim SK, Li S, Bernstein ER (1991) Excited-state intermolecular proton transfer in isolated clusters: 1-naphthol/ammonia and water. J Chem Phys 95:3119–3128

12. Breen JJ, Peng LW, Willberg DM, Heikal A, Cong P, Zewail AH (1990) Real-time probing of reactions in clusters. J Chem Phys 92:805–807

13. Hineman MF, Brucker GA, Kelley DF, Bernstein ER (1992) Excited-state proton transfer in 1-naphthol/ammonia clusters. J Chem Phys 97:3341–3347

14. Knochenmuss R, Leutwyler S (1989) Proton transfer from 1-naphthol to water: Small cluster to the bulk. J Chem Phys 91:1268–1278

15. Steadman J, Syage JA (1990) Picosecond mass-selective measurements of phenol–$(NH_3)_n$ acid–base chemistry in clusters. J Chem Phys 92:4630–4633

16. Steadman J, Syage JA (1991) Time-resolved study of phenol proton transfer in clusters. 3. Solvent structure and ion-pair formation. J Phys Chem 95:10326–10331

17. Syage JA, Steadman J (1992) Probing double-minima ion–molecule reaction coordinates by photoelectron spectroscopy of clusters: $PhOH^+ + NH_3 \rightarrow PhO + NH_4^+$. J Phys Chem 96:9606–9608

18. Syage JA (1993) Tunneling mechanism for excited-state proton transfer in phenol–ammonia clusters. J Phys Chem 97:12523–12529

19. Hineman MF, Kelley DF, Bernstein ER (1993) Proton transfer dynamics and cluster ion fragmentation in phenol/ammonia clusters. J Chem Phys 99:4533–4538

20. Volpel R, Hofmann G, Steidl M, Stenke M, Schlapp M, Trassl R, Salzborn E (1993) Ionization and fragmentation of fullerene ions by electron impact. Phys Rev Lett 71:3439–3441

21. Sattler K, Muhlbach J, Echt O, Pfau P, Recknagel E (1981) Evidence for Coulomb explosion of doubly charged microclusters. Phys Rev Lett 47:160–163

22. Ohashi K, Nishi N (1992) Photodissociation spectroscopy on charge resonance band of $(C_6H_6)_2^+$ and $(C_6H_6)_3^+$. J Phys Chem 96:2931–2932

23. Ohashi K, Nakai Y, Shibata T, Nishi N (1992) Photodissociation spectroscopy of $(C_6H_6)_2^+$. Laser Chem 14:3–14

24. Schriver KE, Hahn MY, Whetten RL (1987) Exciton fusion in molecular clusters. Phys Rev Lett 59:1906–1909

25. Schriver KE, Paguia AJ, Hahn MY, Honea EC, Whetten RL (1987) Are clusters of nonpolar molecules icosahedral? J Phys Chem 91:3131

26. Stace AJ (1988) Evidence of two stable forms of doubly and triply charged water clusters. Phys Rev Lett 61:306–309

27. Stace AJ (1990) Possible ion pairs in multiply charged water clusters Chem Phys Lett 174:103–107

28. Stace AJ, Shukla AK (1982) Preferential solvation of hydrogen ions in mixed clusters of water, methanol, and ethanol. J Am Chem Soc 107:5314

29. Nagashima U, Shinohara H, Nishi N, Tanaka H (1986) Enhanced stability of ion–clathrate structure of magic number water clusters. J Chem Phys 84:209

30. Coolbaugh MT, Peifer WR, Garvey JF (1989) Ion–molecule chemistry within doubly charged ammonia clusters. Chem Phys Lett 156:19–23

31. Frank HS, Wen WY (1957) Structural aspects of ion–solvent interaction in aqueous solutions: A suggested picture of water structure. Discuss Faraday Soc 24:133

32. Franks F, Ives DJG (1966) The structural properties of alcohol–water mixtures. Q Rev Chem Soc 20:1–44

33. Nakanishi K, Ikan K, Okazaki S, Touhara H (1984) Computer experiments on aqueous solutions. III. Monte Carlo calculation on the hydration of tertiary butyl alcohol in an infinitely dilute aqueous solution with a new water–butanol pair potential. J Chem Phys 80:1656–1670

34. Ben-Naim A (1974) Water and aqueous solutions. Plenum Press, New York, p 365

35. Ben-Naim A (1980) Hydrophobic interactions. Plenum Press, New York

36. Ben-Naim A (1989) Solvent-induced interactions: Hydrophobic and hydrophilic phenomena. J Chem Phys 90:7412–7425

37. Eisenberg D, Kauzmann W (1969) The structure and properties of water. Oxford University Press, Oxford

38. Tanford C (1976) The hydrophobic effect. Wiley, New York

39. Smith DE, Haymet ADJ (1993) Free energy, entropy, and internal energy of hydrophobic interactions: Computer simulations: J Chem Phys 98:6445–6454

40. Ben-Naim A, Yaacobi M (1974) Effects of solutes on the strength of hydrophobic interaction and its temperature dependence. J Phys Chem 78:170–175

41. Yaacobi M, Ben-Naim A (1974) Solvophobic interaction. J Phys Chem 78:175–178

42. Nishi N, Takahashi S, Matsumoto M, Tanaka A, Muraya K, Takamuku T, Tamaguchi T (1995) Hydrogen-bonding cluster formation and hydrophobic solute association in aqueous solution of ethanol. J Phys Chem 99:462–468

43. Soper AK, Finney JL (1993) Hydration of methanol in aqueous solution. Phys Rev Lett 71:4346–4349

44. Finney JL, Soper AK (1994) Solvent structure and perturbations in solution of chemical and biological importance. Chem Soc Rev 1–10

45. Nishi N, Koga K, Ohshima C, Yamamoto K, Nagashima U, Nagami K (1988) Molecular association in ethanol–water mixtures studied by mass spectrometric analysis of clusters generated through adiabatic expansion of liquid jets. J Am Chem Soc 110:5246–5255

46. Matsumoto M, Nishi N, Takamuku T, Yamaguchi T, Saita M (1995) Structure of clusters in ethanol–water binary solutions studied by mass spectrometry and X-ray diffraction. Bull Chem Soc Jpn 68(7):1775–1783x

47. Nishi N (1990) Aqueous molecular clusters isolated as liquid fragments by adiabatic expansion of liquid jets. Z Phys D-Atoms Mol Clusters 15:239–255

5.3
Polymer Effects

Teizo Kitagawa and Kazuhiko Seki

When several identical chemical groups are incorporated into a single molecule and interact with each other, they exhibit different properties from those they have when they are isolated. These new properties, arising from direct or indirect interactions of identical chemical groups in a single molecule, are generally called "polymer effects," but when the interaction is caused by short-range forces they are really oligomeric effects. The chemical groups in question can be skeletal repeating units of a linear polymer, their side chains, or separate groups buried in a coiled polymer. Although there are many different features of polymer effects, this chapter focuses mainly on energy levels and reactivities. Simple matrix effects by polymer chains and crystallization effects are not discussed.

Changes in the electronic energy levels of monomers upon incorporation into polymers are important in the design of the electronic properties of materials, particularly for conducting polymers. Recent resonance Raman studies on conducting polymers confirmed that so-called "self-localized excitations" such as neutral and charged solitons, polarons, and bipolarons correspond to radicals and ions, radical ions, and divalent ions, respectively, in the corresponding oligomeric compounds. The wavelength of the maximum absorption depends on the length of the oligomeric units, and the apparent absorption of some useful materials appears very broad and without structure owing to the coexistence of a variety of effective oligomeric lengths.

In oligomeric molecules, cooperativity or feedback control can be displayed between the multiple reaction sites. This property is very important in regulating the reactivity of enzymes in biological systems, and is called "allosteric effects," although such phenomena have not so far been reported in general chemistry. The energy-level problems of synthetic polymers and allosteric effects in biological systems are discussed below.

5.3.1 Polymer Effects on Energy Levels

Incorporation of functional groups into a macromolecule can happen in two ways: via a covalent bond or a non-covalent bond. In the latter case, the interactions between the functional group and the macromolecule are generally so weak that the ionization energies obtained from UV photoemission spectroscopy (UPS) can be interpreted on the basis of that of the pure solid. The ionization energy of a molecule embedded in a polymer matrix, as in the case of a pure solid, is generally lower than that of an isolated molecule owing to the stabilization of the generated cation by the electronic polarization of its surroundings. When polymer chains are oriented, the embedded functional groups can also be aligned and can be monitored by polarization measurements. Their orientation-dependent properties can be altered by temperature changes near the glass transition temperature of the polymer.

In some special cases, such as clathrate-forming and transition metal-coordinated polymers, interactions between the functional group and the macromolecule are fairly strong and accordingly are expected to yield interesting polymer effects. Unfortunately, however, the change of energy levels in these systems have not been extensively studied and therefore, these effects will not be discussed. In contrast, when a functional group is incorporated into a macromolecule via a covalent bond, the effects on the functional group are large and some interesting results have been obtained [1, 2]. Details of such cases are discussed below.

5.3.1.1. Incorporation as a Side Chain

The energy levels of the side-chain functional groups are influenced by the main chain. The magnitude of this influence depends on the orientation of the side chain relative to the main chain and the separation of the interacting energy levels. The two main origins of these effects are main-chain–pendant interactions and pendant–pendant interactions.

Some of the most important materials in this class are vinyl polymers $(CH_2—CHX)_n$ in which a functional group, X, is introduced as a pendant of the polyethylene chain $(CH_2)_n$. The electronic structure of polyethylene and its model compounds have been extensively studied [1]. It has a deep highest occupied molecular orbital (HOMO) (solid state ionization threshold energy $I_s^{th} = 8.5\,eV$) and a high lowest unoccupied molecular orbital (LUMO) (even above the vacuum level), making this polymer electrically and optically inactive.

For vinyl polymers with aromatic pendants, the HOMO and LUMO of the pendant are located between those of the polyethylene chain, and dominate the electronic properties of the polymer. The similarity between the electronic structures of $(CH_2CHX)_n$ and XH is illustrated in Fig. 5.3-1, where UPS and UV absorption spectra of polystyrene are compared with those of benzene ($X = C_6H_5$) and ethyl benzene [2].

Figure 5.3-1 shows some polymer effects. The effect of the main-chain–pendant interaction on the HOMO of an aromatic pendant leads to a lowering of the ionization threshold energy I_s^{th}, which appears as the shift of the right-hand onset of the UPS spectrum in Fig. 5.3-1 (left). The HOMO is raised by (i) the inductive effect of the alkyl chain, and (ii) the slight mixing of the HOMO wavefunction with those of the lower-lying electronic state of the main chain.

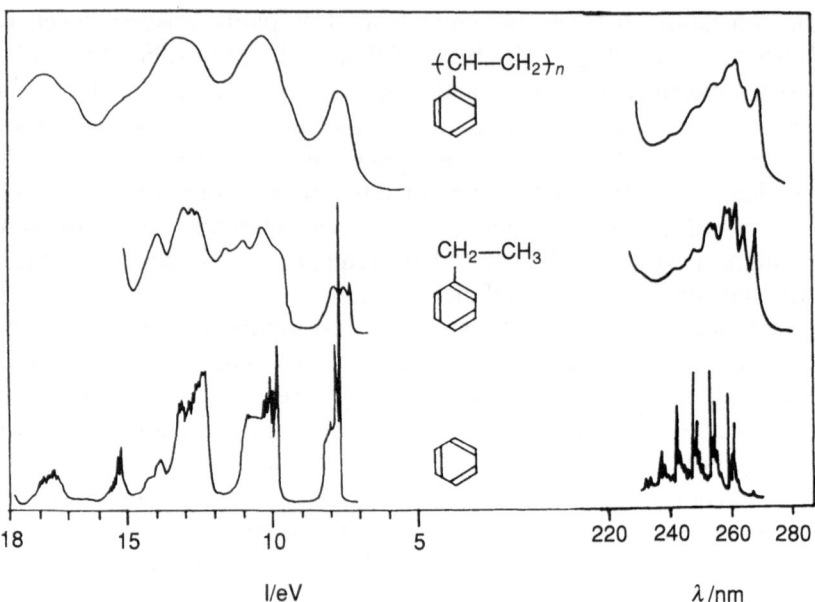

FIG. 5.3-1. Comparison of ultraviolet photoelectron spectroscopy (UPS) spectra (*left*) and optical absorption spectra (*right*) of polystyrene (CH$_2$CH—C$_6$H$_5$)$_n$ [2], ethylbenzene [2], and benzene. (From [1] and [2], with permission)

On the other hand, pendant–pendant interaction does not seem to have a significant effect on the electronic structure, as revealed by the similarity between the UPS spectra of CH$_3$X, XCH$_2$CH$_2$X, and (CH$_2$CHX)$_n$. It should be noted, however, that transport-related properties such as the fluorescence spectrum can be strongly affected even by such a weak pendant–pendant interaction, as shown by the excimer emission in poly(vinyl carbazole) [3]. The insensitivity of an aromatic pendant to the alkyl chain length as well as to the introduction of another pendant may be because of (i) the large HOMO energy difference, or (ii) similar electronegativity of the constituent atoms in the pendant and the main chain.

This situation becomes significantly different if either the pendant or the main chain is changed. Vinyl polymers with nonaromatic pendants, such as —CH$_3$, —F, —Cl, —OH, and —CH=CH$_2$, show different trends from those with aromatic pendants. The ionization energies corresponding to the ionization of either the pendant (—Cl, —OH, and —CH=CH$_2$) or the backbone chain (—CH$_3$ and —F) are dependent on the alkyl chain length, and in some cases also on the introduction of another pendant group. These strong effects are caused by (i) significant mixing of the wavefunctions, and (ii) the strong inductive effects of the substituents because of the large differences in the electronegativity of the constituent atoms of the pendant and the main chain.

Another situation occurs when the backbone chain is changed from polyethylene to polysilane $(SiH_2)_n$, which is a Si analog of polyethylene and is important because of its interesting electronic and electrical properties [4]. These polymers are usually synthesized in the permethylated form in order to avoid hydrolysis in air. The ionization energy of the permethylated polysilane chain (5.9 eV) is much smaller than that of polyethylene (8.5 eV) owing to the smaller electronegativity of Si than of C. Thus the upper part of the occupied states of polysilane overlaps with those of aromatic pendants.

As an example, UPS spectra of poly(methylphenylsilane) (PMPS) $(SiMeC_6H_5)_n$ and related compounds are shown in Fig. 5.3-2 [5]. The spectrum of PMPS can be interpreted as the superimposition of the spectra of the backbone chain (b) and the pendant (c–e), with the HOMO being mainly derived from the polysilane main chain. Comparison of PMPS with alkyl-substituted polysilanes

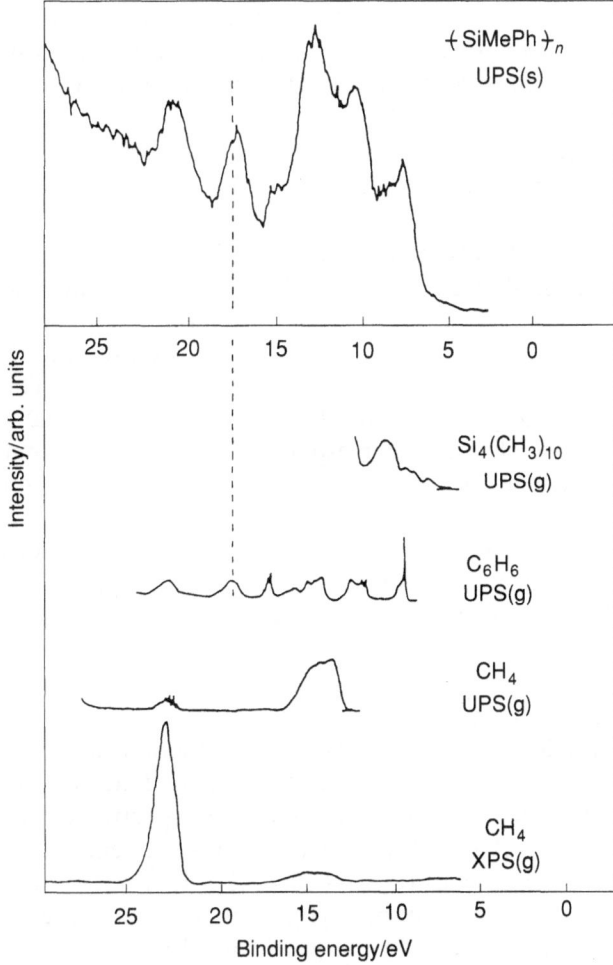

FIG. 5.3-2. UPS or X-ray photoelectron spectroscopy (XPS) spectra of poly(methylphenylsilane) (PMPS) $(SiMeC_6H_5)_n$ and related compounds. Note that the topmost valence states are mainly derived from the states of the main Si chain. (Redrawn from [5], with permission)

revealed that the HOMO energy of the chain in PMPS is lowered by 0.6 eV owing to the main-chain–pendant interaction, indicating an effective mixing of the wavefunctions. This demonstrates the unique ability of polysilanes to produce a strongly delocalized HOMO with low ionization energy, which contains almost equal contributions from the pendant and the main chain.

5.3.1.2 Incorporation in the Main Chain

In discussing the incorporation of a functional group into the main chain, we will focus our attention on aromatic functional groups. Incorporation of aromatic groups often leads to significant rearrangements of the electronic structure, the extent of which depends on the degree of delocalization of the electrons. The effect is small when the aromatic groups are separated by an alkyl group whose electronic states mix very little with those of the aromatics, as seen in the case of aromatic side chains. When the degree of mixing of electronic states increases, the delocalization of the electrons increases, leading to a large rearrangement of the electronic structure. The largest delocalization of electrons often arises from the direct conjugation of aromatic groups, as seen in many conducting polymers.

Such a rearrangement is well demonstrated in Fig. 5.3-3, where the change of the uppermost occupied states from benzene to p-sexiphenyl is shown by the UPS spectra observed and the orbital patterns (inset) [1, 5]. One of the doubly degenerate HOMOs of benzene (b_1 in Fig. 5.3-3) strongly interacts with the other one through direct bonding of phenyl rings, yielding widely split n levels in a n-mer which are delocalized through the molecule. However, the other HOMO in each ring (a_2 in Fig. 5.3-3) remains unaffected owing to the lack of electron density at two carbon atoms forming inter-ring CC bonds. They form a strong peak at the same energy as the HOMOs of benzene. This change in the electronic structure of these oligomers corresponds to the π-band formation in the polymer, poly (p-phenylene) $(C_6H_4)_n$ (at the top in Fig. 5.3-3). A similar band evolution was also traced in polyenes and oligothiophenes.

It should be noted that the formation of delocalized states is essential for achieving high conductivity by doping acceptors. The reason for this is firstly that the band formation raises the HOMO, leading to easier doping because of the resultant low ionization energy. A similar situation also occurs for the unoccupied states, resulting in easier doping of donors. Secondly, the formation of a wide band means a strong interaction among constituent units, which leads to large intra-chain mobility in the carriers formed by doping.

Figure 5.3-4 demonstrates the dependence of the degree of delocalization on the connecting unit, revealed by UPS spectra, for polymers with phenyl groups in the main chain [7]. The spectrum of solid benzene is also shown. The spectrum of poly(p-xylylene) $(C_6H_4—CH_2CH_2)_n$ is similar to that of benzene, indicating the localized nature of phenyl groups when they are separated by alkyl groups. Its ionization threshold energy ($I_s^{th} = 6.9$ eV, indicated by T in Fig. 5.3-4(b)) is also close to that of benzene (7.1 eV). This indicates that the levels derived from the

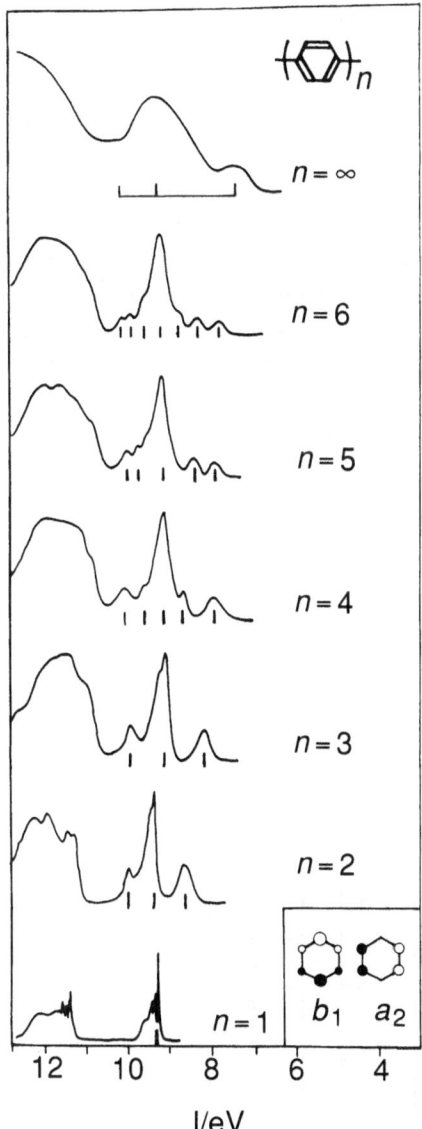

Fɪɢ. 5.3-3. UPS spectra of oligo-*p*-phenylenes and poly(*p*-phenylene) $(C_6H_4)_n$. (Redrawn from [6], with permission)

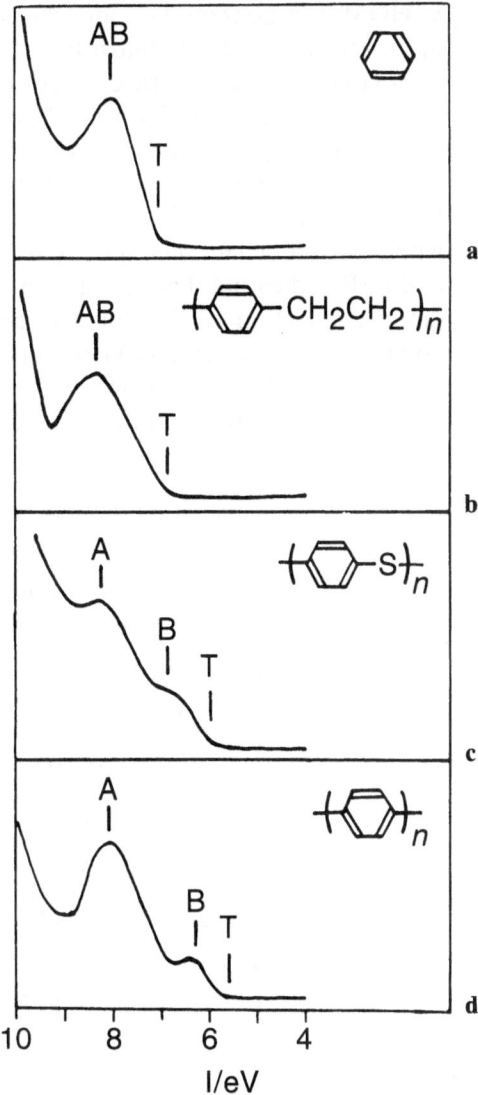

Fɪɢ. 5.3-4. UPS spectra of **a** benzene, **b** poly(*p*-xylylene) $(C_6H_4$—$CH_2CH_2)_n$, **c** poly(*p*-phenylene sulfide) (PPS) $(C_6H_4$—$S)_n$, and **d** poly(*p*-phenylene) (PPP). *T* denotes the onset corresponding to the ionization threshold. The feature *A* corresponds to the levels derived from the noninteracting a_2 orbital of benzene, while *B* corresponds to the uppermost level derived from the interaction of the b_1 orbital of benzene (see inset, Fig. 5.3-3). (Data from [7], with permission)

a_2 HOMO of benzene (A) and those from the b_1 HOMO of benzene (B) (see inset, Fig. 5.3-3) are at almost the same energy.

In contrast, the spectra of poly(p-phenylene sulfide) (PPS) $(C_6H_4—S)_n$ and poly(p-phenylene) (PPP) are different from that of benzene, with ionization threshold energies of 6.0 eV for PPS and 5.65 eV for PPP (marked T in Figs. 5.3-4(c) and (d), respectively) being much low than that of benzene because of the delocalization of the HOMO of benzene (b_1 in Fig. 5.3-3) throughout the whole polymer chain. The spectrum of PPS is intermediate between those of PPX and PPP, indicating that interaction through sulfur is slightly less effective than direct connection of the benzene rings. On the other hand, the spectrum of poly(p-phenylene vinylene) (PPV) $(C_6H_4—CH=CH)_n$ at the bottom of Fig. 5.3-5(a) shows delocalization to an extent similar to that seen in PPP, indicating effective delocalization throughout the vinylene group.

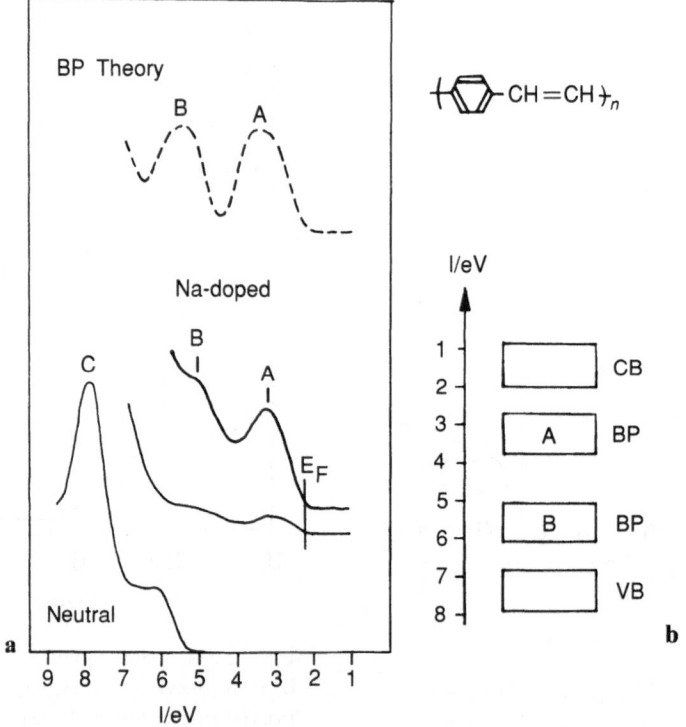

FIG. 5.3-5. **a** UPS spectra of poly(p-phenylene vinylene) (PPV) $(C_6H_4—CH=CH)_n$ in the neutral and Na-doped (ca. one Na per phenyl ring) states compared with the theoretically simulated spectrum for the bipolaron (*BP*) states. *A* and *B* are new peaks induced by doping, which are assigned to the two kinds of bipolaron states. The feature *C* corresponds to the levels derived from the noninteracting a_2 orbital of benzene (see inset, Fig. 5.3-3). E_F denotes the position of the Fermi level of the metal substrate [8]. **b** Energy diagram by valence effective Hamiltonian (VEH) calculation [8] for Na-doped PPV. *I* is the ionization energy and *CB* and *VB* denote the conductance band and the valence band, respectively. (From [8], with permission)

As mentioned at the beginning of this chapter, the formation of self-localized states such as solitons, polarons, and bipolarons is expected for conjugated polymers. These states are believed to form in the bandgap (between the HOMO and the LUMO) of the polymer. Figure 5.3-5(a) demonstrates the formation of such states in the UPS spectra of PPV for the neutral and Na-doped states [8]. By doping Na, two new states appear in the originally flat bandgap region of the spectrum. These can be interpreted as bipolarons formed by Na doping, for which theoretical calculations provide the energy diagram illustrated in Fig. 5.3-5(b). The spectrum simulated on the basis of these calculations shows good agreement with experimental results as seen in Fig. 5.3-5(a) [8]. The existence of such states is a unique aspect of conducting polymers, and can be regarded as a special kind of polymer effect.

5.3.2 Polymer Effects in Biology

The most characteristic feature of biological systems is a hierarchy of molecular organization. Monomer units such as amino acids, nucleotides, sugars, and fatty acids, which can be derived from simpler molecules such as carbon dioxide, water, and ammonia, are combined into various polymers such as proteins, nucleic acids, polysaccharides, and lipids. These macromolecules are further noncovalently bound together into supramolecular complexes, which ultimately are assembled into cell organelles such as mitochondrion, nucleus, and lysosome.

Monomer units used in biological macromolecules are generally versatile as a single molecule, but when they are incorporated into macromolecules their properties are modified. For instance, while mononucleotides can serve as signal transducers and energy carriers, a linear chain of four kinds of mononucleotides (DNA) serves to store information of biological activities; the sequence of monomers specifies proteins to be synthesized and all activities of living things. In the linear chain of amino acids, the sequence of monomers determines the three-dimensional structures of a whole molecule, which is specific and essential to active enzymes. Thus polymerization of monomers generates quite new properties which cannot be predicted on the basis of the properties of the monomers.

5.3.2.1 Higher-Order Structures

The three-dimensional structure of a linear chain of amino acids can be categorized into two major classes, fibrous and globular (Fig. 5.3-6). The fibrous proteins consist of polypeptide chains arranged in parallel along a single axis to yield a long fiber or sheet. Fibrous proteins are physically tough and insoluble in water or dilute salt solutions. They are basic structural elements in the connective tissue of higher animals, as seen for collagen, α-keratin, and elastin. In globular proteins, however, the polypeptide chains are tightly folded into compact spherical

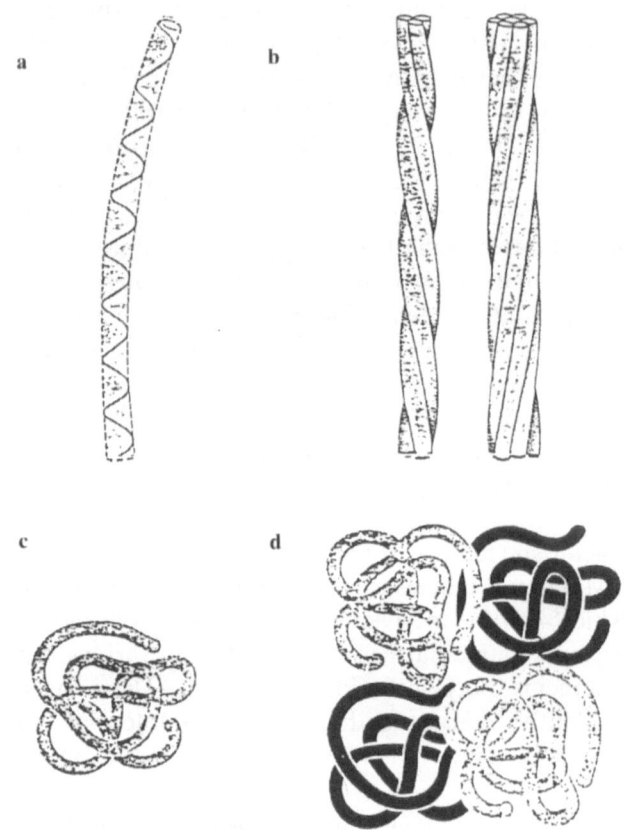

FIG. 5.3-6. **a,b** Fibrous and **c,d** globular proteins. **a** and **c** consist of a single polypeptide, while **b** and **d** are composed of two or more peptides

or globular shapes. Most globular proteins are soluble in aqueous systems and are usually mobile or dynamic in the cell. A protein structure is characterized by higher-order structures.

The sequence of amino acid residues along the covalent backbone of a polypeptide chain is known as its primary structure. The secondary structure refers to regular recurring arrangements in space of the polypeptide chain, and these are broadly classified as helices, planar sheets, turns, and randomly coiled structures. The tertiary structure refers to the arrangement of the secondary structural elements of a polypeptide chain in three dimensions to form the compact, tightly folded structure which is typical of globular proteins. Searches for a given amino acid residue in each secondary structure in various proteins by X-ray crystallographic analysis have made it possible to deduce qualitatively which secondary structure each amino acid favors and partially to infer a three-dimensional structure from the primary structure.

When a molecule consists of more than two peptide chains (Fig. 5.3-6, (b) and (d)) which are not covalently bound but are strongly associated, the relative arrangements of the constituent subunits (or protomers) are easily altered. The quaternary structure refers to the way in which individual subunits are arranged in relation to each other. The energy necessary to change the structure of proteins becomes less as the order of structure becomes higher. The quaternary structure of a globular protein, which controls its function, can easily be changed by simple protonation of an amino acid residue and is therefore physiologically very important. This kind of effect is called an "allosteric effect."

5.3.2.2. Multiple Ligand Binding and Allosteric Effects

The simplest case of polymer effects on reactivity is the binding of ligands to a macromolecule with multiple equivalent binding sites. The binding of the ith ligand is represented by

$$ML_{i-1} + L \leftrightarrow ML_i \qquad (5.3\text{-}1)$$

where M, L, and ML_i denote a macromolecule with q equivalent binding sites, a ligand, and a macromolecule with i bound ligands, respectively. The macroscopic equilibrium constant for the reaction in Eq. (5.3-1) is given by

$$K_i' = [ML_i] / \{[ML_{i-1}][L]\} \qquad (5.3\text{-}2)$$

Since the number of vacant sites of ML_{i-1} is $q-(i-1)$ and the number of occupied sites of ML_i is i, the microscopic binding constant of individual sites, K_i, is related to the macroscopic binding constant by Eq. (5.3-3).

$$K_i = [i/(q-i+1)]K_i' \qquad (5.3\text{-}3)$$

Then the concentration of ML_i is given by Eq. (5.3-4).

$$\begin{aligned}[ML_i] &= [(q-i+1)/i][ML_{i-1}][L]K_i \\ &= [q!/(q-i)!i!][L]^i[M_0]\prod_{j=1}^{i} K_j \end{aligned} \qquad (5.3\text{-}4)$$

Since the total number of polymer molecules, $[M]_{total}$, is given by

$$[M]_{total} = [M_0] + \sum_{i=1}^{q}[ML_i] = [M_0]P([L]) \qquad (5.3\text{-}5)$$

where

$$P(x) = 1 + \sum_{i=1}^{q}[q!/(q-i)!i!]x^i \prod_{j=1}^{i} K_j \qquad (5.3\text{-}6)$$

and the average number of bound ligands per macromolecule, I, is represented by

$$I = \sum_{i=1}^{q} i[ML_i] \Big/ [M]_{total} = [L]P'([L]) \Big/ P([L])$$
(5.3-7)

where P' is the first derivative of P. The extent of fractional saturation is derived from Eqs. (5.3-5) to (5.3-7) as

$$X = I/q = [L]P'([L]) \Big/ qP([L]) = \{d\ln P([L]) \Big/ d\ln[L]\} \Big/ q$$
(5.3-8)

If all K_js are equal to K irrespective of the value of j, X is represented by Eq. (5.3-9).

$$X = K[L] \Big/ \{1 + K[L]\}$$
(5.3-9)

This has the same form as the Michaelis–Menten equation for a one-substrate enzyme-catalyzed reaction rate provided that X and L are replaced with the rate and substrate, respectively. Using this approximation, the ratio of the number of occupied sites to that of unoccupied sites is given by

$$Y = X/(1-X) = K[L]$$
(5.3-10)

When K_j increases with j, there is positive cooperativity for ligand binding. In the reverse case there is a kind of feedback inhibition of the reaction.

The qualitative features of X vs [L] and $\log Y$ vs $\log$[L] are illustrated in Fig. 5.3-7, where the solid and broken lines represent the curves for positive allosteric and non-allosteric cases, respectively. Since the effective binding constants for small and large [L] should be K_1 and K_q, respectively, the two asymptotes could be represented by $\log Y = \log[L] + \log K_1$ and $\log Y = \log[L] + \log K_q$. In this figure $K_q > K_1$ is postulated, and the corresponding difference in the binding free energy ($\Delta G_j = -RT \ln K_j$)

$$\Delta\Delta G_{q1} = \Delta G_q - \Delta G_1$$
(5.3-11)

can be evaluated from the ordinate values of the asymptotes at $\log$[L] = 0, as shown in Fig. 5.3-7(b). When $\Delta\Delta G_{q1}$ is not zero there would be a transition area between the two asymptotes. The largest slope of the $\log Y$ vs $\log$[L] curve, $n = \{d(\log Y)/d(\log[L])\}_{max}$, which is called the Hill coefficient and serves as a measure of cooperativity, never exceeds the number of cooperative binding sites.

When two different kinds of ligands, A and B, bind to a macromolecule which has q equivalent sites for A and r equivalent sites for B, the binding equations are

$$MA_{i-1}B_j + A \leftrightarrow MA_iB_j \qquad (i = 1 \sim q)$$
$$MA_iB_{j-1} + B \leftrightarrow MA_iB_j \qquad (i = 1 \sim r)$$
(5.3-12)

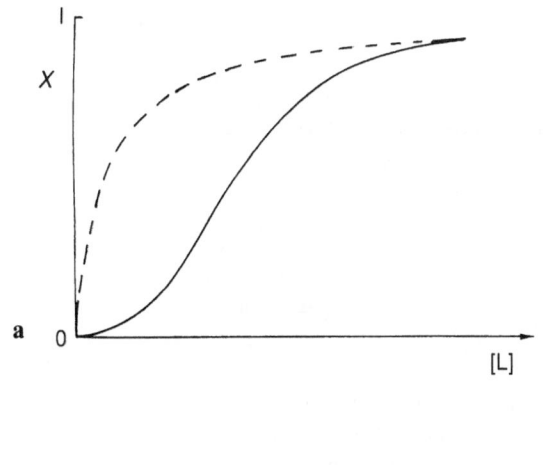

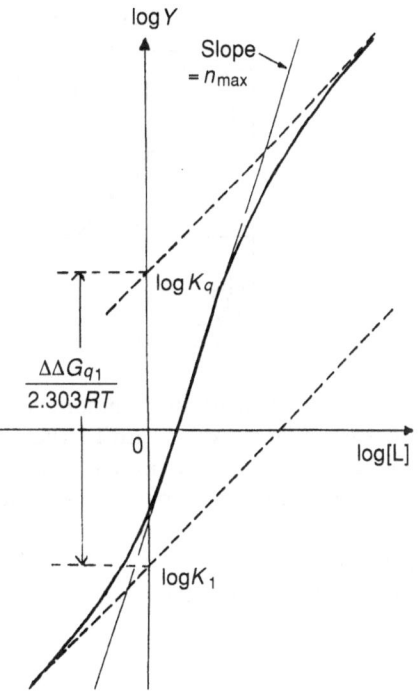

FIG. 5.3-7. **a** X vs [L] curve: *solid* and *broken lines* denote the allosteric and nonallosteric cases, respectively. **b** $\log[X/(1-X)]$ vs. $\log[L]$ curve corresponding to the solid line in **a**. *Broken lines* represent the asymptotes at the limit of low and high concentrations of ligands

In this case the function P Eq. (5.3-6) is represented as a double sum with respect to x and y. The fractional saturations for the A and B ligands, X_A and X_B, are found in the same way as in Eq. (5.3-8), and it follows that

$$q\left\{\delta X_A / \delta \ln[B]\right\}_{[A]} = r\left\{\delta X_B / \delta \ln[A]\right\}_{[B]} \qquad (5.3\text{-}13)$$

Equation (5.3-13) is a linkage relation between two different kinds of ligands bound to the same macromolecule, and means that if the binding of one ligand influences the binding of the other, the binding of the latter must influence the binding of the former. This kind of regulation and control is made possible by oligomerization of functional monomer units, and has been noticed in various enzymes [9]. When K_j for the binding of L changes with j it is called a homotropic allosteric effect, and when K_j is altered by the binding of other types of ligands Eq. (5.3-12), it is called a heterotropic allosteric effect.

5.3.2.3 Two-State Model for Allosteric Effects

Multiple substrate binding to an oligomeric enzyme and accompanied cooperative effects are common and essential to regulative enzymes. To interpret this a two-state model has been proposed [10], and as a typical substance for studying general allostery hemoglobin (Hb) has been extensively investigated [11]. Pro-

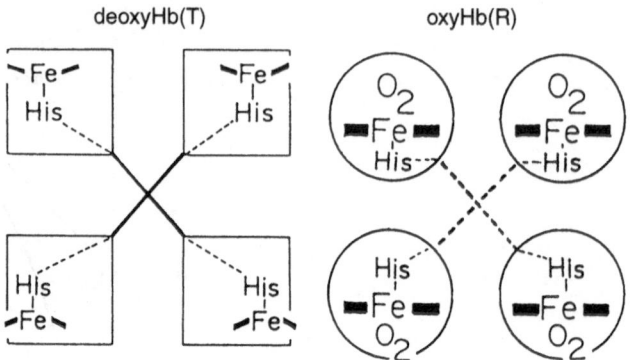

deoxyHb(T)   oxyHb(R)

FIG. 5.3-8. Schematic diagram for two state models of hemoglobin (*Hb*). The left and right sides stand for deoxyHb with T structure and oxyHb with R structure, respectively. *Circles* and *squares* denote the protein tertiary structures of each subunit in the R and T quaternary structures, respectively. Each subunit contains one heme group to which histidyl imidazole is coordinated at the axial position of the heme iron and O_2 is bound to its *trans* position. The intersubunit interactions are stronger in the T than the R state (represented by *solid* and *broken lines*)

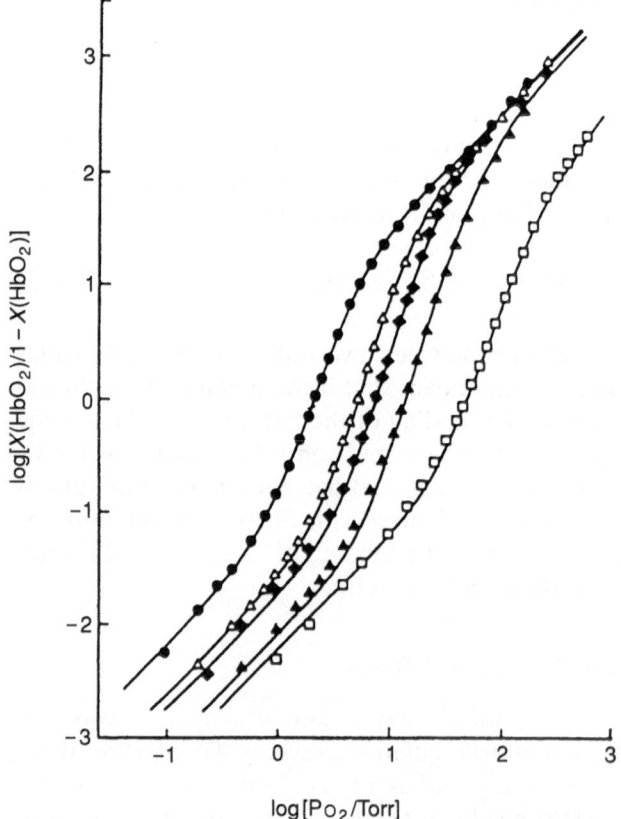

FIG. 5.3-9. Fitting of oxygen-binding equilibrium curves of hemoglobin (*Hb*) by the two-state model. The points are experimental and the lines are calculated for best-fit. *Circles*, pH 9.1, 0.1 M Cl⁻; *open triangles*, pH 7.4, 0.1 M Cl⁻; *diamonds*, pH 7.4, 0.1 M Cl⁻, 5% CO_2; *solid triangles*, pH 7.4, 0.1 M Cl⁻, 2 mM DPG; *squares*, pH 6.5, 0.1 M Cl⁻, 2 mM inositol hexaphosphate (an allosteric effecter). (Redrawn from [12], with permission)

vided that there are two structures for macromolecules with i bound ligands, R_i and T_i, with different binding constants, K_R and K_T, respectively, and that R_i and T_i are in chemical equilibrium and their microscopic binding constants do not depend on the number of bound ligands, then

$$T_{i-1} + L \leftrightarrow T_i \quad : \quad K_T$$
$$R_{i-1} + L \leftrightarrow R_i \quad : \quad K_R \qquad (5.3\text{-}14)$$

$$[T_i]/[R_i] = ([R_0]/[T_0])(K_T/K_R)^i = Lc^i \qquad (5.3\text{-}15)$$

In this case it is possible to describe the system using the three constants L $(=[R_0]/[T_0])$, K_T, and K_R.

A practical image of this model for Hb is illustrated in Fig. 5.3-8, where the circles and squares denote the protein structures corresponding to the R and T states, respectively, and each subunit contains one heme group co-ordinated by histidine. Human Hb is a tetramer protein to which four oxygen molecules can be bound ($q = 4$). As well as oxygen, other physiological ligands including protons, carbon dioxide, chloride ions, and diphosphoglycerols (DPG) are also bound, and serve as allosteric effectors although their binding sites are not represented in Fig. 5.3-8. The experimental oxygen binding equi-libriums observed for human Hb under various pH and salt concentrations, and theoretical curves fitted using the two-state model, are depicted in Fig. 5.3-9 [12].

5.3.2.4 Structural Mechanism of Cooperativity

Hb is a protein of the type shown in Fig. 5.3-6(d). Although there is no covalent bond between any two subunits, there are many weak interactions such as hydro-gen bonds and salt bridges. The cooperative oxygen binding of Hb can be ex-plained in terms of reversible transition between two quaternary structures T and R, as illustrated in Fig. 5.3-8. However, this raises the question: What is the reality of the T and R structures? Almost all spectroscopic data exhibit no distinguish-able difference between oxygenated T and R structures. The porphyrin vibra-tions and even the Fe—O_2 stretching frequency are practically the same between the T and R states. However, the Fe–histidine stretching frequency, which reflects the strength of the sole covalent bond between the heme group and a polypeptide chain in each subunit (see Fig. 5.3-8), is found to be lower in deoxyHb (T) than in deoxyHb (R), as illustrated in Fig. 5.3-10(a) [13]. This is consistent with the presumption from X-ray crystallographic analysis [11] that the stronger intersubunit interactions in the T state let the globin take a tensed structure and pull the iron-bound histidine (His) away from the heme, thus weakening the Fe–His bond.

It might be thought that when one residue in the polypeptide chain ($\sim$140 residues) is replaced by another amino acid the communication between the subunit interfaces and the Fe–His bond is altered, and accordingly the strain

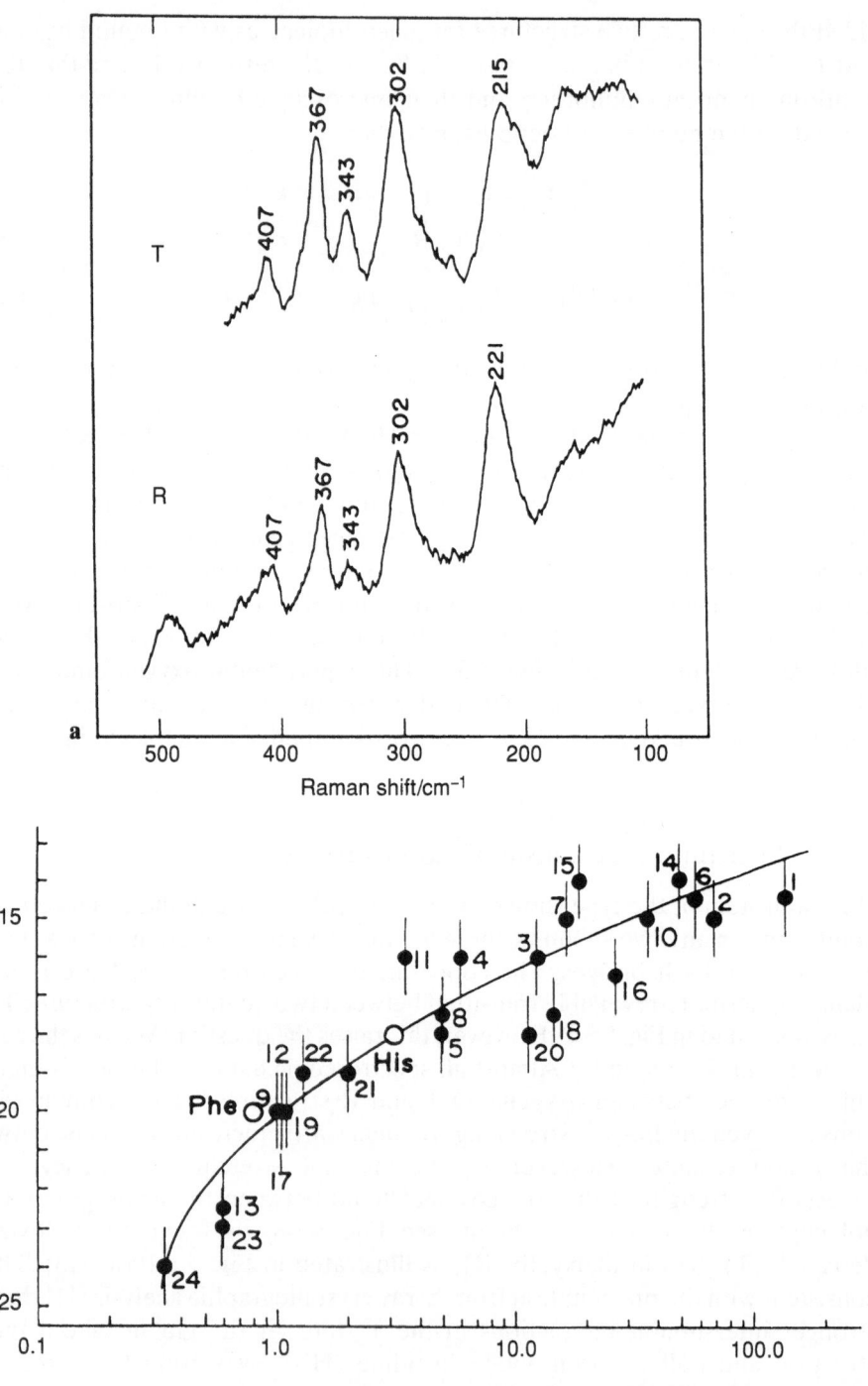

FIG. 5.3-10. **a** Typical low-frequency resonance Raman spectra of deoxyHb with T and R structures. The band around 215–221 cm^{-1} is assigned to the Fe–His stretching mode on the basis of the Fe isotopic frequency shift. **b** The relation between the Fe–His stretching frequency and the equlibrium constant for the first oxygen to bind to deoxyHb. The equilibrium constant is represented in terms of a dissociation constant. *Closed* and *open circles* indicate the experimental values observed for natural and artificial mutants, respectively, and the *solid line* denotes the theoretical curve calculated by assuming that the strain imosed on the Fe–His bond is proportional to the stabilization energy at the subunit interface

imposed on the Fe–His bond by the globin would be affected. In fact, the magnitude of the strain changes upon mutation of amino acid residues and the oxygen affinity is also changed. Figure 5.3-10(b) shows the observed correlation between the Fe–His stretching frequency and the equilibrium constant for the first oxygen to bind to deoxyHb. The solid line represents a theoretical curve derived with the assumption that the destabilization of the Fe–His bond at the heme is proportional to the stabilization at the subunit interfaces. This curve reproduces the observed trend very well [13], suggesting that the quaternary structure change is triggered by mechanical strain. In this way the allosteric effects of Hb, which are characteristic of oligomeric proteins, can reasonably be interpreted as a conformational change accompanied by a change in the quaternary structure. In other words, a previously mysterious feature of biological reactions can be understood by applying the principles of chemistry.

References

1. Seki K (1989) Photoelectron spectroscopy of polymers. In: Baessler H (ed.) Optical techniques to characterize polymer systems. Studies in Polymer Science vol 5, Elsevier, Amsterdam, p 115
2. Salaneck WR (1985) Photoelectron spectroscopy of the valence electronic structure of polymers. CRC Crit Rev Solid State Mat Sci 12:267–296
3. Kloepper W (1969) Transfer of electronic excitation energy in polyvinyl carbazole. J Chem Phys 50:2337–2343
4. Zeigler JM, Fearon FWG (eds) (1990) Silicon-based polymer science. A comprehensive resource. ACS, Washington, DC
5. Ishii H, Yuyama A, Narioka S, Seki K, Hasegawa S, Fujino M, Isaka H, Fujika M, Matsumoto N (1995) Photoelectron spectroscopy of polysilanes, polygermanes and related compounds. Syn Metals 69:595–596
6. Seki K, Karlsson UO, Engelhardt R, Koch EE, Schmidt W (1984) Intramolecular band mapping of poly (*p*-phenylene) via UV photoelectron spectroscopy of finite polyphenyls. Chem Phys 91:459–470
7. Seki K, Asada S, Mori T, Inokuchi H, Murase I, Karlsson UO, Engelhard R, Koch EE (1987) UV photoelectron spectroscopy of conducting polymers and their model compounds. Syn Metals 17:629–634

8. Fahlman M, Beljonne D, Loeglund M, Friend RH, Holmes AB, Bredas JL, Salaneck WR (1993) Experimental and theoretical studies of the electronic structure of Na-doped poly(para-phenylenevinylene). Chem Phys Lett 214:327–332
9. Monod J, Changeux J-P, Jacob F (1963) Allosteric proteins and cellular control systems. J Mol Biol 6:306–329
10. Monod J, Wyman J, Changeux J-P (1965) On the nature of allosteric transitions: A plausible model. J Mol Biol 12:88–118
11. Perutz MF (1979) Regulation of oxygen affinity of hemoglobin: Influence of structure of the globin on the heme iron. Annu Rev Biochem 48:327–386
12. Imai K (1982) Allosteric effects in haemoglobin. Cambridge University Press, London, p 213
13. Kitagawa T (1988) The heme protein structure and the iron–histidine stretching mode. In: Spiro TG (ed.) Biological applications of Raman spectroscopy. Wiley, New York, pp 97–131

5.4
Electric and Magnetic Field Effects

Yoshifumi Tanimoto

All substances are composed of electrons and nuclei. The electrons have electric charges and their spin motions bring about magnetism. Therefore, it is to be expected that the chemical and physical behaviors of substances will be affected by factors in the physical environment such as electric and magnetic fields.

The effects of an electric field are known as the Stark effect, the Pockels effect, the Kerr effect, dielectric alignment, electric field effects upon molecular aggregation, and electric field effects upon electron–hole pairs. In Sect. 5.4.1, the Stark and Pockels effects and electric field effects upon the chemical and physical behavior of molecular systems will be described.

The effects of a magnetic field are known as the Zeeman effect, the Faraday effect, magnetic birefringence, the magnetic Kerr effect, magnetic field effects upon chemical reactions, magnetic field effects upon emission, magnetic field effects upon chemical equilibrium, magnetic orientation, and others. Many chemical and physical processes are known to be affected by an external magnetic field. In particular, magnetic field effects on chemical reactions via radical pairs in solution have been widely observed. Their mechanisms are primarily interpreted in terms of the radical-pair model. It should be noted that a magnetic field can affect the spin dynamics of a molecule even though the magnetic-field-induced energy change in the molecule is several orders of magnitude smaller than the thermal energy of the molecule at room temperature. Discussion of the magnetic field effects on chemical reactions and related phenomena appear in Sect 5.4.2.

5.4.1 Electric Field Effects

5.4.1.1 Stark Effect

When atoms and molecules are placed in a static electric field, their energies change with a concomitant mixing of their wavefunctions. The splitting and shifts of atomic and molecular states induced by an electric field are called the Stark effect. The interaction of atoms and molecules with an electric field is given by the following Hamiltonian [1]:

$$H = -\mu \cdot E \tag{5.4-1}$$

where μ is the electric dipole moment and E is the external electric field. The energy shift caused by Eq. (5.4-1) can be calculated using perturbation theory. If n and m are eigenfunctions of the nth and mth nondegenerate or degenerate states, respectively, at zero electric field, and E_n and E_m are the respective energies, the electric field changes E_n by amount E', where

$$E' = -\langle n|\mu|n \rangle \cdot E + \sum_{m \neq n} \left| \langle m|\mu|n \rangle \cdot E \right|^2 \Big/ \left(E_n - E_m \right) \tag{5.4-2}$$

The first term of the right-hand side of Eq. (5.4-2) represents the interaction between a permanent electric dipole and the electric field and is called the first-order Stark effect. The ground state (1s state) of the hydrogen atom shows no first-order Stark effect since it has no permanent dipole moment. The $2s$, $2p_0$, $2p_1$, and $2p_{-1}$ states are degenerate at zero field, where 0, 1, and -1 are magnetic quantum numbers. When an electric field is applied in the z direction, the wavefunctions of $2s$ and $2p_0$ mix with each other and the four degenerate states at zero field split into three sublevels. The first-order Stark effect is actually observed in the emission spectra from these states.

The second term of Eq. (5.4-2) arises from the interaction of n with other states and shows the second-order perturbation energy $-(1/2\alpha E) \cdot E$, where α is the polarizability.

An electric field also affects degenerate rotational levels of ground-state molecules in a gas. The Stark effect is widely used in microwave spectroscopy to determine electric dipole moments of polyatomic molecules.

5.4.1.2 Pockels Effect

The refractive index ellipsoid of a substance is given by the three principal refractive indices n_x, n_y, and n_z, where x, y, and z are its principal dielectric axes. A potassium dihydrogen phosphate (KDP) crystal is a uniaxial crystal in which z is the optical axis, and its refractive index ellipsoid is given by the following equation when the electric field E_z is applied in the z direction [2]:

$$\left(x^2 + y^2 \right) \Big/ n_o^2 + z^2 / n_e^2 + 2r_{63} E_z xy = 1 \tag{5.4-3}$$

where n_0 and n_e are the indices for ordinary and extraordinary light, and r_{63} is the first-order electrooptical constant. In the absence of an electric field ($E_z = 0$), light propagating to the z-axis does not exhibit birefringence since refractive indices in the direction of x and y are the same.

Choosing a new coordinate generated by a 45° rotation of the x- and y-axes around the z-axis, Eq. (5.4-3) is expressed as

$$\left(x' \right)^2 \Big/ n_{x'}^2 + \left(y' \right)^2 \Big/ n_{y'}^2 + z^2 / n_e^2 = 1 \tag{5.4-4}$$

where

$$n_{x'} \sim n_0\left(1 - n_0^2 r_{63} E_z / 2\right), \qquad n_{y'} \sim n_0\left(1 + n_0^2 r_{63} E_z / 2\right) \qquad (5.4\text{-}5)$$

and x' and y' are the new coordinate axes. Equation (5.4-5) means that this KDP crystal becomes birefringent for the light propagating to the z axis when an electric field E_z is applied, since $n_{x'} \neq n_{y'}$. These changes in refractive indices in proportion to an applied field are called Pockels effect. Inorganic crystals such as KDP and LiTaO$_3$ and organic liquid crystals like N-(4-methoxybenzylidene)-4'-n-butylaniline exhibit this effect.

A refractive index change proportional to the square of an applied electric field is also observed for CS$_2$ and nitrobenzene. This is called the Kerr effect.

The phase difference between the two lights polarized to x' and y' is given as

$$
\begin{aligned}
\delta &= \left(2\pi/\lambda\right)\left(n_{x'} - n_{y'}\right)D \\
&= \left(2\pi/\lambda\right)n_0^3 r_{63} E_z D = \left(2\pi/\lambda\right)n_0^3 r_{63} V
\end{aligned}
\qquad (5.4\text{-}6)
$$

where λ is the wavelenth of light, D is the length of the crystal, and V is the applied voltage. When the crystal is placed between crossed Nicol prisms and V is applied to the crystal, light can pass through the crossed Nicol prisms, and its intensity I is

$$I = I_0 \sin^2\left(\left(\pi/\lambda\right)n_0^3 r_{63} V\right) \qquad (5.4\text{-}7)$$

where I_0 is the intensity of the light incident to the crystal.

Figure 5.4-1 shows schematically a device that can control light intensity using an electric field. It is composed of a polarizer, an analyzer, and a KDP crystal. The two crystal surfaces must be perpendicular to the optical axis z, and transparent electrodes are attached to them. Light can pass through the cell only when a voltage V is applied to the crystal. The optical parameters of KDP for 546.1 nm light are $n_0 = 1.5095$, $n_e = 1.4684$, and $r_{63} = -1.07 \times 10^{-11}\,\mathrm{m\,V^{-1}}$. When $D = 1\,\mathrm{cm}$, δ becomes $\pi/2$ by applying a voltage of around 3.7 kV. The duration of light transmittance is controlled by the duration of the applied voltage. This device is called a Pockels cell and is used as a fast electric shutter.

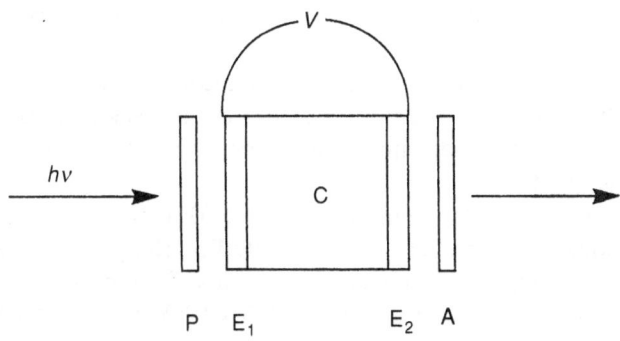

FIG. 5.4-1. Structure of a Pockels cell. P, polarizer; A, analyzer; C, potassium dihydrogen phosphate crystal; E_1, E_2, transparent electrodes; V, applied electric field

5.4.1.3 Dielectric Alignment of Liquid Crystals

The electric alignment of liquid crystals has been studied in great detail [3, 4], and a simple example of alignment follows.

The free energy of a dielectric liquid crystal per unit volume is $-\varepsilon_a E^2/8\pi$, where ε_a is the anisotropic dielectric constant and E is the applied electric field. The energy for a single molecule is several orders of magnitude smaller than the thermal energy kT, where k is the Boltzmann constant and T is the absolute temperature. However, when N molecules form an aggregate in which all molecules are orientated in the same direction, the free energy becomes $-N\varepsilon_a E^2/8\pi$. Liquid crystal molecules form a molecular aggregate with $N \sim 10^8$. The free energy becomes larger than the thermal energy and electric alignment of the crystal takes place in the presence of an electric field.

A nematics liquid crystal with $\varepsilon_\parallel < \varepsilon_\perp$, for example, is uniformly orientated at zero field and its orientation is deformed by application of an electric field, generating a tunable birefringence. The threshold voltage (V_{th}) for deformation [3] is given by

$$V_{th}^2 = \pi^2 K_{33}\Big/\Big(\big(\varepsilon_\perp - \varepsilon_\parallel\big)\varepsilon_o\Big) \qquad (5.4\text{-}8)$$

where K_{33} is the curvature-elastic modulus for the bending of the orientation lines. Since liquid crystals are electrically anisotropic, this deformation in the orientation of a crystal results in a change in its optical properties. Nowadays, many liquid crystals are used commercially as liquid crystal displays in electronic devices.

Optical nonlinear films can be prepared by using the effect mentioned above. It is possible to orient a polar polymer film by heating it above its glass-transition temperature under an electric field to align the molecules, then cooling it below the transition temperature under the electric field to fix their orientation. This technique is called "poling," and is used to prepare optical nonlinear films [5].

5.4.1.4 Electric Field Effects on Aggregation of Polar Molecules

When photochromic spiropyran in nonpolar solvents is photoirradiated at low temperature ($\sim 240\,K$), dimers (AB) are formed (Fig. 5.4-2) [6, 7]. Here A is the original less-polar and colorless form and B is the polar and photocolored form of the photochrome. Solvation of AB by n molecules of A results in the formation of charge transfer complexes ($A_{n+1}^+B^-$, CTC), where $n = 2$–3. The CTC and dimers finally form an aggregate in which about 10^6 molecules are associated.

When a voltage larger than $6 \times 10^5\,V\,m^{-1}$ is applied to the photocolored solutions, the aggregates are deposited on the positive wall of the condenser and the remaining solution becomes colorless. When the spiropyran solution is irradiated in the presence of a voltage (2–$3 \times 10^6\,V\,m^{-1}$), colored crystals form. The crystals

$$\mathrm{A} \xrightarrow{h\nu} \mathrm{B}$$

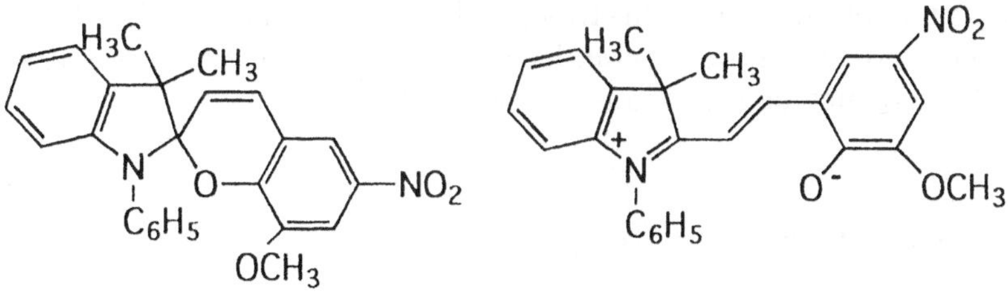

$$\mathrm{A} + \mathrm{B} \longrightarrow \mathrm{AB}$$

$$\mathrm{AB} + n\mathrm{A} \longrightarrow \mathrm{A}_{n+1}^{+}\mathrm{B}^{-}\ (\mathrm{CTC})$$

$$\mathrm{CTC} + \mathrm{AB} \longrightarrow \text{Aggregate}$$

FIG. 5.4-2. Reaction scheme of spiropyran. *CTC*, charge transfer complex

extend along the lines of electric force and exhibit optical dichromism, which results from the orientation of the CTC in the field. The mechanism of the effect is not well understood. It seems to arise from the electrostatic interaction between the external field and the electric dipole moment of the aggregate.

5.4.1.5 Electric Field Effects on Electron-Hole Pairs

An example of electric field effects on reaction intermediates is described briefly below. The intensity of exciplex fluorescence from a poly (*N*-vinylcarbazole) (PVCz) film doped with 1,2,4,5-tetracyanobenzene (TCB) is quenched by about 3% in the presence of a voltage of $2 \times 10^7\,\mathrm{V\,m^{-1}}$[8].

$$\underset{\text{ion pair}}{(\mathrm{PVCz^{+}} \ldots \mathrm{TCB^{-}})} \longrightarrow \underset{\text{free carrier}}{\mathrm{PVCz^{+}} + \mathrm{TCB^{-}}} \longrightarrow$$

$$\searrow \underset{\text{exciplex}}{(\mathrm{PVCz^{\delta+}}\ \mathrm{TCB^{\delta-}})} \longrightarrow$$

In this reaction, an ion pair composed of a PVCz cation radical and a TCB anion radical is photogenerated. A fraction of the pair undergoes geminate recombination to the relaxed fluorescent exciplex state, whereas the rest undergoes electric-field-assisted dissociation into free carriers. The field is believed to reduce the

Coulombic interaction between two radicals in a pair. Thus, the fluorescence intensity of the exciplex is affected by an external electric field. Very recently, electric field effects on the fluorescence of organic molecules in poly (vinyl alcohol) thin films and Langmuir–Blodgett layers were reported [9]. The mechanism of these effects on the photoinduced dynamics of organic molecules in thin films and layers may be clarified in the near future.

5.4.2 Magnetic Field Effects

5.4.2.1 Magnetic Field Effects on Chemical Reactions in Solution: Radical-Pair Mechanism

In chemical reactions such as decomposition, electron transfer, and hydrogen transfer, a pair of radicals is often generated as a short-lived reaction intermediate in a solvent cage and called a radical pair [10–17]. This pair has two electron spin states, singlet (S) and triplet (T), where T is composed of three sublevels T_+, T_0, and T_-. Figure 5.4-3 shows the pathways of a radical pair in solution. An energy gap between the two states arises from the electron exchange energy $2J$, which is usually assumed to be negative and an exponential function of the interradical distance. The gap in the pair immediately after its formation is large and no transition occurs between the S and T states. After several diffusive motions the energy gap between the two states becomes nearly degenerate ($2J\sim0$) and transition then takes place. By further diffusion, the radical pair becomes two escape radicals. A singlet pair undergoes cage reaction, resulting in the formation of stable cage products. The escape of radicals from the solvent cage takes place mainly from the triplet state, leading to the formation of escape products.

Magnetic interaction in a pair induces spin transitions when the energies of the S and T states are nearly degenerate ($2J\sim0$). At zero field, S, T_0, T_+, and T_- are degenerate. When a magnetic field is applied, T_+ and T_- split away from each other because of the electronic Zeeman effect. Isotropic magnetic interaction induces intersystem crossing (ISC) among spin states, whereas anisotropic interaction induces spin relaxation (SR). At zero field, ISC between S and the three triplet sublevels occurs via isotropic electron-nuclear hyperfine (hf) interaction. In a magnetic field, the S–T_+ and S–T_- ISC rates are reduced (hf mechanism), whereas a difference in the isotropic electron g values of the two radicals accelerates the S–T_0 ISC rate (Δg mechanism). Furthermore, in a magnetic field the SR of $T_\pm$–T_0 and $T_\pm$–S takes place via anisotropic hf (δhf mechanism) and g (δg mechanism) interactions at component radicals as well as dipole interaction between two radicals (dipole mechanism).

The processes mentioned above are all magnetic-field-dependent. An external magnetic field can affect the decay-rate constants of singlet and triplet radical pairs, and therefore yields of products derived from them are influenced by the magnetic field. The magnetic field effects (MFEs) induced by the mechanisms,

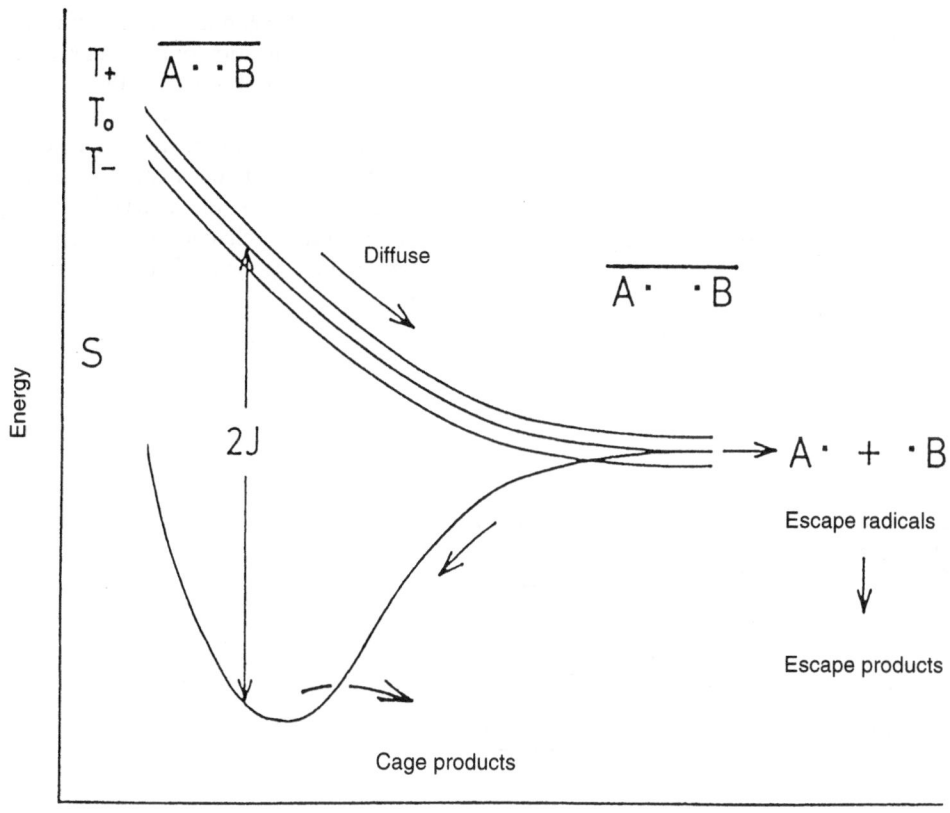

FIG. 5.4-3. Pathways of a radical pair in solution

details of which will be described later, are schematically shown in Fig. 5.4-4. The MFE based on the hf mechanism appears in a low magnetic field (<0.1 T) (1 T = 10000 G) [18]. The effect based on the δhf and dipole mechanisms appears in the magnetic fields of 0.1~2 T. The effect based on the Δg mechanism appears in a magnetic field larger than 1 T [18], and the effects due to the δg mechanism seem to occur in a magnetic field higher than 2 T [16]. Generally speaking, experimentally observed MFEs are explained by

In the case of a radical pair whose interradical distance is restricted by a short methylene chain, the situation is slightly different from that mentioned above. S and T are nondegenerate at zero field and the S–T ISC rate is slow. By application of a magnetic field, T_ shifts to a lower energy. When S and T_ become degenerate, the S–T_ ISC rate is enhanced (S–T_ level crossing). By further increasing the magnetic field, the ISC rate is again reduced because of the large energy gap. Generally speaking, experimentally observed MFEs are explained by a combination of the mechanisms mentioned above.

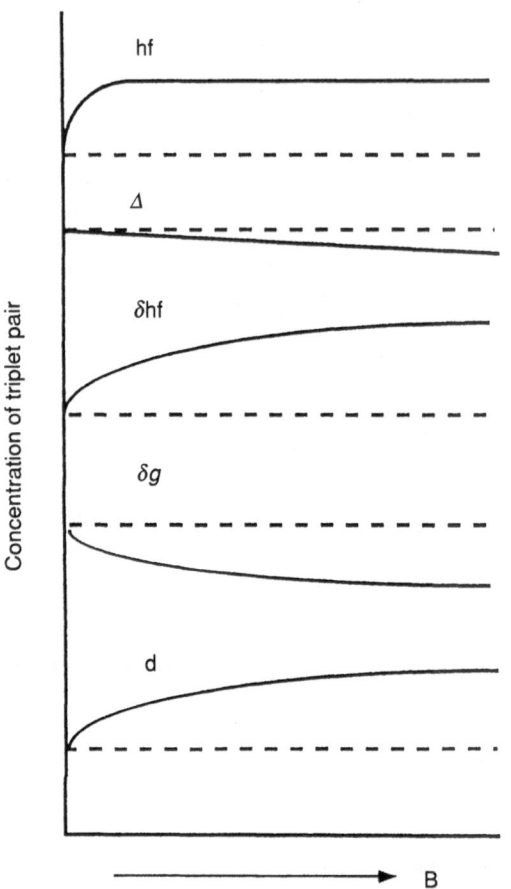

FIG. 5.4-4. Magnetic-field-dependence of the concentration of a triplet radical pair. The initial state of the pair is assumed to be the triplet state. *hf*, hyperfine mechanism; Δg, Δg mechanism; δhf, δhf mechanism; δg, δg mechanism; *d*, dipole mechanism. The concentration change of the singlet pair is the reverse of that of the triplet pair

The magnitude of an MFE is related to the time the pair reside in the energy region $2J\sim0$. In nonviscous solutions the effects are small (typically a few percent) because of fast diffusion of the radicals. Conversely, the effects on a radical pair in a micellar solution and a chain-linked biradical in a nonviscous solution are significant (typically more than 100%).

Consider the following example. When a benzophenone derivative, BP-8, is photolysed in a polyoxyethylene dodecylether (Brij 35, ICI Americas, West Warwick, RI, USA) micellar solution, a triplet radical pair is generated composed of a ketyl radical derived from BP-8 and an alkyl radical derived from the Brij 35 surfactant [16]. Figure 5.4-5 shows MFEs on the lifetime of a triplet pair in a micellar solution. The steep increase in the lifetime between zero and about 0.1 T is due to the hf mechanism. The gradual change between 0.1 and about 2 T is attributable to a reduction in relaxation induced by δhf and dipole mechanisms, and the decrease in the lifetime in a magnetic field higher than 2 T may result from an enhancement of relaxation due to the δg mechanism. The yield of the escape radical from the micelle cage exhibits a magnetic-field dependence

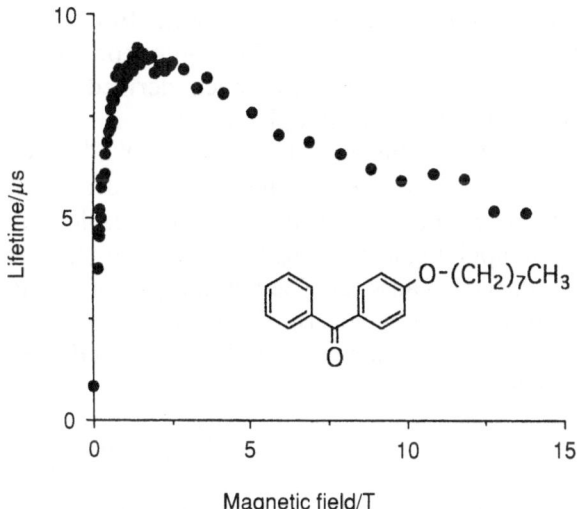

FIG. 5.4-5. Magnetic field effects on the lifetime of a triplet radical pair composed of a ketyl radical derived from BP-8 and an alkyl radical derived from Brij 35 surfactant in Brij 35 micellar solution

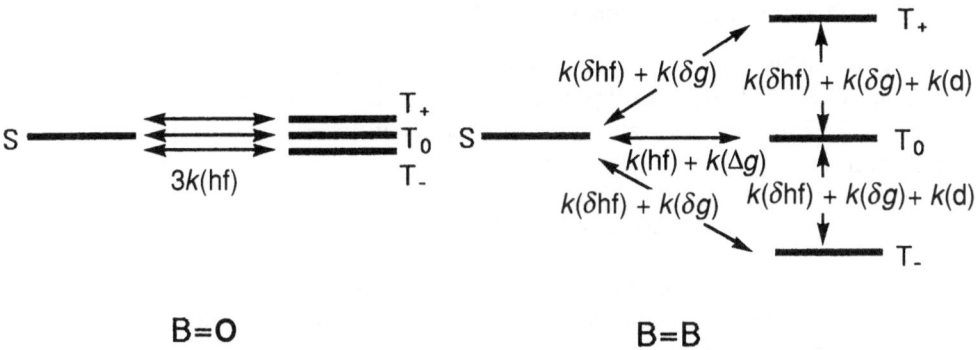

FIG. 5.4-6. Energy levels and pathways relevant to a radical pair with $2J \sim 0$

analogous to that shown in Fig. 5.4-5. Spin relaxation among spin states is enhanced by the application of microwaves when its energy matches the energy difference of two spin states. Thus by using magnetic and microwave fields simultaneously, one can further control the pathways of chemical reactions via radical pairs [19].

The mechanisms of MFEs caused by isotropic and anisotropic magnetic interactions are now described in detail.

Isotropic Magnetic Interaction. Figure 5.4-6 shows the energy levels and pathways relevant to a radical pair with $2J \sim 0$. The $S-T_0$ ISC rate constant $k(\text{hf})$ for the hf interaction is given by the following equation [20, 21]:

$$k\left(\text{hf}\right) \sim g\beta B_{\text{av}}/\left(\pi\hbar\right) \tag{5.4-9}$$

where g is the electron g value, β is the Bohr magneton, and $\hbar$ is Planck's constant. B_{av} is the average of the hf coupling constants of component radicals 1 and 2. Analogous equations have been derived for the rate constants of S–T$_+$ and S–T$_-$ ISC, which are not given here.

The k(hf) value is of the order of 10^7 s^{-1}, since B_{av} of a typical organic radical pair is about 10 G. At zero field, S and T are nearly degenerate and the hf-induced ISC occurs between S and the three triplet sublevels (T$_+$, T$_0$, T$_-$) (Fig. 5.4-6). Upon applying a magnetic field much larger than B_{av}, the ISC between S and T$_\pm$ is reduced because the Zeeman splitting of T$_+$ and T$_-$ lifts the degeneracy of S and of these two sublevels (hf mechanism).

The S–T$_0$ ISC rate constant $k(\Delta g)$ induced by the difference in electron g values is as follows:

$$k(\Delta g) \sim \Delta g \beta B_{ex} / (\pi \hbar) \tag{5.4-10}$$

where $\Delta g = g_1 - g_2$ and B_{ex} is the external magnetic field. g_i is the isotropic electron g value of radical i. The $k(\Delta g)$ value is zero at zero field. Application of an external magnetic field enhances the ISC rate. In a typical organic radical pair, Δg is about 0.001 and $k(\Delta g)$ becomes about 10^7 s^{-1} at 1 T.

Anisotropic Magnetic Interaction. Spin relaxation among S, T$_0$, T$_+$, and T$_-$ takes place via δhf, δg, and dipole interactions in a magnetic field (Fig. 5.4-4) [14, 15, 22, 23]. The SR rate constant $k(\delta$hf$)$ induced by the anisotropic hf interaction is

$$k(\delta hf) = (1/4)(k_{\delta hf}(1) + k_{\delta hf}(2)) \tag{5.4-11}$$

where $k_{\delta hf}(i)$ is the SR rate constant $k_{\delta hf}$ induced by the anisotropic hf interaction of radical i. When one magnetic nucleus with spin I is at a distance r from the radical electron, $k_{\delta hf}$ is

$$k_{\delta hf} = (4/3)I(I+1)\gamma_i^2 g^2 \beta^2 \langle r^{-6} \rangle \tau_c / (1 + \omega_0^2 \tau_c^2) \tag{5.4-12}$$

where γ_i is the magnetogyric ratio of nucleus I, $\omega_0 = g\beta B_{ex}/\hbar$, and τ_c is its correlation time.

The SR rate constant $k(\delta g)$ induced by anisotropic g values is

$$k(\delta g) = (1/4)(k_{\delta g}(1) + k_{\delta g}(2)) \tag{5.4-13}$$

where $k_{\delta g}(i)$ is the SR rate constant $k_{\delta g}$ induced by the anisotropic g value of radical i. $k_{\delta g}$ is also given as

$$k_{\delta g} = (1/5)\delta g^2 \beta^2 \hbar^{-2} B_{ex}^2 \tau_c / (1 + \omega_0^2 \tau_c^2) \tag{5.4-14}$$

where $\delta g^2 = g_{xx}^2 + g_{yy}^2 + g_{zz}^2 - 3g^2$ and $g = (g_{xx} + g_{yy} + g_{zz})/3$.

T$_\pm$–T$_0$ spin relaxation also occurs via dipolar spin–spin interaction between two radical electrons. The SR rate constant k(d) induced by this interaction is

$$k(d) = (3/10)g^4\beta^4 r_{12}^{-6}\hbar^{-2}\tau_{12}/(1+\omega_0^2\tau_{12}^2) \qquad (5.4\text{-}15)$$

where r_{12} is the distance between two radical centers and τ_{12} is the correlation time of the vector r_{12}.

The $k(\delta\text{hf})$ value decreases with increasing external magnetic field (δhf mechanism). The $k(d)$ value also decreases with increasing magnetic field (dipole mechanism). These rate constants are estimated to be roughly of the order of $10^5\,s^{-1}$ at 1 T. The $k(\delta g)$ value increases with increasing magnetic field (δg mechanism). Its value is also estimated to be of the order of $10^5\,s^{-1}$ at 1 T.

5.4.2.2 Magnetic Isotope Effects

According to the hf mechanism, singlet–triplet ISC rates at zero field are determined by the electron–nuclear hf coupling constants of component radicals. The coupling constants of atoms such as ^{1}H, ^{13}C, and ^{17}O, which have nuclear magnetic moments, are nonzero, whereas those of atoms such as ^{12}C and ^{16}O, which have no nuclear magnetic spin, are zero. Therefore hf-induced singlet–triplet ISC is a process dependent on the nuclear magnetic isotope of atoms in a radical pair.

For example, in photolysis of dibenzyl ketone (DBK) in micellar solution, triplet radical pairs composed of a benzyl radical $Ph\text{–}\dot{C}H_2$ and an acyl radical $\dot{C}OCH_2\text{–}Ph$ are generated [24].

$$DBK \underset{ISC}{\overset{h\nu}{\longleftrightarrow}} \,^3(Ph\text{—}\dot{C}H_2 \quad \dot{C}O\text{—}CH_2\text{—}Ph) \xrightarrow{escape} Ph\text{—}\dot{C}H_2 + \dot{C}O\text{—}CH_2\text{—}Ph$$

$$\downarrow ISC$$

$$\,^1(Ph\text{—}\dot{C}H_2 \quad \dot{C}O\text{—}CH_2\text{—}Ph)$$

When DBK containing 48% ^{13}C at the CO position is photolyzed at zero field, the concentration of the recovered DBK having ^{13}C at CO becomes 62% after 90% conversion. The hf coupling constant of ^{13}C at the CO position is 125 G, whereas that of ^{12}C is zero. Since the cage product from the singlet pair is DBK, the yield of recovered DBK is higher in the radical pair having ^{13}C at the CO position than in that having ^{12}C at the CO position. This effect is simple, and useful for the separation of nuclear magnetic isotopes.

5.4.2.3 Magnetic Field Effects on Emission in the Gas Phase

Magnetic quenching of the fluorescence from the nonmagnetic excited singlet state ($^1A_2$) of gaseous CS_2 was first studied in 1974 using an N_2 nanosecond laser [25]. The fluorescence intensity decreases to about 50% by the application of a magnetic field of around 1.3 T. Since then, MFEs on the dynamic behavior of the excited singlet CS_2 and other excited molecules in the gas phase have been extensively studied to clarify the details of the mechanisms [13, 14]. Fluorescence from the $^1B_2$ state of gaseous CS_2, which exhibits magnetic-field-dependence similar to that of $^1A_2$, was examined using a picosecond laser [26]. Figure 5.4-7 shows the MFE on the fluorescence decay. This clearly demonstrates that fast energy redistribution processes, which cannot be detected by

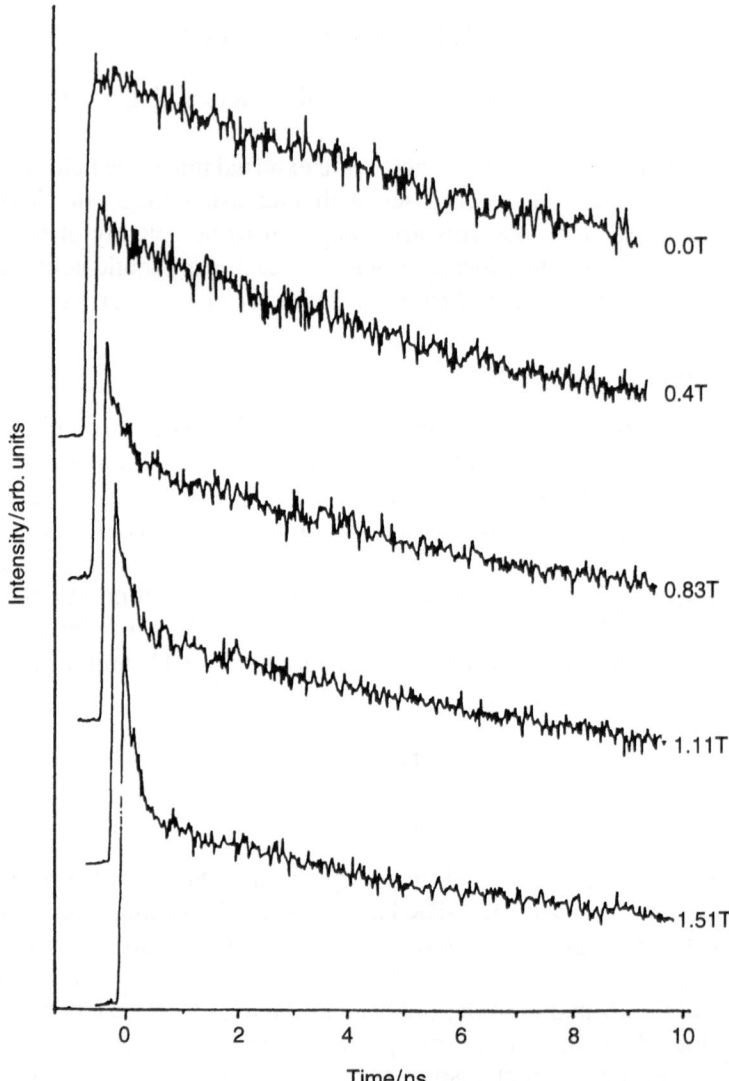

FIG. 5.4-7. Magnetic-field-dependence of the fluorescence decay of CS_2 vapor following excitation at the band head of the 6V band using a picosecond dye laser. From [26], with permission of S. Nagakura; Elsevier Science Publishers B.V. (1987)

measurements in the nano- and microsecond regions, are enhanced by magnetic fields.

The MFEs on emission from gaseous molecules can be explained by the theory of radiationless transition in a magnetic field, as originally proposed by Stannard [27] and Matsuzaki and Nagakura [28]. The Zeeman Hamiltonian $\gamma(L \cdot B_{ex} + 2S \cdot B_{ex})$ induces nonradiative transitions, where γ is the magnetogyric ratio, L is the orbital angular momentum, and S is the spin angular momentum. There are two mechanisms responsible for the phenomena mentioned above, i.e., the direct mechanism (DM) and the indirect mechanism (IM).

The direct mechanism arises from the $L \cdot B_{ex}$ term. The rate constant $k_{nr}(DM)$ for the nonradiative transition due to the DM can be given as

$$k_{nr}(DM) = \alpha B_{ex}^2 \tag{5.4-16}$$

where α is the matrix element which directly connects the initially prepared state (ϕ_i) by photoexcitation to the levels of an isoenergetic dark state ($\{\phi_d\}$). A magnetic field induces mixing between the *different* electronic states possessing the *same* spin state. Internal conversion from ϕ_i to $\{\phi_d\}$ is induced *directly* by the presence of a magnetic field.

The indirect mechanism arises from the $2S \cdot B_{ex}$ term. The $k_{nr}(IM)$ value for the nonradiative transition due to the IM is

$$k_{nr}(IM) = \begin{cases} \beta B_{ex}^2 & (\text{low magnetic field}) \\ \text{constant} & (\text{high magnetic field}) \end{cases} \tag{5.4-17}$$

where β is the constant of proportionality. Zeeman effects take place between *different* spin states of the *same* electronic state when its electron spin multiplicity is nonzero. A magnetic field cannot directly induce nonradiative decay from ϕ_i to $\{\phi_d\}$. In the case where ϕ_i is the excited singlet state, $\{\phi_d\}$ is the excited triplet state. The two states are coupled by the spin–orbit interaction, which is magnetic-field-independent. When a magnetic field is applied, the triplet sublevels undergo Zeeman shift and Zeeman mixing. Under a higher magnetic field, electron spins are quantized along the laboratory-fixed axes, though they are quantized along the molecular axes at zero and low magnetic fields (spin decoupling). This results in enhancement of singlet-to-triplet nonradiative decay.

The magnetic quenching ratio $Q(B_{ex})$ for the steady-state emission intensity I_{ss}, which is defined by Eq. (5.4-18), is a useful parameter when considering MFE.

$$Q(B_{ex}) = I_{ss}(0)/I_{ss}(B_{ex}) - 1 \tag{5.4-18}$$

where $I_{ss}(0)$ and $I_{ss}(B_{ex})$ are the intensities in the absence and presence of a magnetic field B_{ex}.

When the mechanism of the MFE is the DM and $\{\phi_d\}$ is the statistical limit case, the fluorescence decay is an exponential and its rate constant $k_f(B_{ex})$ in a magnetic field B_{ex} is given by

$$k_f(B_{ex}) = k_{nr} + k_{nr}(DM) \tag{5.4-19}$$

where k_{nr} is the decay rate constant which is magnetic-field-independent. Then the $Q(B_{ex})$ value becomes

$$Q(B_{ex}) = \left[k_f(B_{ex}) - k_f(0)\right]/k_f(0) = \alpha B_{ex}^2/k_{nr} \tag{5.4-20}$$

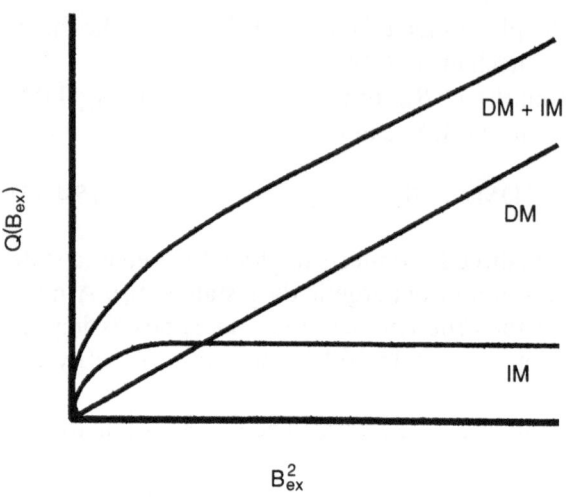

FIG. 5.4-8. Magnetic-field-dependence of $Q(B_{ex})$ on B_{ex}^2. DM, direct mechanism; IM, indirect mechanism; $DM + IM$, hybrid mechanism

$Q(B_{ex})$ increases in proportion to B_{ex}^2 as shown in Fig. 5.4-8. Similarly, when the mechanism is the IM and $\{\phi_d\}$ is the intermediate case, $Q(B_{ex})$ increases asymptotically to a saturation value at a higher magnetic field. When both DM and IM act simultaneously, the magnetic field dependence of $Q(B_{ex})$ can be expressed by the combination of DM and IM (hybrid mechanism). The dependence of $Q(B_{ex})$ on the fluorescence from the $^1A_2$ and $^1B_2$ states of gaseous CS_2 is explained in terms of the hybrid mechanism. From the MFE on the fluorescence decay shown in Fig. 5.4-7, it is concluded that $\{\phi_d\}$ in the $^1B_2$ fluorescence belongs to the intermediate case. MFEs on excited molecules in the gas phase are useful to investigate radiationless processes and energy levels of excited states in gaseous molecules, because fluorescence intensity and its decay profile are strongly related to the mechanisms of radiationless transition.

5.4.2.4 Thermodynamic Effects of Magnetic Fields: Magnetic Field Effects on Chemical Equilibrium

Usually, the magnetic free energy is much smaller than the thermal energy and consequently has little influence on chemical equilibrium. However, when a chemical reaction accompanies a large change of isotropic magnetic susceptibility, the magnetic field can also affect the chemical equilibrium.

Ferromagnetic hydrides $LaCo_5H_{3.4}$ adsorb hydrogen molecules at room temperature [29].

$$2.22\ LaCo_5H_{3.4} + H_2 \rightleftharpoons 2.22\ LaCo_5H_{4.3}$$

The saturation moment of the solids $LaCo_5H_{3.4}$ and $LaCo_5H_{4.3}$ are 45 and $11\ emu\,g^{-1}$, respectively. As hydrogen desorbs, the moment of solids increases at a rate of $\Delta M_s = 1.64 \times 10^4\ emu$ for one mole of hydrogen atom. Therefore, by applying a magnetic field, the equilibrium of the above reaction shifts to the left in favor of the magnetic free energy and the hydrogen pressure increases when the reaction is carried out under a constant volume. When $P_{B_{ex}}$ and P_0 are the hydrogen pressure in the presence and absence of a magnetic field B_{ex}, the logarithmic pressure change $\ln(P_{B_{ex}}/P_0)$ is

$$\ln\left(P_{B_{ex}}/P_0\right) = 2\Delta M_s B_{ex}/RT \tag{5.4-21}$$

where R is the gas constant and T is the absolute temperature. The change $\ln(P_{B_{ex}}/P_0)$ becomes 0.19 at 14 T, which corresponds to a 21% change in the equilibrium constant.

5.4.2.5 Mechanical Effects of Magnetic Fields

Magnetic Orientation of Macromolecular Systems: Anisotropy of Magnetic Susceptibility. The magnetic anisotropic energy of a diamagnetic molecule, $(x_\parallel - x_\perp)B_{ex}^2$, where $x_\parallel$ and $x_\perp$ are the magnetic susceptibilities parallel and perpendicular to the molecular axis, respectively, is much smaller than the thermal energy. However, when large numbers of molecules associate to form a giant molecular aggregate with their principal diamagnetic axes in the same direction, the magnetic energy becomes larger than the thermal energy and magnetic orientation of this aggregate takes place.

Fibrinognen (molecular weight 340000) is a soluble plasma protein which polymerizes in the presence of the appropriate enzyme to form fibrin gel [30]. The Δx of fibrinogen is estimated to be $2.5 \times 10^{-27}\,\text{erg G}^{-2}$, and fibrin fibers (1 μm in diameter and 100 μm in length), of which the gel is composed, contain about 10^7 fibrinogen monomers. Figure 5.4-9 shows scanning electron microscope images of fibrin gel polymerized in the presence and absence of a magnetic field of 8 T. The magnetic orientation of the fiber can be clearly seen.

Analogously, orientation of liquid crystalline polyester in a magnetic field of 6 T has been reported [31]. The magnetic orientation of macromolecular systems is useful in the preparation of orientated functional molecular systems.

Magnetohydrodynamics Mechanism. When a charged particle e moves with a velocity v in a magnetic field B_{ex}, the particle is acted on by an electromagnetic force F.

$$F = ev \times B_{ex} \tag{5.4-22}$$

Since the direction of F is perpendicular to that of v, the particle motion is disturbed by the magnetic field [magnetohydrodynamics (MHD) mechanism]. In electrochemical reactions, substrates in a bulk solution are transported to electrodes on which oxidation/reduction of the substrates takes place. In the presence of a magnetic field, this mass-transport process is affected by the flow of solution induced by the force F (MHD-induced flow).

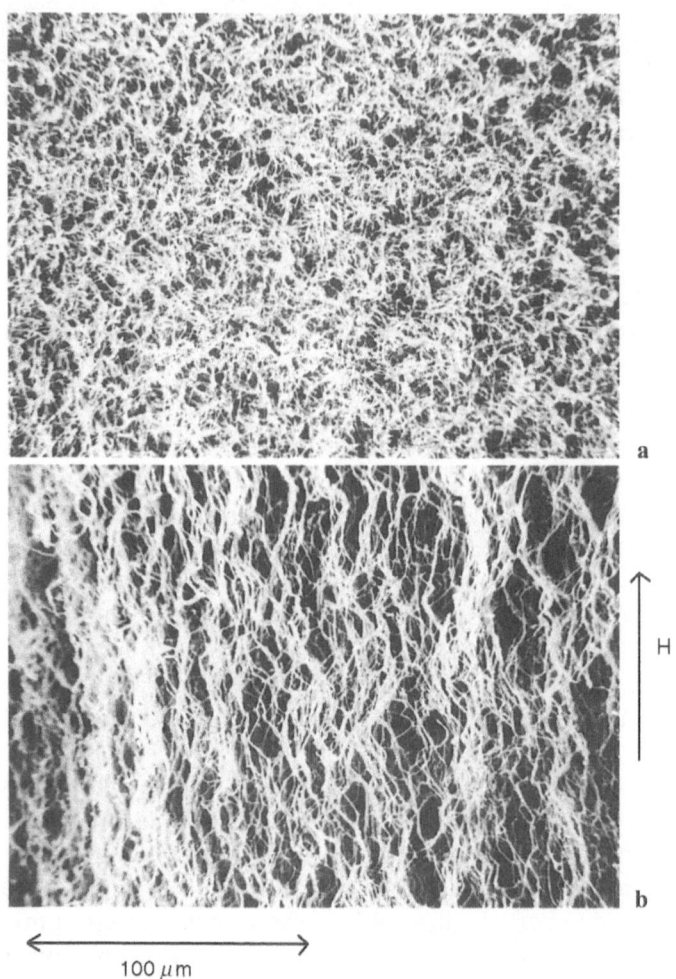

FIG. 5.4-9. Scanning electron microscopic images of fibrin gels, polymerized **a** at zero field and **b** at 8 T. The field direction is indicated by an *arrow*. Courtesy of Prof. T. Higashi of Osaka University

In anodic oxidation of the phenylacetate ion, the formation yield of benzaldehyde increases by about 40% in the presence of a magnetic field of 0.67 T, whereas a main product, 1,2-diphenylethane, is insensitive to that field [32].

$$PhCH_2COO^- \longrightarrow Ph\dot{C}H_2 \longrightarrow PhCH_2-CH_2Ph$$
$$\text{(adsorbed)} \qquad \text{(adsorbed)}$$
$$O_2 \searrow$$
$$PhCHO$$

Two-electron oxidation of a phenylacetate ion adsorbed on the electrode results in the formation of a benzyl radical. 1,2-Diphenylethane results from recombination of two adsorbed benzyl radicals. On the other hand, benzaldehyde is generated from the reaction of an adsorbed benzyl radical and dissolved oxygen. MHD-induced flow of the solution enhances the transport of oxygen to the electrode, leading to an increase in the yield of benzaldehyde. The MHD effect is extremely useful in controling chemical reactions on electrode surfaces.

References

1. Nagakura S, Nakajima T (1979) Kagaku to ryoshiron (Chemistry and quantum theory). Iwanami, Tokyo
2. Fujiwara K, Yamaguchi S (1965) Kogaku densikogaku II (Optics and physical optics). Asakura, Tokyo
3. Helfrich W (1973) Electric alignment of liquid crystals. Mol Cryst Liq Cryst 21:187–209
4. Iwayanagi S (1984) Ekisyo (Liquid cystals). Kyoritu, Tokyo
5. Williams DJ (1984) Organic polymeric and non-polymeric materials with large optical nonlinearities. Angew Chem Int Ed Engl 23:690–703
6. Krongauz VA, Parshutkin AA (1972) The effects of the electric field on photochromism in spiropyran: The dipole crystallization of a dye along the lines of force. Photochem Photobiol 15:503–507
7. Parshutkin AA, Krongauz VA (1974) Spectral and electrical properties of associates of photochromic spiropyrans in solution. Crystallization of dye molecules in a constant electric field. Mol Photochem 6:437–462
8. Yokoyama M, Shimokihara S, Matsubara A, Mikawa H (1982) Extrinsic carrier photogeneration in poly-N-vinylcarbazole. III. CT fluorescence quenching by an electric field. J Chem Phys 76:724–728
9. Ohta N (1995) Electric field effects on photoinduced dynamics. Annual Meeting of Chemical Society of Japan, 27–30 March 1995, Kyoto
10. Turro NJ, Kraeutler B (1980) Magnetic field and magnetic isotope effects in organic photochemical reactions. A novel probe of reaction mechanisms and a method for enrichment of magnetic isotopes. Acc Chem Res 13:369–377
11. Nagakura S, Hayashi H (1984) External magnetic effects upon photochemical reactions. Int J Quant Chem Quant Chem Symp 18:571–578
12. Molin YN (ed.) (1984) Spin polarization and magnetic effects in radical reactions. Elsevier, Amsterdam
13. Steiner UE, Ulrich T (1989) Magnetic field effects in chemical kinetics and related phenomena. Chem Rev 89:51–147
14. Hayashi (1990) Magnetic field effects on dynamic behavior and chemical reactions of excited molecules. In: Rabek JF (ed.) Photochemistry and photophysics, vol 1. CRC Press, Boca Raton, FL, pp 59–136
15. Steiner UE, Wolf HJ (1991) Magnetic field effects in photochemistry. In: Rabek JF, Scott GW (eds) Photochemistry and photophysics, vol 4. CRC Press, Boca Raton, FL, pp 1–130
16. Tanimoto Y, Fujiwara Y (1995) Effects of high magnetic field on organic reactions. J Synth Org Chem Jpn 53:413–422

17. Hayashi H, Nagakura S (1978) The theoretical study of external magnetic field effect on chemical reactions in solution. Bull Chem Soc Jpn 51:2862–2866
18. Sakaguchi Y, Hayashi H, Nagakura S (1980) Classification of the external magnetic field effects on the photodecomposition reaction of dibenzoyl peroxide. Bull Chem Soc Jpn 53:39–42
19. Mukai M, Fujiwara Y, Tanimoto Y, Konishi Y, Okazaki M (1993) Product yield-detected ESR studies of photochemical reaction of a bifunctional chain molceule. Magentic field and microwave effects. Z Phys Chem 180:223–233
20. Shulten K, Wolynes PG (1978) Semiclassical description of electron spin motion in radicals including the effect of electron hopping. J Chem Phys 68:3292–3297
21. Weller A, Nolting F, Staerk H (1983) A quantitative interpretation of the magnetic field effect on hyperfine-coupling-induced triplet formation from radical ion pairs. Chem Phys Lett 96:24–27
22. Hayashi H, Nagakura S (1984) Theoretical study of the relaxation mechanism in magnetic field effects on chemical reactions. Bull Chem Soc Jpn 57:322–328
23. Luders K, Salikhov KM (1987) Theoretical treatment of the recombination probability of radical pairs with consideration of singlet–triplet transitions induced by paramagnetic relaxation. Chem Phys 117:113–131
24. Turro NJ, Chow MF, Chung CJ, Kraeutler B (1981) Magnetic and micellar effects on photoreactions. 1. ^{13}C isotope-enrichment of dibenzyl ketone via photolysis in aqueous detergent solution. J Am Chem Soc 103:3886–3891
25. Matsuzaki A, Nagakura S (1974) Magnetic quenching of fluorescence observed with carbon disulfide excited by a nitrogen laser. Chem Lett 675–678
26. Imamura T, Tamai N, Fukuda Y, Yamzaki I, Nagakura S, Abe H, Hayashi H (1987) External magnetic field effect on the fluorescence of CS_2 excited to the V^1B_2 state with nanosecond and picosecond dye lasers. Chem Phys Lett 135:208–212
27. Stannard PR (1978) Radiationless transitions in a magnetic field. J Chem Phys 68:3932–3939
28. Matsuzaki A, Nagakura S (1978) On the mechanism of magnetic quenching of fluorescence in gaseous state. Helv Chim Acta 61:675–684
29. Yamaguchi M, Yamamoto I, Goto T, Miura S (1989) Shift in the chemical equilibrium of the $LaCo_5$–H system by strong magnetic fields. Phys Lett A 134:504–506
30. Torbet J, Freyssinet JM, Hudry-Clergeon G (1981) Oriented fibrin gels formed by polymerization in strong magnetic fields. Nature 289:91–93
31. Ito E, Sata H, Yamato M (1993) Orientation behavior of liquid crystalline polyester in a magentic field. Mem Fac Tech Tokyo Metropolitan Univ 43:4677–4682
32. Watanabe T, Tanimoto Y, Nakagaki R, Hiramatsu M, Nagakura S (1987) The magnetic field effects on electrolysis. III. The anodic oxidation of phenylacetate ion. Bull Chem Soc Jpn 60:4166–4168

Part VI
Molecular Design and Functionality of Molecular Systems

6.1
Molecular Design and Functionality of Molecular Systems

Hiroo Inokuchi

The creation of new molecules is one of the major goals of chemistry; chemists are constantly synthesizing new molecules. At present the number of molecules identified exceeds 12 million. According to an International Union of Pure and Applied Chemistry (IUPAC) report, the number of registered molecules in 1954 was only 600 000. Nowadays 600 000–700 000 new molecules are synthesized or extracted and identified every year. At the beginning of the 21st century, the total number of registered molecules will easily exceed 20 million (Fig. 6.1-1).

From this vast number of molecules, useful molecules are chosen depending on which molecular characteristics are required. For example, colored molecules are used as dyes, fragrant molecules as cosmetics, and the physiologically active molecules as medicines. These applications have made use of the original functionality that each molecule possesses. Color, fragrance, and physiological activity constitute some facets of molecular functionality. Molecular design will play a vital role in future progress in achieving the required molecular functionality.

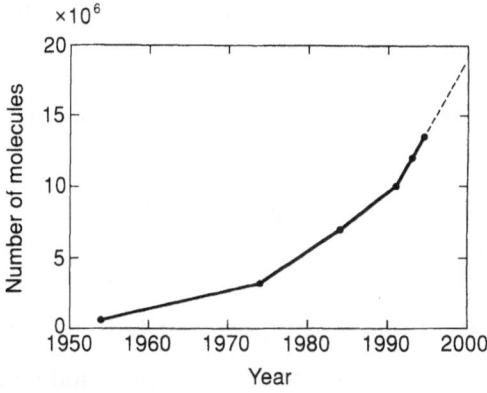

Fig. 6.1-1. Increase in the number of molecules registered since 1950

TABLE 6.1-1. Binding energy of typical solids

Substance	Binding energy ($KJ\,mol^{-1}$)	Classification
He	0.10	Molecular crystal
Benzene	44.7	
Anthracene	89.1	
LiF	1013.8	Ionic crystal
NaCl	764.0	
KBr	663.2	
Diamond	718.4	Covalent crystal
B	517	
SiC	592	
Pt	447	Metal
Fe	354	
Al	291	

6.1.1 Molecules and Molecular Environments

The molecular environment that surrounds each molecule determines its properties. Typical examples are the behavior of molecules dissolved in a solvent, and the changes in molecular properties caused by electric and magnetic fields. Molecular properties vary enormously, and in extreme cases completely new molecules can be formed, if the environment is suitable. Some examples are described in Part 5. When a molecule approaches another molecule one of two kinds of assemblies is formed: a molecular assembly composed of identical molecules or one composed of different molecules.

6.1.1.1 Molecular Assemblies Composed of Identical Molecules

In a molecule formed by chemical bonds between constituent atoms, interatomic bonds are completed and thus the molecule is naturally stable. The force exerted between molecules is the van der Waals force, which is composed of the dispersion force, the orientation effect, and the induction effect. However, these forces are usually very weak compared with chemical bonds, as shown in Table 6.1-1. Therefore thin films can be formed by vapor deposition from molecular assemblies. Molecular solids can easily be deformed by a weak external force. Furthermore, by application of an external field, the molecular orientation can be changed and sometimes an amorphous state can result, a typical example of which is a liquid crystal. Owing to the weak intermolecular interaction in molecular assemblies, their durability to heat is low. This is one of the drawbacks of molecular assemblies.

6.1.1.2 Molecular Assemblies Composed of Different Molecules

In 1949, it was reported that the color of a solution of iodine in benzene is quite different from that of a solution in chloroform. This is due to the formation of a new complex by the charge transfer force in addition to the van der Waals force. Since the pioneering work by Mulliken [1] and Nagakura and Tanaka [2, 3], extensive studies have been carried out both theoretically and experimentally. These are described in more detail in Part 3. The solid crystals of the perylene–bromine complex are also formed by the charge transfer force. The formation of this complex confirmed that organic solids can also be electrical conductors (see Sect. 4.3.1).

The force that binds elements consists of chemical bonds and the van der Waals bond (Fig. 6.1-2). The chemical bonds can be classified into three types: covalent bonds, metallic bonds, and ionic bonds. A typical substance formed by covalent bonds is diamond. The metals Au and Ag are formed by metallic bonds, and NaCl by ionic bonds. Solid Ar is formed by the van der Waals force. It is rare that a substance is formed by only one type of bond. In general, a solid is formed by a combination of these bonds. For example, Ge and Si crystals are formed by a combination of covalent and metallic bonds, and AgI and SiC by a combination of covalent and ionic bonds.

Let us apply this concept to molecular assemblies. The perylene–bromine complex is formed by a combination of ionic and van der Waals bonds. A large number of charge transfer complexes belong to the same group. The molecular assemblies formed by a combination of van der Waals and covalent bonds have also been synthesized. They have attracted great attention in the fields of solid-state chemistry and physics. In general, a functional molecular system may be selected from a series of possible substances by considering the relative strengths of the van der Waals bond, which is the most important factor, and one of the three chemical bonds.

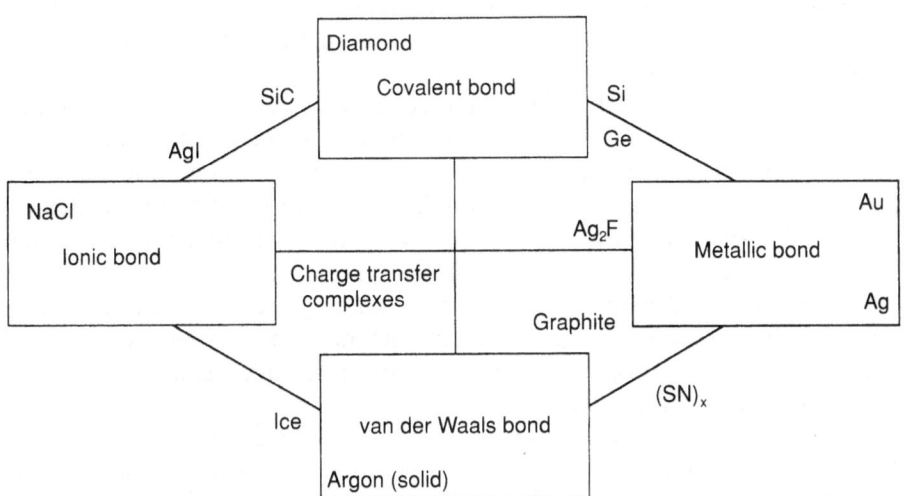

Fig. 6.1-2. The classification of chemical compounds by bonding characteristics

6.1.1.3 Molecular Crystals

The van der Waals force plays the most important role in the formation of molecular crystals. This force holds the molecules together, and the molecules combine three-dimensionally to form molecular crystals. van der Waals forces are weak, but they are additive. Molecules with high molecular weights can form crystals that can withstand the processes used to obtain measurements that are necessary to clarify their molecular functionality. For example, coronene fused with seven benzene rings forms a monoclinic single crystal with a melting point of 442°C. Molecular crystals are essentially formed by weak bonds, and thus are susceptible to external forces such as high pressure, electric fields, and magnetic fields. The way their properties change in response to external forces is the subject of much on-going research.

6.1.2 Studies of the Functionality of Molecular Systems

The functionalities of materials may be understood by examining the six solid-state properties, i.e., electrical, optical, magnetic, thermal, dynamic, and dielectric properties. Each property has its representative characteristics. The electrical conduction of a substance is the result of both its electronic and ionic conduction. Electronic conduction depends on conducting, semiconducting, and insulating characteristics. Recently superconductivity has become a very important subject. The magnetic properties of a substance include diamagnetic, paramagnetic, ferromagnetic, and anti-ferromagnetic characteristics. The optical properties include reflection, absorption, fluorescence, and phosphorescence. The thermal properties include specific heat and thermal conductivity. The mechanical properties include elasticity and compressibility, and viscosity is important in relation to the handling of molecular crystals. The dielectric properties include the all-important polarization and also dielectric dispersion. Ferroelectric characteristics are also of great interest.

The study of the functionalities of molecular systems is further complicated because combinations of some of the six solid-state properties must also be examined. For example, photoelectric properties appear when optical and electrical properties are combined. To examine the photoelectric properties of a substance, the following three phenomena are utilized: photoconduction, photoemission, and the photovoltaic effect (Table 6.1-2). The experimental study of these properties is indispensable to an understanding the electronic structure of a material. Information on electrical and magnetic properties as well as on electrical and thermal properties is very important to understand the full functionality of a molecular system.

Figure 6.1-3 shows the combinations of the six solid-state properties that have recently attracted attention. Even property combinations which are unimportant at present may be of interest in the future as a result of on-going research.

TABLE 6.1-2. Photoelectric phenomena of solids

Classification	Figure	Phenomena	Application
Photoconduction	*hv* ⊖ electron ⊕ hole	Conductivity of semiconductors is increased under illumination	Photocell
Photoemission	*hv*	Electron is emitted from a solid surface under illumination	Phototube
Photovoltaic effect	*hv*	Electricity is produced by illuminating the *p–n* junction of a semiconductor	Solar cell

	E	O	M	T	D	De
E		◎	◎	○	○	○
O	Photo-electricity		◎		○	
M	Magneto-resistance	Magnetic resonance				
T	Thermo-electricity	Photo-acoustic spectroscopy	Magnetic transition		○	◎
D	Motor effect	Mössbauer effect	Magnetic balance	Rheology		○
De	Piezo-electricity	Nonlinear optical effect	Magnetic liquid crystal	Liquid crystal	Electret	

FIG. 6.1-3. Combinations of pairs of the six solid-state properties. *Single circles*, of current interest; *double circles*, of great current interest. Letters represent the following properties: *E*, electrical; *O*, optical; *M*, magnetic; *T*, thermal; *D*, dynamic; *De*, dielectric

6.1.3 Molecular Functionality and Design

The ultimate goal of molecular design is to predict the structure of a molecular system with a desired functionality without carrying out experiments. Out of the many millions of molecules known to science, chemists have selected those with functionalities useful to human society. How have they selected these molecules? Let us take dyes as an example.

Over thousands of years, human beings have gained experience in using flower petals and plant leaves for dyeing cotton and linen. Indigo is one of the dyestuffs that was found in this way. Indigo is obtained from plants of the genus *Indigofera* and has long been used for dyeing cotton. The isolation and structural determination of indigo has opened up the possibility of making dyestuffs artificially. In 1880 indigo was first synthesized by A. Baeyer, and since then natural dyestuffs have been replaced by synthetic dyes. Chemists have synthesized a variety of similar dyes based on the chemical structure of indigo. This is an example in which the basic structure of a dye molecule was first learned from nature and then other dye molecules were designed and synthesized.

Consider the example of another well-known pigment, phthalocyanine. This compound was discovered by chance. In 1929 a beautiful blue material was found on the inside wall of a dye reaction vessel at a dyestuff factory. The structure of the blue material was determined by physical chemists, and it was found to be a metal coordination compound including iron from the vessel. For the last 60 years, a series of phthalocyanine complexes have been synthesized by replacing the complexed iron by other metals and have been used as pigments and materials in electronics. This is an example in which an accidental discovery of a compound with unique properties has taught chemists, aided with a vast accumulated knowledge of chemistry, to design similar molecules.

However, these examples should not be considered cases of true molecular design, because chemists started with the known molecular structures of natural compounds or accidentally discovered compounds with unique properties. Chemists then searched for useful molecules based on their experience and knowledge of chemistry.

To achieve real molecular design, the behavior of the electrons in a molecular system must be studied theoretically as described in Part 2, and at the same time the relation between the structure and the functionality must be clarified experimentally. It is then necessary to compare the theoretical and experimental results for many molecular systems. Although much research will be needed to develop the necessary techniques, this field is sure to become increasingly important in the future.

The functionality of molecular systems, backed by molecular design, will become the science best able to predict the potential functionalities of a vast number of artificially created molecules.

References

1. Mulliken RS (1952) Molecular compounds and their spectra II. J Am Chem Soc 74:811–824
2. Nagakura S, Tanaka J (1954) The relation between energy levels of substituent groups and electron migration effects in some monosubstituted benzenes. J Chem Phys 22:236–240
3. Nagakura S, Tanaka J (1954) On the relation between the chemical reactivity and energy levels of chemical reagents. J Chem Phys 22:563–563

Index